MATHEMATICAL SURVEYS AND MONOGRAPHS SERIES LIST

Volume

1 The problem of moments, J. A. Shohat and J. D. Tamarkin

2 The theory of rings, N. Jacobson

3 Geometry of polynomials, M. Marden

4 The theory of valuations, O. F. G. Schilling

5 The kernel function and conformal mapping, S. Bergman

6 Introduction to the theory of algebraic functions of one variable, C. C. Chevalley

7.1 The algebraic theory of semigroups, Volume I, A. H. Clifford and G. B. Preston

7.2 The algebraic theory of semigroups, Volume II, A. H. Clifford and G. B. Preston

8 Discontinuous groups and automorphic functions, J. Lehner

9 Linear approximation, Arthur Sard

10 An introduction to the analytic theory of numbers, R. Ayoub

11 Fixed points and topological degree in nonlinear analysis, J. Cronin

12 Uniform spaces, J. R. Isbell

13 Topics in operator theory, A. Brown, R. G. Douglas, C. Pearcy, D. Sarason, A. L. Shields; C. Pearcy, Editor

14 Geometric asymptotics, V. Guillemin and S. Sternberg

15 Vector measures, J. Diestel and J. J. Uhl, Jr.

16 Symplectic groups, O. Timothy O'Meara

17 Approximation by polynomials with integral coefficients, Le Baron O. Ferguson

18 Essentials of Brownian motion and diffusion, Frank B. Knight

19 Contributions to the theory of transcendental numbers, Gregory V. Chudnovsky

20 Partially ordered abelian groups with interpolation, Kenneth R. Goodearl

21 The Bieberbach conjecture: Proceedings of the symposium on the occasion of the proof, Albert Baernstein, David Drasin, Peter Duren, and Albert Marden, Editors

22 Noncommutative harmonic analysis, Michael E. Taylor

23 Introduction to various aspects of degree theory in Banach spaces, E. H. Rothe

24 Noetherian rings and their applications, Lance W. Small, Editor

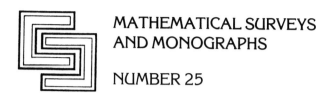

MATHEMATICAL SURVEYS
AND MONOGRAPHS

NUMBER 25

ASYMPTOTIC BEHAVIOR OF DISSIPATIVE SYSTEMS

JACK K. HALE

American Mathematical Society
Providence, Rhode Island

1980 Mathematics Subject Classification (1985 *Revision*). 34-02, 34C35, 34D30, 35-02, 35B32, 35B40, 58F09, 58F10, 58F12

Library of Congress Cataloging-in-Publication Data
Hale, Jack K.
  Asymptotic behavior of dissipative systems.
  (Mathematical surveys and monographs, ISSN 0076-5376; no. 25)
  Bibliography: p.
  1. Differentiable dynamical systems. 2. Differential equations, Partial–Asymptotic theory. 3. Stability. 4. Differential equations–Asymptotic theory. I. Title. II. Series.
  QA614.8.H35   1988   003   87-33495
  ISBN 0-8218-1527-X (alk. paper)

Copying and reprinting. Individual readers of this publication, and nonprofit libraries acting for them, are permitted to make fair use of the material, such as to copy a chapter for use in teaching or research. Permission is granted to quote brief passages from this publication in reviews, provided the customary acknowledgment of the source is given.

Republication, systematic copying, or multiple reproduction of any material in this publication (including abstracts) is permitted only under license from the American Mathematical Society. Requests for such permission should be addressed to the Executive Director, American Mathematical Society, P.O. Box 6248, Providence, Rhode Island 02940.

The owner consents to copying beyond that permitted by Sections 107 or 108 of the U.S. Copyright Law, provided that a fee of $1.00 plus $.25 per page for each copy be paid directly to the Copyright Clearance Center, Inc., 21 Congress Street, Salem, Massachusetts 01970. When paying this fee please use the code 0076-5376/88 to refer to this publication. This consent does not extend to other kinds of copying, such as copying for general distribution, for advertising or promotion purposes, for creating new collective works, or for resale.

Copyright ©1988 by the American Mathematical Society. All rights reserved.
Printed in the United States of America
The American Mathematical Society retains all rights
except those granted to the United States Government.
The paper used in this book is acid-free and falls within the guidelines
established to ensure permanence and durability. ∞

Contents

Acknowledgment	ix
Chapter 1. Introduction	1
Chapter 2. Discrete Dynamical Systems	8
2.1. Limit sets	8
2.2. Stability of invariant sets and asymptotically smooth maps	10
2.3. Examples of asymptotically smooth maps	13
2.4. Dissipativeness and global attractors	16
2.5. Dependence on parameters	21
2.6. Fixed point theorems	23
2.7. Stability relative to the global attractor and Morse-Smale maps	25
2.8. Dimension of the global attractor	26
2.9. Dissipativeness in two spaces	28
Notes and Remarks	33
Chapter 3. Continuous Dynamical Systems	35
3.1. Limit sets	35
3.2. Asymptotically smooth and α-contracting semigroups	36
3.3. Stability of invariant sets	38
3.4. Dissipativeness and global attractors	38
3.5. Dependence on parameters	40
3.6. Periodic processes	41
3.7. Skew product flows	43
3.8. Gradient flows	49
3.9. Dissipativeness in two spaces	54
3.10. Properties of the flow on the global attractor	56
Notes and Remarks	60

Chapter 4. Applications	61
4.1. Retarded functional differential equations (RFDE's)	61
4.1.1. Properties of the semigroup	61
4.1.2. Global attractor	63
4.1.3. An example	63
4.1.4. A gradient system	65
4.1.5. Equations with negative feedback	67
4.1.6. Periodic equations	70
4.2. Sectorial evolutionary equations	71
4.3. A scalar parabolic equation	75
4.3.1. Existence and gradient	75
4.3.2. Qualitative properties of the flow on the attractor	79
4.3.3. Stability properties of equilibria	84
4.3.4. A bifurcation problem—Dirichlet conditions	87
4.3.5. A bifurcation problem—other boundary conditions	92
4.3.6. Equations whose flow is equivalent to an ODE.	94
4.3.7. A method for determining stability	97
4.3.8. Stable solutions in a singularly perturbed equation	99
4.3.9. Quenching as a dynamic problem	105
4.4. The Navier-Stokes equation	107
4.5. Neutral functional differential equations	113
4.5.1. Properties of the semigroup	113
4.5.2. Global attractor in the space of continuous functions	116
4.5.3. Global attractor in $W^{1,\infty}$	117
4.6. Some abstract evolutionary equations	120
4.7. A one dimensional damped wave equation	125
4.7.1. Linear damping	125
4.7.2. A bifurcation problem	129
4.7.3. Nonlinear damping	130
4.7.4. Periodic forcing	132
4.8. A three dimensional damped wave equation	134
4.8.1. Nonlinear damping	134
4.8.2. Nonlinear damping, periodic forcing	138
4.8.3. Linear damping	139
4.8.4. Linear damping, periodic forcing	145
4.9. Remarks on other applications	145
4.9.1. Retarded equations with infinite delays	145
4.9.2. Strongly damped quasilinear evolutionary equations	146
4.9.3. A Beam equation	148
4.9.4. Other hyperbolic systems	151
4.9.5. Kuramoto-Sivashinsky equation	153
4.9.6. A Nonlinear diffusion problem	154
4.9.7. Age-dependent populations	155

4.10. Dependence on parameters and approximation of the attractor   160
    4.10.1. Reaction diffusion equations   161
    4.10.2. Singular perturbations   165
    4.10.3. Approximation of attractors   170
    4.10.4. Lower semicontinuity of the attractor   171
    4.10.5. Remarks on inertial manifolds   176

Appendix. Stable and Unstable Manifolds   179

References   187

Index   197

Acknowledgment

The author is indebted to many students and colleagues who have helped to develop the ideas in these notes over the years. In particular, I owe a special acknowledgment to Genevieve Raugel who read and constructively criticized almost all of the original manuscript. Thanks to Ezoura Fonseca, Kate MacDougall, Jeri Murgo, and the professional staff of the American Mathematical Society, especially Janet Scappini, for their assistance in the preparation of the text. Finally, the author gratefully appreciates financial support from the National Science Foundation, the Air Force Office of Scientific Research, the Army Research Office, and a Carnegie Fellowship from Heriot-Watt University.

CHAPTER 1

Introduction

In the last thirty years, many of the concepts of dynamical systems on locally compact spaces have been adapted to infinite dimensional systems, especially to delay differential equations and partial differential equations. These latter systems define dynamical systems on spaces which are not locally compact. Even when the base space is locally compact but not bounded, there is not a very general theory of dynamical systems. In this case, one can and often does restrict the class to one which has some type of dissipative property in order to reduce the essential part of the flow to a compact set.

A natural concept of dissipation (which we refer to as *point dissipative* or *ultimately bounded*) is to assume that there is a bounded set into which every orbit eventually enters and remains. For finite dimensional spaces and motivated by his studies of the periodically forced van der Pol equation, Levinson [1944] was led to study this formal concept. In this case, the dynamical system was discrete where the mapping $T\colon R^n \to R^n$ is the mapping taking the initial data for the differential equation to the value of the solution at time w, where w is the period of the forcing function. Because the space is locally compact, point dissipative implies *bounded dissipative* or *uniformly ultimately bounded*; that is, there is a bounded set into which the orbit of any bounded set eventually enters and remains (see, for example, Yoshizawa [1966] or Pliss [1966]). Using this fact, one easily sees that there is a maximal compact invariant set A such that the ω-limit set $\omega(U)$ of any bounded set U belongs to A; that is, A is a *global attractor*. The ω-limit set of U is defined as $\omega(U) = \bigcap_{m \geq 0} \mathrm{Cl} \bigcup_{n \geq m} T^n U$. In particular, the set $\omega(U)$ consists of all of the limit points of the orbit of U. However, $\omega(U)$ is generally much larger than this latter set.

In the applications of dynamical systems to infinite dimensional problems, the base space is usually not locally compact and other ideas must come into play.

If we have a dynamical system on some Banach space X and it is known that the orbits are precompact, then it has an ω-limit set which is compact and invariant. At first glance, it would appear that the set B consisting of the union of all ω-limit sets of points of the space would contain all of the relevant information concerning the asymptotic behavior of the orbits of the flow. However, this is not the case even in finite dimensions. The global attractor A mentioned above is not the same as B. The reason for this is that, in general, the set of nonwandering points in the flow may consist of many disjoint invariant sets. In particular, there may be several minimal invariant sets in the flow, some of which will be stable and some of which will be unstable. The structure of the unstable sets plays a fundamental role in the determination of the basins of attractions of the stable minimal invariant sets. None of this information is contained in the set B. Due to this fact, the set B also may not be stable for the flow. It can change drastically if the dynamical system is subjected to a small perturbation. This can even occur in R^2. As an example, consider the flow in R^2 defined by the family of curves shown in Figure 1.1. The ω-limit set of every point is an equilibrium point.

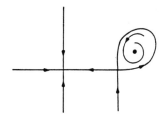

FIGURE 1.1

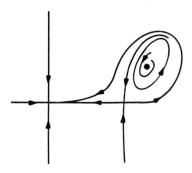

FIGURE 1.2

Under a small perturbation, it is possible for the new flow to possess an unstable periodic orbit as shown in Figure 1.2. The set B originally consisted of the three equilibrium points. After perturbation, it consists of the three equilibrium points and the large unstable periodic orbit. There is an "explosion" of B under perturbations! The homoclinic orbit consisting of part of the unstable set of an equilibrium caused the trouble. One could not hope to develop a reasonable qualitative theory of dynamical systems without considering such homoclinic orbits.

It is clear from the above example that the manner in which the points in B are connected by orbits plays a crucial role in the dynamics. How do we find a set containing B which will reflect this behavior? In the finite dimensional case, if the dynamical system is point dissipative, then we have remarked that the local compactness of the space allows one to prove that the dynamical system has a global attractor A. It is the maximal compact invariant set of the dynamical system, and the ω-limit set of any bounded set belongs to A. In particular, the set A is uniformly asymptotically stable. The set A will contain all of the unstable sets of any compact invariant set. One can also show that the set A cannot "explode" under perturbations. For the previous example, the corresponding sets A are shown in Figure 1.3. In the first case, it is homeomorphic to a disk with a line attached. In the second case, it is homeomorphic to a disk.

FIGURE 1.3

The set A may "shrink" under perturbations. For the above example, it is possible to perturb the dynamical system so that the flow and the attractor are shown in Figure 1.4. The attractor is homeomorphic to a line.

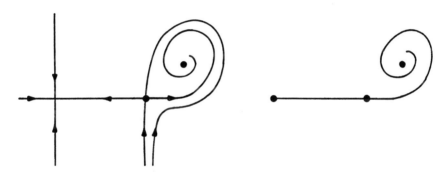

FIGURE 1.4

A change in the "size" of the attractor always will correspond to some type of bifurcation in the flow. However, there may be a bifurcation in the flow without any appreciable change in the "size" of the attractor. For example, see Figure 1.2.

The above simple examples are given to illustrate that the global attractor A contains much of the relevant information about the flow. It is reasonable therefore to take as a first objective the understanding of the flow restricted to A. The other part of the flow consists of transient behavior. This does not mean that the transient behavior is unimportant. It contains all of the separatrices which determine the basins of attraction of the minimal nonwandering sets of the flow. However, these separatrices change very little under perturbation if the flow restricted to A does not change its qualitative properties under perturbation.

Our objective in these notes is to present a theory of dissipative systems in infinite dimensions which will have the same general flavor as the one mentioned above for finite dimensions. More precisely, we give a general class of systems for which there is a global attractor A; that is, A will be compact, invariant, maximal with respect to this property and the ω-limit set of any bounded set will belong to A.

We discover the class (called *asymptotically smooth*) by investigating in detail the relationships between various types of stability for compact invariant sets. The class seems to be the appropriate one to have asymptotic stability and uniform asymptotic stability to be equivalent concepts.

To see why some hypotheses must be imposed on the class of dynamical systems in order to have the above property, it is instructive to consider some linear equations. Consider the neutral delay equation

$$\dot{x}(t) - \dot{x}(t-1) + cx(t) = 0$$

where $c > 0$ is a constant. Let us consider the weak form of this equation

$$(d/dt)[x(t) - x(t-1)] + cx(t) = 0$$

and consider initial data in $C[-1, 0]$. The characteristic equation is

$$\lambda(1 - e^{-\lambda}) + c = 0.$$

One easily shows that each solution λ of this equation is simple, has real part less than zero, and $\operatorname{Re} \lambda \to 0$ as $|\lambda| \to \infty$. This is enough to show that every solution approaches zero as $t \to \infty$. The principle of uniform boundedness implies the zero solution is stable. However, the zero solution is not uniformly asymptotically stable, because, from the linearity of the equation, this would imply that every solution approaches zero at a uniform exponential rate. Of course, this is not true since $\operatorname{Re} \lambda \to 0$ as $|\lambda| \to \infty$.

The above example of a linear neutral delay equation for which asymptotic stability and uniform asymptotic stability are not equivalent is somewhat artificial. However, neutral equations of similar type can be constructed in the theory of lossless transmission lines from the telegraph equation. If there is no resistance in the lines, the same phenomenon can occur. This occurs also in linear hyperbolic equations with some types of boundary conditions (see the text for references).

The reason for the difficulty in the above examples is that the radius of the essential spectrum of the semigroup generated by the equation is equal to one. If we require this radius to be less than one, then asymptotic stability and uniform asymptotic stability are equivalent. Linear systems with this property are equivalent to α-*contractions* where α is the measure of noncompactness in the sense of Kuratowski.

For the nonlinear case, the following concept is a generalization of the above linear examples. A continuous map $T: X \to X$, X a Banach space, is called *asymptotically smooth* if, for any bounded set B for which $TB \subset B$, there exists a compact set K such that $\text{dist}(T^n B, K) \to 0$ as $n \to \infty$. This implies, in particular, that the ω-limit set $\omega(B)$ of B is compact, $\omega(B) \subset K$, and $\text{dist}(T^n B, \omega(B)) \to 0$ as $n \to \infty$. Similar definitions are given for C^0-semigroups $T(t)$ on X.

Showing that a dynamical system defined by an evolutionary equation is asymptotically smooth depends on general properties of the equation. In fact, in many important cases, it follows from a variation of constants formula.

To obtain a global attractor, one must impose other properties of dissipation. It is somewhat surprising that the following general result is true.

THEOREM. *If $T(t)\colon X \to X$ is an asymptotically smooth C^0-semigroup which is point dissipative and has orbits of bounded sets bounded, then $T(t)$ has a global attractor.*

This result is very convenient for the applications since the hypotheses are reasonable to verify. We have already remarked about how to verify the hypothesis of being asymptotically smooth. Having orbits of bounded sets bounded involves a priori estimates on the solutions. Point dissipativeness involves obtaining a pointwise estimate on the orbits. Even with this weak form of dissipation, the theorem implies uniform convergence of the orbits of bounded sets to a compact set.

Chapters 2 and 3 of the text are devoted to a rather detailed discussion of abstract properties and examples of asymptotically smooth maps and semigroups. There is also a discussion of some of the continuity properties of the global attractor under perturbation, its capacity and Hausdorff dimension, and the stability of the flow on the global attractor under perturbation, in particular, Morse-Smale systems. For semigroups, we discuss properties of asymptotically smooth gradient flows and asymptotically smooth skew product flows. When the map or semigroup is defined on two spaces with one compactly imbedded in the other and has some additional properties, it is possible to determine very strong limiting behavior of the orbits in the small space from very weak limiting behavior in the large space. This topic is discussed under the heading "Dissipativeness in two spaces."

Even though the abstract theory has interest in itself, the final test is the relevance of the theory in applications to evolutionary equations. Chapter 4 is devoted to these applications. The elements of the theory and most of the fundamental results had their origin in our efforts in the late 1960s (see references in the text) to understand the asymptotic behavior of functional differential equations of both retarded and neutral type. Parabolic systems of partial differential equations fall naturally into our class of asymptotically smooth maps. In the last few years, it has been observed that the general theory applies as well to many other classes of problems, including some dissipative hyperbolic partial differential equations and age dependent populations. These problems are addressed in Chapter 4.

As one will see from the examples in Chapter 4, it is not an easy exercise to show that an equation has a global attractor; that is, the verification of the hypotheses in the above theorem requires a rather detailed analysis of the equation. Even when the evolutionary equation is considered as a perturbation of some linear equation, there are many problems in obtaining a semigroup. In this case, one would like to make extensive use of the variation of constants formula. This implies first of all that one must show that the linear equation generates a C^0-semigroup, and one must obtain excellent bounds on the essential spectrum of the semigroup, hopefully from the spectrum of the generator. This is a nontrivial exercise. The next step is to determine global existence of solutions of the complete equation to obtain a C^0-semigroup—another very difficult exercise. This is often enough to obtain the property of being asymptotically smooth. The dissipative properties always require special methods for each problem.

In each of the examples in Chapter 4, we have tried to show how one verifies the hypotheses of the above theorem to obtain the existence of a global attractor. Some properties of the flow on the global attractor also are discussed for several examples.

The subject of dissipative systems as presented here is still in its infancy. From the number of examples in Chapter 4 that have been shown to define asymptotically smooth semigroups, it is clear that they are important. Restricting the flow to the compact attractor can be considered as a necessary first step in discussing these infinite dimensional dissipative systems. It not only gives a way of discussing the qualitative changes in the flow as the semigroup is perturbed, but it also leads in a natural way to the posing of important problems that have received little attention up to this time.

Progress will be slow because the problems are difficult. On the other hand, once it is accepted that much of the qualitative information about the infinite dimensional flow is contained in the structure of the flow on the global attractor, many new important questions can be formulated and unexpected results occur. For example, we now have the following theorem of Oliva (see Hale, Magalhães, and Oliva [1984]):

> *When flows are restricted to the attractor, Morse-Smale systems are stable under smooth perturbations of the semigroup.*

Such a result is very important in the applications.

Another very surprising success story is the following result of Henry [1985]:

> *For a scalar parabolic equation*
> $$u_t = u_{xx} + f(x, u, u_x)$$
> *with boundary conditions specified at $x = 0$ and 1, the stable and unstable manifolds of hyperbolic equilibria are always transversal.*

This result says that global dynamical behavior can change only by local bifurcation of equilibria. Its discovery was a consequence of trying to determine the nature of the flow on the global attractor for the above equation with $f(u) = \lambda(u - u^3)$ and Dirichlet conditions on the boundary. The above theorem allows one to give a complete solution to this as well as other problems.

Another important result for a differential difference equation

$$\dot{x}(t) = f(x(t), \, x(t-1))$$

with negative feedback is due to Mallet-Paret [1985] who has shown that *there is a Morse decomposition on the global attractor with the basic invariant sets in the decomposition being determined by the number of zeros in (0,1)*.

Results of the above type would not have been possible if the interest had not been in the properties of the flow on the global attractor. This attitude also suggests a way to formulate meaningful questions concerning the changes in the flow on the global attractor when certain physical parameters are subjected to variations. Such problems could be difficult or perhaps impossible to formulate on the whole space. For example, in reaction-diffusion equations, what are the effects on the dynamics of the boundary conditions, the region of definition of the equations, the diffusion coefficients, etc.? What are the effects of singular perturbation terms on the dynamics? How are the global attractors related when an hyperbolic equation degenerates to a parabolic one? How does the flow of a neutral delay equation degenerate to a retarded one as the singular parameter approaches zero? One also can consider the approximation of the global attractor and the flow on the attractor by numerical procedures. Some of these topics are discussed in §4.10.

Throughout the notes, most of the basic material on the theory of dissipative systems is presented in detail. In the applications, many of the more complicated proofs were omitted with appropriate references given. For each application, a more complete presentation could be the basis of a book. My hope is that the attitude will prove useful to the reader in his own research.

CHAPTER 2

Discrete Dynamical Systems

2.1. Limit sets.

Suppose X is a complete metric space and $T\colon X \to X$ is a continuous mapping. For any $x \in X$, the *positive orbit* $\gamma^+(x)$ through x is defined as $\gamma^+(x) = \bigcup_{n\geq 0} T^n x$. A *negative orbit* through x is a sequence $\{x_j, j = 0, -1, -2, \ldots\}$ such that $x_0 = x$, $T x_{j-1} = x_j$ for all j. A *complete orbit* through x is a sequence $\{x_j, j = 0, \pm 1, \pm 2, \ldots\}$ such that $x_0 = x$ and $T x_{j-1} = x_j$ for all j.

Since the range of T need not be all of X, to say that there is a negative or complete orbit through x may impose restrictions on x. In spite of this fact, it will be clear from later discussions that some of the most interesting orbits are complete orbits. Also, since T need not be one-to-one, it is not necessary for a negative orbit through x to be unique if one exists. Let *the negative orbit* $\gamma^-(x)$ through x be defined to be the union of all negative orbits through x. Then

$$\gamma^-(x) = \bigcup_{n\geq 0} H(n,x),$$

$$H(n,x) = \{y \in X \colon \text{ there is a negative orbit } \{x_{-j}, j = 0, 1, 2, \ldots\}$$
$$\text{through } x \text{ with } x_{-n} = y\}$$

The complete orbit $\gamma(x)$ through x is $\gamma(X) = \gamma^-(x) \cup \gamma^+(x)$. For any subset $B \subset X$, let $\gamma^+(B) = \bigcup_{x \in B} \gamma^+(x)$, $\gamma^-(B) = \bigcup_{x \in B} \gamma^-(x)$, $\gamma(B) = \bigcup_{x \in B} \gamma(x)$ be respectively the positive orbit, negative orbit, complete orbit through B if the latter exist.

For any set $B \subset X$ define the ω-*limit set* $\omega(B)$ *of* B and the α-*limit set* $\alpha(B)$ *of* B as

$$\omega(B) = \bigcap_{n \geq 0} \mathrm{Cl} \bigcup_{k \geq n} T^k B,$$

$$\alpha(B) = \bigcap_{n \geq 0} \mathrm{Cl} \bigcup_{k \geq n} H(k, B).$$

For a point x, the ω-limit set $\omega(x)$ can be characterized by saying that $y \in \omega(x)$ if and only if there is a sequence of integers $n_j \to \infty$ such that $T^{n_j} x \to y$ as $j \to \infty$. For a set B, $y \in \omega(B)$ if and only if there is a sequence $x_j \in B$ and a sequence of integers $n_j \to \infty$ such that $T^{n_j} x_j \to y$ as $j \to \infty$. Similar characterizations hold for $\alpha(x)$ and $\alpha(B)$.

A set $S \subset X$ is said to be *invariant* under T if, for any $x \in S$, a complete orbit through x exists and $\gamma(x) \subset S$. It is not difficult to show that S is invariant if and only if $TS = S$.

Notice that the map T is not assumed to be invertible. Thus, it is possible that a point x has more than one complete orbit. As a consequence, there may exist an invariant set S which does not contain all of the complete orbits through each point in S. In fact, suppose $T: [0,1] \to [0,1]$, $Tx = 4x(1-x)$, $0 \le x \le 1$. The map T is not invertible and has a fixed point $x_0 = 3/4$. If $S = \{x_0\}$, then S is invariant. On the other hand, if $y = 1/4$, then $Ty = x_0$ and $\omega(y) = S$. The iterates $T^{-n}y$, $n = 0, 1, 2, \ldots$, are well defined and $\{T^{-n}y, n = 0, \pm 1, \pm 2, \ldots\}$ is a complete orbit through x_0 which does not belong to S. Note also that $T^{-n}y \to x_0$ as $n \to +\infty$, that is, this is an orbit homoclinic to x_0 (see Block [1978]). For a similar example in delay differential equations, see Walther [1981], an der Heiden and Walther [1983], and Hale and Lin [1986].

A set A is said to *attract* a set B under T if, for any $\varepsilon > 0$, there exists an $n_0 = n_0(\varepsilon, A, B)$ such that $T^n B$ belongs to the ε-neighborhood of A for $n \ge n_0$.

The following lemmas are fundamental.

LEMMA 2.1.1. *If $\omega(B)$ is compact and attracts B, then $\omega(B)$ is invariant. If, in addition, $\omega(B) \subset B$, then $\omega(B) = \bigcap_{n \ge 0} T^n B$. If $\gamma^-(B) = \bigcup_{k \ge 0} H(k,B)$, $\alpha(B)$ is compact, $\alpha(B)$ attracts $H(k,B)$ for each k, then $\alpha(B)$ is invariant.*

PROOF. Suppose $\omega(B)$ is compact and attracts B. If $\omega(B) = \varnothing$, there is nothing to prove. Suppose $\omega(B) \ne \varnothing$. To prove $\omega(B)$ is invariant, observe first that T being continuous implies $T\omega(B) \subset \omega(B)$. To prove that $\omega(B) \subset T\omega(B)$, suppose $y \in \omega(B)$. Then there are sequences $x_j \in B$ and $k_j \to \infty$ as $j \to \infty$ such that $T^{k_j} x_j \to y$ as $j \to \infty$. Consider the set $H = \{T^{k_j - 1} x_j, j \ge 1\}$. Since $\omega(B)$ is compact, attracts B and $k_j \to \infty$ as $j \to \infty$, we can assume there is a subsequence (which we label the same) which converges to z. Then $z \in \omega(B)$ and $Tz = y \in T\omega(B)$. Thus, $\omega(B) \subset T\omega(B)$. This fact together with the previous assertion implies that $\omega(B) = T\omega(B)$ and $\omega(B)$ is invariant. If $\omega(B) \subset B$, then it is clear that $\omega(B) \subset \bigcap_{n \ge 0} T^n(B)$ since $\omega(B)$ is invariant. Since $T^k B \subset \operatorname{Cl} \bigcup_{n \ge k} T^n B$ for all k, we have $\bigcap_{k \ge 0} T^k B \subset \omega(B)$. Thus, $\omega(B) = \bigcap_{k \ge 0} T^k B$.

The same type of argument proves the assertions for $\alpha(B)$ and the lemma is proved.

LEMMA 2.1.2. *If B is a nonempty subset of X such that $\operatorname{Cl} \gamma^+(B)$ is compact, then $\omega(B)$ is nonempty, compact, invariant, and $\omega(B)$ attracts B. If $\gamma^-(B)$ is nonempty and $\operatorname{Cl} \gamma^-(B)$ is compact, then $\alpha(B)$ is nonempty, compact, and invariant.*

PROOF. Suppose $\operatorname{Cl} \gamma^+(B)$ is compact and B is not empty. Then $\omega(B)$ is nonempty and compact.

We now prove that $\omega(B)$ attracts B. If not, then there is an $\varepsilon > 0$, a sequence of integers $k_j \to \infty$ as $j \to \infty$, and a sequence $y_j \in B$ such that $\operatorname{dist}(T^{k_j} y_j, \omega(B)) > \varepsilon$ for $j = 1, 2, \ldots$. Since the set $\{T^{k_j} B, j \ge 1\}$ belongs to a compact set there is a convergent subsequence of the $\{T^{k_j}y_j\}$ which we label the same. Since the limit of this sequence must belong to $\omega(B)$, we obtain a contradiction. Therefore, $\omega(B)$ attracts B. Lemma 2.1.1 implies $\omega(B)$ is invariant. This completes the proofs of the assertions concerning $\omega(B)$.

The same type of argument proves the assertions for $\alpha(B)$ and the lemma is proved.

2.2. Stability of invariant sets and asymptotically smooth maps.

Our objective in this section is to study the relationships between asymptotic stability and uniform asymptotic stability of compact invariant sets. This will motivate the introduction of an important class of mappings, namely, asymptotically smooth maps.

Suppose $T: X \to X$ is continuous and J is an invariant set. The set J is *stable* if, for any neighborhood V of J, there is a neighborhood U of J such that $T^n U \subset V$ for all $n \geq 0$. The set J *attracts points locally* if there is a neighborhood W of J such that J attracts points of W. The set J is *asymptotically stable* (a.s.) if J is stable and attracts points locally. The set J is *uniformly asymptotically stable* (u.a.s.) if J is stable and attracts a neighborhood of J.

If J attracts points locally, then it need not be that J is stable. This is easily seen by choosing T to be the time one map of an appropriate flow in the plane.

Clearly, J u.a.s. implies J is a.s. Without further hypotheses on T, J being a.s. may not imply that J is u.a.s. It is not trivial to give an example for which a.s. does not imply u.a.s. However, there are examples in Cooperman [1978] for linear maps and Brumley [1970] (see also Hale [1974]) for linear neutral delay differential equations. We content ourselves with remarking that these examples have the property that the ω-limit set of every point is zero and yet the rate of approach to zero is not uniform in t for every point in X. For a linear map T, it is not difficult to show that u.a.s. implies $|T^n| \leq k\lambda^n$, for all n, where $\lambda \in [0,1)$ and k is a constant. Therefore, for such maps, a.s. does not imply u.a.s.

We now study the stability concepts in more detail. An equivalent definition of stability is contained in

LEMMA 2.2.1. *An invariant set J is stable if and only if, for any neighborhood V of J, there is a neighborhood $V' \subseteq V$ of J such that $TV' \subset V'$.*

PROOF. If J is stable and V is a neighborhood of J, then there is a neighborhood W of J such that $T^n W \subseteq V$. Let $V' = \bigcup_{n \geq 0} T^n W$. Then V' satisfies the stated properties. The converse is clear.

The following result relates various types of attractivity properties for stable J.

LEMMA 2.2.2. *If J is a compact invariant set which is stable, then the following statements are equivalent:*
  (i) *J attracts points locally.*
  (ii) *There is a bounded neighborhood W of J such that $TW \subset W$ and J attracts compact sets of W.*

PROOF. Suppose (i) is satisfied; that is, there is neighborhood W of J such that J attracts points of W. From Lemma 2.2.1 we may assume $TW \subseteq W$. We may also suppose that W is bounded. Let U be a fixed open neighborhood of J, $U \subset$ interior of W. Let H be an arbitrary compact set in W. From Lemma 2.1.2, $\omega(H)$ is compact and attracts H. For any $x \in H$, there is an integer $n_0 = n_0(x, U)$ such that $T^n x \in U$ for $n \geq n_0$. Since T is continuous, there is a neighborhood O_x of x such that $T^{n_0} O_x \subset U$. Let $\{O_{x_i}, i = 1, 2, \ldots, p\}$ be a finite cover of H, let $N(H, U) = \max_i n_0(x_i, U)$, and let $V(H, U) = \bigcup_i O_{x_i}$. Then $T^n V(H, U) \subset U$ for $n \geq N(H, U)$. If we now take a sequence $\{U_i\}$ of open sets $U_1 \supset U_2 \supset \cdots$ with $\bigcap_{i \geq 1} U_i = J$, we see that $\omega(H) \subset J$; that is, J attracts H. This shows that (i) implies (ii).

The fact that (ii) implies (i) is obvious and the proof is complete.

An invariant set J is said to be *isolated* if there is a neighborhood V of J such that, if K is an invariant set in V, then $K \subseteq J$. Note that an isolated invariant set J is closed since the continuity of T implies that the closure of J is invariant. In the case where X is finite dimensional, Lemma 2.2.2 implies that a.s. of a compact invariant set J and u.a.s. of J are equivalent. Therefore, if J is a.s. and X is finite dimensional, then J is isolated. If X is infinite dimensional, this assertion is not implied by Lemma 2.2.2. Under additional hypothesis on T, we will show that a.s. of a compact invariant set J is equivalent to u.a.s. and therefore J will be isolated.

A motivation for this class of maps comes from Lemma 2.2.2 although historically this is not the way these maps were introduced. We would obtain u.a.s. if, in (ii) of Lemma 2.2.2, the compact set J would attract W. Therefore, we introduce a class of maps which will have this property.

A continuous mapping $T: X \to X$ is *asymptotically smooth* if, for any nonempty closed bounded set $B \subset X$ for which $TB \subset B$, there is a compact set $J \subset B$ such that J attracts B. Examples of interesting asymptotically smooth maps will be given in the next section. The following characterization of asymptotically smooth maps is interesting.

LEMMA 2.2.3. *The mapping $T: X \to X$ is asymptotically smooth if and only if, for any nonempty closed bounded set $B \subset X$, there is a nonempty compact set $J = J(B) \subset X$ such that $J(B)$ attracts the set $L(B) = \{x \in B: T^n x \in B$ for $n \geq 0\}$.*

PROOF. Let $J(B)$ be defined as in the lemma. If $B \subset X$ is closed and bounded with $TB \subset B$, then clearly $L(B) = B$ and $J(B)$ attracts B.

If B is any closed bounded set and $L = L(B)$, $\overline{L} = \mathrm{Cl}\,L$, then $TL \subset L$, $T\overline{L} \subset \overline{L}$. Thus, T asymptotically smooth implies there is a compact set J in $\overline{L}$ that attracts $\overline{L}$. This completes the proof.

COROLLARY 2.2.4. *If T is asymptotically smooth and B is a nonempty bounded set in X such that $\gamma^+(B)$ is bounded, then $\omega(B)$ is nonempty, compact, and invariant and $\omega(B)$ attracts B. In particular, if, for some $x \in X$,*

$\gamma^+(x)$ is bounded, then $\gamma^+(x)$ is precompact and $\omega(x)$ is nonempty, compact, and invariant.

PROOF. Since $T\gamma^+(B) \subset \gamma^+(B)$ and T is continuous, we have $T\operatorname{Cl}\gamma^+(B) \subset \operatorname{Cl}\gamma^+(B)$. Since T is asymptotically smooth, there is a compact set $J \subset \operatorname{Cl}\gamma^+(B)$ such that J attracts B. Thus, there are sequences $\varepsilon_j \to 0, k_j \to \infty$ such that $T^n B \subset N(J, \varepsilon_j)$, the ε_j-neighborhood of J, for $n \geq k_j$. This implies $\omega(B) \subset J$. Since $\omega(B)$ is closed and J is compact, the set $\omega(B)$ is compact. It remains to prove that $\omega(B)$ attracts B. If not, then there is an $\varepsilon > 0$ and sequences of integers $n_j \to \infty$ as $j \to \infty$, and of points $y_j \in B$ such that $T^{n_j} y_j \notin N(\omega(B), \varepsilon)$. Since J attracts B and J is compact, we can assume $T^{n_j} y_j$ converges as $j \to \infty$ to a $z \in J$. But then $z \in \omega(B)$, which is a contradiction. Lemma 2.1.1 completes the proof.

For asymptotically smooth maps, we have the following result.

THEOREM 2.2.5. *If T is asymptotically smooth and J is a compact invariant set that attracts points locally, then the following statements are equivalent:*

(i) *There is a bounded neighborhood W of J such that $TW \subset W$ and J attracts compact sets of W;*

(ii) *J is stable;*

(iii) *J is u.a.s.;*

and each condition implies that J is isolated.

PROOF. Let us first prove part of the last statement of the theorem, namely, that (i) implies J is isolated. Since T is asymptotically smooth, Corollary 2.2.4 implies $\omega(W)$ is compact. Every invariant set in W therefore must be in $\omega(W)$ since $\omega(W) = \bigcap_{n\geq 0} T^n W$ and $TW \subset W$. Since J attracts compact sets and $\omega(W)$ is invariant, we have $\omega(W) \subset J$. It is clear that (iii) implies (ii). Condition (ii) implies (i) from Lemma 2.2.2. We now prove that (i) implies (ii). In this part of the proof, we do not use the fact that T is asymptotically smooth, but only that J attracts compact sets. Suppose (i) does not imply (ii). The compactness of J implies there are an $\varepsilon > 0$ (as small as desired), a sequence of integers $n_j \to \infty$, a sequence of points $y_j \to y \in J$ as $j \to \infty$ such that $T^n y_j \in N(J, \varepsilon)$, $0 \leq n \leq n_j$, and $T^{n_j+1} y_j \notin N(J, \varepsilon)$, where $N(J, \varepsilon)$ is the ε-neighborhood of J. The set $\{y, y^j, j \geq 1\} = H$ is compact, J attracts H, $\operatorname{Cl}\gamma^+(H)$ is compact, and $\omega(H) \subset J$. Thus, we may assume without loss of generality that $T^{n_j} y_j \to z$ as $j \to \infty$. Since $z \in \omega(H) \subset J$, this is a contradiction since $Tz \in \omega(H) \subset J$ and $Tz \notin N(J, \varepsilon)$. this shows that (i) implies (ii).

To show that (ii) implies (iii), we will use the fact that (ii) implies (i) from Lemma 2.2.2. Let U be a fixed neighborhood of J, $U \subset W$. From Lemma 2.2.1, we may assume without loss of generality that $TU \subset U$. For any compact set $H \subset W$, there is an integer $n_0 = n_0(H, U)$ such that $T^n H \subset U$ for $n \geq n_0$. Since T is continuous and H is compact, there is a neighborhood H_1 of H such that $T^{n_0} H_1 \subset U$. Thus $T^n H_1 \subset U$ for $n \geq n_0$. Since T is asymptotically smooth, there is a compact set $K \subset U$ such that K attracts U. Since J attracts K by (i), it follows that J attracts U and, in particular, J attracts H_1. By taking $H = J$, we have that J attracts a neighborhood J_1 of J and J is u.a.s. This completes the proof of the theorem.

COROLLARY 2.2.6. *If T is asymptotically smooth, then a compact invariant set J is a.s. if and only if it is u.a.s.*

2.3. Examples of asymptotically smooth maps.

A continuous map $T: X \to X$ is *conditionally completely continuous* if $A \subset X$ bounded and TA bounded imply $\text{Cl}(TA)$ is compact. The map T is *completely continuous* if it is conditionally completely continuous and also a bounded map. The following observation is immediate.

LEMMA 2.3.1. *A conditionally completely continuous map is asymptotically smooth.*

To introduce other important examples of asymptotically smooth maps, we need some additional terminology. A *measure of noncompactness* β on a complete metric space X is a function β from the bounded sets of X to the nonnegative real numbers satisfying
  (i) $\beta(A) = 0$ for $A \subset X$ if and only if A is precompact.
  (ii) $\beta(A \cup B) = \max[\beta(A), \beta(B)]$.
  (iii) $\beta(A + B) \leq \beta(A) + \beta(B)$.
The *Kuratowski measure of noncompactness*, α, is defined by

$$\alpha(A) = \inf\{d: A \text{ has a finite cover of diameter } < d\}.$$

The α-measure of noncompactness satisfies the additional conditions
  (iv) $\alpha(\overline{\text{co}}A) = \alpha(A)$
where $\overline{\text{co}}A$ is the closed convex hull of A.
  (v) If $A_1 \supset A_2 \supset \cdots$ are nonempty closed sets of X such that $\alpha(A_j) \to 0$ as $j \to \infty$, then $\bigcap_{j \geq 1} A_j$ is nonempty and compact.

The motivation for the hypothesis in the following case of an asymptotically smooth map arose from the situation in which T is the time one map of a C^0-semigroup $V(t): X \to X$, $t \geq 0$; that is, $T = V(1)$. Then $T^n = V(n)$.

LEMMA 2.3.2. *For each integer n, suppose $T^n = S_n + U_n: X \to X$ has the property that U_n is completely continuous and, for any $r > 0$, there is a constant $k(n, r)$, $k(n, r) \to 0$ as $n \to \infty$ such that $|S_n x| \leq k(n, r)$ if $|x| \leq r$. Then T is asymptotically smooth.*

PROOF. Suppose $B_r = \{x \in X: |x| \leq r\}$ and $L_r = \{x \in B_r: T^n x \in B_r \text{ for } n \geq 0\}$. Let α be the Kuratowski measure of noncompactness on X. Since $\gamma^+(T^n L_r) = T^n \gamma^+(L_r)$ and $\gamma^+(L_r) \subset B_r$, it follows that $\alpha(\gamma^+(T^n L_r)) \leq k(n, r)$ for all n. Since $\alpha(\text{Cl}\, \gamma^+(T^n L_r)) = \alpha(\gamma^+(T^n L_r))$ and $k(n, r) \to 0$ as $n \to \infty$, it follows, from the definition of $\omega(L_r)$, that $\omega(L_r)$ is nonempty and compact. We are using here the following property of α-contractions. If $A_1 \subset A_2 \subset \cdots$ are nonempty closed subsets of X such that $\alpha(A_j) \to 0$ as $j \to \infty$, then $\bigcap_j A_j$ is nonempty and compact.

We now prove that $\omega(L_r)$ attracts L_r. Suppose this is not the case. Then there are an $\varepsilon > 0$ and a sequence $\{T^{k_j} x_j\}$, $j \geq 1$, in L_r, $k_1 < k_2 < \cdots, k_j \to \infty$ as $j \to \infty$ such that $\{T^{k_j} x_j\} \notin N_\varepsilon(\omega(L_r))$. If $U_l = \{T^{k_j} x_j : k_j \geq l\}$, $l \geq 1$, then $U_1 \supset U_2 \supset \cdots$ and $\alpha(\operatorname{Cl} U_l) = \alpha(T^l \{T^{k_j - l} x_j, k_j \geq l\}) \leq \alpha(T^l(L_r)) \leq k(l, r) \to 0$ as $l \to \infty$. Thus, $\bigcap_{l \geq 1} \operatorname{Cl} U_l$ is compact. Thus, without loss of generality, we may assume $T^{k_j} x_j \to y$ as $j \to \infty$. Then $y \in \omega(L_r)$, but this is a contradiction since $y \notin N_\varepsilon(\omega(L_r))$. Therefore, $\omega(L_r)$ attracts L_r and we have proved there is a compact set in B that attracts L_r. If B is an arbitrary bounded set in X, then there is an $r > 0$ such that $B \subseteq B_r$. Then $L(B) = \{x \in X : T^n x \in B, n \geq 0\} \subset L_r$. Since $\omega(L_r)$ is compact, so is $\omega(L(B))$ and we have proved that T is asymptotically smooth. This completes the proof of the lemma.

A continuous map $T: X \to X$ is a *conditional β-contraction* of order k, $0 \leq k < 1$, with respect to the measure of noncompactness β if $\beta(TA) \leq k\beta(A)$ for all bounded sets $A \subset X$ for which TA is bounded. The map T is a *β-contraction* if it is a conditional β-contraction which is also a bounded map, that is, takes bounded sets to bounded sets.

A continuous map $T: X \to X$ is *conditionally β-condensing* if $\beta(TA) < \beta(A)$ for bounded sets $A \subset X$ for which TA is bounded and $\beta(A) > 0$. The map T is *β-condensing* if it is a conditional β-condensing map which is also a bounded map.

We now suppose that X is a Banach space and summarize some properties of bounded linear operators and their relation to α-contractions. If T is a bounded linear operator on a Banach space X and α is the Kuratowski measure of noncompactness on X, let $\alpha(T) = \inf\{k : \alpha(TB) \leq k\alpha(B) \text{ for all } B \subset X, \text{ bounded}\}$. The number $\alpha(T)$ is related to the radius of the essential spectrum of T. There are several ways to define the essential spectrum of an operator. We give the two most common ones.

For a bounded linear operator $T: X \to X$, the *essential spectrum* $\mathrm{E}\sigma(T)$ of T according to Browder [1961] is the set of $\lambda \in \sigma(T)$, the spectrum of T, such that at least one of the following holds:

(i) $R(\lambda I - T)$ is not dense,

(ii) λ is a limit point of $\sigma(T)$,

(iii) $N_\lambda(T)$ is infinite dimensional,

where $N_\lambda(T)$ is the generalized null space of $\lambda I - T$.

Following Schechter [1971], the *essential spectrum*

$$\mathrm{Es}(T) = \bigcap \{\sigma(T + K) : K \text{ is a compact linear operator}\}.$$

If $r_{\mathrm{E}\sigma}(T)$ and $r_{\mathrm{Es}}(T)$ denote the radii of the essential spectra defined above, then the following interesting properties hold.

LEMMA 2.3.3. *If T is a bounded linear operator on a Banach space X, then*

(i) $r_{\mathrm{E}\sigma}(T) = r_{\mathrm{Es}}(T)$;

(ii) $r_{\mathrm{E}\sigma}(T) = \lim_{n \to \infty} (\alpha(T^n))^{1/n}$;

(iii) *If $\lambda_0 \in \sigma(T)$, $|\lambda_0| > r_{\mathrm{E}\sigma}(T)$, then λ_0 is a pole of finite order of $(\lambda I - T)^{-1}$;*

(iv) *For any $\varepsilon > 0$, there is an equivalent norm in X such that*
$$T = S + U,$$
where U has finite dimensional range and $|S| < \alpha(T) + \varepsilon$.

Property (iii) is due to Browder [1961], property (ii) to Nussbaum [1970], and the proofs of all of the results can be found in Deimling [1985].

Property (i) shows that the radii of the essential spectra are the same for each of the above definitions. Property (ii) relates the radius of the essential spectrum to $\alpha(T)$ in the same way that $\lim_{n\to\infty} |T^n|^{1/n}$ is the radius of the spectrum of T. Property (iv) shows that every bounded linear operator with $\alpha(T) < 1$ (that is, linear α-contraction) can be written in the form $T = S + U$ where $|S| < 1$ in an equivalent norm.

An important example of a nonlinear operator which is a conditional β-contraction of order $k = 0$ is a map T which is conditionally completely continuous. This is a consequence of property (i) in the definition of the β-measure of noncompactness. A more interesting example of a conditional β-contraction is $T = S + U$ where S is globally Lipschitz with Lipschitz constant $k \in [0, 1)$ and U is conditionally completely continuous. This is a consequence of properties (i) and (iii) of the β-measure of noncompactness.

The next results relate these maps to asymptotically smooth maps.

LEMMA 2.3.4. *Conditional β-contractions are asymptotically smooth.*

A proof of this lemma could be given following the lines of the proof of Lemma 2.3.2. However, it is also a consequence of the following more general result.

LEMMA 2.3.5. *Conditional β-condensing maps are asymptotically smooth.*

PROOF. Let B be a closed bounded set in X such that $B \subset X$, $TB \subset B$. We want to show first that every sequence $\{T^{k_j} x_j\}$ with $k_j \to \infty$ as $j \to \infty$ has compact closure. To do this, we introduce some notation. Let C be the set of all such sequences in B and let $\eta = \sup\{\beta(h): h \in C\}$. If $h \in C$, then $h \in B$ and we have $\eta < \beta(B)$. We claim there is an $h^* \in C$ such that $\beta(h^*) = \eta$. Let $\{h_l\}$ be a sequence of elements of C such that $\beta(h_l) \to \eta$. Define $\tilde{h}_l = \{T^{k_j} x_j : T^{k_j} x_j \in h_l, k_j > l\}$. Then $\beta(\tilde{h}_l) = \beta(h_l)$. If $h^* = \bigcup_l \tilde{h}_l$ ordered in any way, then $h^* \in C$ and $\beta(h^*) \geq \beta(h_l)$ for each l. So $\beta(h^*) = \eta$. Let $\tilde{h}^* = \{T^{k_j} x_j : T^{k_j-1} x_j \in h^*\}$. Then $\tilde{h}^* \in C$ and $\beta(\tilde{h}^*) \leq \eta$. Since $T(\tilde{h}^*) = h^*$ and $\beta(h^*) = \eta$, we have $\beta(T\tilde{h}^*) \geq \beta(\tilde{h}^*)$. If $\beta(\tilde{h}^*) > 0$, this contradicts the fact that T is conditional β-condensing. Thus, $\beta(\tilde{h}^*) = 0$, $\tilde{h}^*$ and h^* are precompact, and $\eta = 0$. Thus, every sequence in C is precompact.

This shows that $\omega(B)$ is nonempty. Also, $\omega(B)$ attracts B. In fact, if this is not the case, then there is a sequence $\{T^{k_j} x_j\}$, $j \geq 1$, and an $\varepsilon > 0$ such that $T^{k_j} x_j \notin N_\varepsilon(\omega(B))$ for all $j \geq 1$. But $\{T^{k_j} x_j\}$ is precompact by the above argument and contains a subsequence which converges to a point z which is outside $N_\varepsilon(\omega(B))$. This is obviously a contradiction since $z \in \omega(B)$. Thus, $\omega(B)$ attracts B.

We now prove $\omega(B)$ is invariant. Clearly $T\omega(B) \subset B$. To prove that $\omega(B) \subset T\omega(B)$, suppose $y \in \omega(B)$, $y = \lim_{k\to\infty} T^{k_j} x_j$ where $x_j \in B$ and $k_j \to \infty$. Let $H = \{T^{k_j-1} x_j, j \geq 1\}$. By the argument of the first part of this proof $\text{Cl}\, H$ is compact. Thus, we may assume without loss in generality that $T^{k_j-1} x_j \to z$. Then $z \in \omega(B)$ and $Tz = y \in T\omega(B)$. Thus, $\omega(B) \subset T\omega(B)$. As a consequence of the fact that $T\omega(B) \subset \omega(B)$, we have $\omega(B)$ invariant.

Finally, $\alpha(\omega(B)) = \alpha(T\omega(B)) < \alpha(\omega(B))$ if $\alpha(\omega(B)) > 0$. Thus, $\alpha(\omega(B)) = 0$ and $\omega(B)$ is compact. Since $\omega(B)$ attracts B from the argument above, we have that T is asymptotically smooth and the result is proved.

Let us give another example of an α-contraction which will play a role in the discussion of hyperbolic partial differential equations.

A pseudometric ρ on a Banach space X is said to be *precompact* (with respect to the norm of X) if any bounded sequence (in norm) has a subsequence which is Cauchy with respect to ρ.

LEMMA 2.3.6. *If $T\colon X \to X$ is a continuous map such that*

$$|Tx - Ty| \leq q|x - y| + \rho(x,y) \quad \text{for all } x, y \in X,$$

where $0 \leq q$ is a constant and ρ is a precompact pseudometric, then $\alpha(T(B)) \leq q\alpha(B)$ for any bounded set B. In particular, if $0 \leq q < 1$, then T is an α-contraction.

PROOF. For any $\varepsilon > 0$, there are sets $F_1, \ldots, F_n$ such that $B = F_1 \cup \cdots \cup F_n$ and $\text{diam}\, F_i < \alpha(B) + \varepsilon$, $i = 1, 2, \ldots, n$. Since ρ is a precompact pseudometric, there are sets $C_1, \ldots, C_m$ such that $B = C_1 \cup \cdots \cup C_m$ and $\rho(x,y) < \varepsilon$ if $x, y \in C_i$, $i = 1, 2, \ldots, m$. Since $B = \bigcup\{F_i \cap C_j, i = 1, \ldots, n, j = 1, \ldots, m\}$, it follows that $T(B) = \bigcup\{T(F_i \cap C_j), i = 1, \ldots, n, j = 1, \ldots, m\}$. Also, for $x, y \in F_i \cap C_j$, we have

$$|Tx - Ty| \leq q|x - y| + \rho(x,y) \leq q[\alpha(B) + \varepsilon] + \varepsilon.$$

Thus, $\text{diam}(F_i \cap C_j) \leq q\alpha(B) + \varepsilon(q+1)$. Since ε is arbitrary, the lemma is proved.

2.4. Dissipativeness and global attractors.

In the previous section, we have seen how asymptotically smooth maps play a role in determining the equivalence of a.s. and u.a.s. for a compact invariant set. In this section, we want to explore similar questions on a more global level. More precisely, we determine conditions which will ensure that there is a compact invariant set which attracts bounded sets of X. These conditions are expressed in terms of dissipative and boundedness properties of the mapping T.

We need some definitions. A continuous mapping T on a complete metric space X is said to be *point dissipative (compact dissipative) (locally compact dissipative) (bounded dissipative)* on X if there is a bounded set B in X such that B attracts each point of X (each compact set of X) (a neighborhood of each compact set of X) (each bounded set of X) under T. In the classical theory of differential equations, point (bounded) dissipativeness is often referred to as ultimate (uniform ultimate) boundedness.

For a given continuous map $T\colon X \to X$, a compact invariant set A is said to be a *maximal compact invariant set* if every compact invariant set of T belongs to A. An invariant set A is said to be a *global attractor* if A is a maximal compact invariant set which attracts each bounded set $B \subset X$. Notice that this implies that $\omega(B) \subset A$. A global attractor A has very strong stability properties since $\delta(T^n B, A) \to 0$ as $n \to \infty$, where the distance $\delta(B, A)$ from B to A is defined by

$$(2.1) \qquad \delta(B, A) = \sup_{y \in B} \inf_{x \in A} \operatorname{dist}(y, x).$$

The number $\delta(B, A)$ is denoted sometimes by $\operatorname{dist}(B, A)$. Note that $\delta(B, A) \neq \delta(A, B)$ in general. We say that an invariant set A is a *local attractor* if A is compact, invariant, and there exists a bounded neighborhood B of A such that A attracts B. Notice that A is a local attractor if and only if it is uniformly asymptotically stable.

The major portion of this book is concerned with the existence and flow on the global attractor. For this reason, we will omit the word global and sometimes use attractor to refer to global attractor. However, we always will keep the word local when dealing with local attractors. Global attractors must always be connected as the following result shows.

LEMMA 2.4.1. *If X is a Banach space and J is a compact invariant set which attracts compact sets, then J is connected.*

PROOF. If $\overline{co}\, J$ is the closed convex hull of J, then $\overline{co}\, J$ is compact, connected, and J attracts $\overline{co}\, J$. If J is not connected, then there are open sets U, V with $U \cap J \neq \emptyset$, $V \cap J \neq \emptyset$, $J \subset U \cup V$, and $U \cap V = \emptyset$. By the continuity of T, the set $T^n(\overline{co}\, J)$ is connected for each $n \geq 0$. Since $J \subset T^n(\overline{co}\, J)$ for all $n \geq 0$, we have $U \cap T^n(\overline{co}\, J) \neq \emptyset$, $V \cap T^n(\overline{co}\, J) \neq \emptyset$ for all $n \geq 0$. Since $T^n(\overline{co}\, J)$ is connected there is an $x_n \in T^n(\overline{co}\, J) \setminus (U \cup V)$. Since J attracts $\{x_n, n \geq 0\}$, this set must be compact and we may assume that $x_n \to x \in J$. Clearly $x \notin U \cup V$, which is a contradiction.

One of the basic results on the existence of a maximal compact invariant set is the following.

THEOREM 2.4.2. *If $T\colon X \to X$ is continuous and there is a nonempty compact set K that attracts compact sets of X and $A = \bigcap_{n \geq 0} T^n K$, then*

(i) *A is independent of K and, if X is a Banach space, A is connected.*
(ii) *A is maximal, compact, and invariant.*
(iii) *A is stable and attracts compact sets of X.*

If, in addition, T is asymptotically smooth, then

(iv) *for any compact set H in X, there is a neighborhood H_1 of H such that $\gamma^+(H_1)$ is bounded and A attracts H_1. In particular, A is u.a.s.*
(v) *If C is any subset of X such that $\gamma^+(C)$ is bounded, then A attracts C.*

PROOF. The ideas in the proof will follow very closely the ones in the proof of Theorem 2.2.5. Suppose H is a compact set in X. Since $T^n H \to K$, the set $\mathrm{Cl}\,\gamma^+(H)$ is compact and $\omega(H) \subset K$. In particular $\omega(K) \subset K$, $\omega(K) = \bigcap_{n \geq 0} T^n K$, is a nonempty, compact invariant set which attracts compact sets of X. Also, $\omega(K)$ is connected by Lemma 4.1 if X is a Banach space. If K_1 is any other compact set that attracts compact sets of X, then clearly $\omega(K_1) \subset \omega(K) \subset \omega(K_1)$ since $\omega(K)$ and $\omega(K_1)$ attract compact sets of X. Thus, $A = \omega(K)$ is independent of K and maximal. This proves (i) and (ii). Also, this proof shows that A attracts compact sets.

The proof that A is stable is verbatim the same as the proof that (i) implies (ii) in Theorem 2.2.5. This part of the proof of Theorem 2.2.5 used only the fact that J attracts compact sets and did not use either the fact that T was asymptotically smooth or that W was bounded and $TW \subseteq W$. With this remark (iii) is proved.

We now prove (iv). The proof is very similar to the proof that (ii) implies (iii) in Theorem 2.2.5. Since A is stable, Lemma 2.2.1 implies there is a bounded neighborhood U of A such that $T\overline{U} \subset \overline{U}$, $\overline{U} = \mathrm{Cl}\,U$. Since A attracts compact sets of X, for any compact set H of X, there is an integer $n_0 = n_0(H, U)$ such that $T^n H \subset U$ for $n \geq n_0$. Since T is continuous and H is compact, there is a neighborhood H_1 of H such that $T^{n_0} H_1 \subset U$. Since $TU \subset U$, it follows that $T^n H_1 \subset U$ for $n \geq n_0$. Since T is asymptotically smooth, there is a compact set $K \subset U$ such that K attracts U. Since A attracts K by (iii), then A attracts U and, in particular, A attracts H_1. This proves (iv).

To prove (v), suppose C is a subset of X such that $\gamma^+(C)$ is bounded. Since $T\,\mathrm{Cl}\,\gamma^+(C) \subset \mathrm{Cl}\,\gamma^+(C)$ and T is asymptotically smooth, there is a compact set K such that K attracts $\gamma^+(C)$. Then A attracts $\gamma^+(C)$ by (iii). This completes the proof of the theorem.

An important consequence of Theorem 2.4.2 is

COROLLARY 2.4.3. *If T is asymptotically smooth, then the following statements are equivalent:*

(i) *There is a compact set which attracts compact sets of X.*

(ii) *There is a compact set which attracts a neighborhood of each compact set of X.*

To apply Theorem 2.4.2, one must have some practical procedure for the verification of the hypotheses. A step in this direction is the following result.

LEMMA 2.4.4. *If T is asymptotically smooth and T is compact dissipative, then there exists a compact invariant set which attracts compact sets and the conlcusions of Theorem 2.4.2 hold.*

PROOF. Let C be a bounded set which attracts compact sets. Let $B = \{x \in C : T^n x \in C, n \geq 0\}$. Then $T\overline{B} \subseteq \overline{B}$, $\overline{B} = \mathrm{Cl}\,B$, and there is a compact set K which attracts $\overline{B}$. This set K attracts compact sets. Now use Theorem 2.4.2.

The following result relates the different types of dissipativeness.

LEMMA 2.4.5. *Let T be asymptotically smooth and point dissipative. If the orbit of any compact set is bounded, then T is locally compact dissipative. If the orbit of any bounded set is bounded, then T is bounded dissipative.*

PROOF. Let B be a nonempty closed bounded set which attracts points of X and let $U = \{x \in B : \gamma^+(x) \subset B\}$. Then U attracts points of X and $\gamma^+(U)$ is bounded. Since $T\gamma^+(U) \subset \gamma^+(U)$ and T is asymptotically smooth, there is a compact set $K \subset \text{Cl}\,U$ such that K attracts U and also attracts points of X. The set K also attracts itself and so $\gamma^+(K)$ is precompact. If $J = \omega(K)$, then J is a compact invariant set that attracts points of X.

Now suppose that the orbit of any compact set is bounded. We first show there is a neighborhood V of J such that $\gamma^+(V)$ is bounded. If this is not the case, then there is a sequence $x_j \in X$ and a sequence of integers $k_j \to \infty$ and a y in J such that $\lim_{j\to\infty} x_j = y$ and $|T^{k_j} x_j| \to \infty$ as $j \to \infty$. But then $\text{Cl}\{x_j, j \geq 1\}$ is a compact set with $\gamma^+\{\text{Cl}\{x_j, j \geq 1\}\}$ unbounded. This is a contradiction. Thus, there is a neighborhood V of J such that $\gamma^+(V)$ is bounded. Since J attracts points of X and T is continuous, for any $x \in X$, there is a neighborhood O_x of x and an integer n_0 such that $T^n O_x \subset \gamma^+(V)$ for $n \geq n_0$; that is, $\gamma^+(V)$ attracts O_x. For any compact set H in X, one can find a finite covering of H and thus an open set H_1 containing H such that $\gamma^+(V)$ attracts H_1. Thus, T is locally compact dissipative.

Now suppose that orbits of bounded sets are bounded. Then orbits of compact sets are bounded and T is compact dissipative from the above proof. Lemma 2.4.4 and Theorem 2.4.2 imply the existence of a maximal compact invariant set A that attracts bounded sets. Thus, T is bounded dissipative and the lemma is proved.

As a consequence of Lemmas 2.4.5, 2.4.4, and Theorem 2.4.2, we have the following result which is very useful in the applications.

THEOREM 2.4.6. *If T is asymptotically smooth, point dissipative, and orbits of bounded sets are bounded, then there exists a connected global attractor A.*

From Lemma 2.3.4, β-contractions are asymptotically smooth. As we shall see later, this case occurs often in applications. The special case in which T is completely continuous also occurs in several applications. For such mappings, Theorem 2.4.6 can be improved to

THEOREM 2.4.7. *If $T: X \to X$ is completely continuous and point dissipative, then there is a connected global attractor A.*

PROOF. From Lemma 2.4.5, it is only necessary to show that the orbits of bounded sets are bounded. Since TC is precompact for C bounded, it is only necessary to show that orbits of compact sets are bounded. Suppose H is any compact set. Since T is point dissipative, there is a bounded set B which may be assumed to be open such that, for any $x \in H$, there is an integer $n_0 = n_0(x, B)$ with the property that $T^n x \in B$ for $n \geq n_0$. By continuity of T, there is a neighborhood O_x of x such that $T^{n_0} O_x \in B$. Let $\{O_{x_i}, i = 1, 2, \ldots, p\}$ be a finite cover of H, let $N(H) = \max n_0(x_i, B)$, let $K = \text{Cl}\,T(B)$ and let $\overline{K} = \bigcup_{n=1}^{N(K)} T^n(B)$. Then $T^n B \subset \overline{K}$ for $n \geq 1$. Also, $T^n H \subset \overline{K}$ for $n \geq n_0(H) + 1$ and $\gamma^+(H)$ is bounded. This proves the theorem.

The following result is the global analogue of Theorem 2.2.5.

THEOREM 2.4.8. *If T is conditionally α-condensing and J is a compact invariant set that attracts points, then the following statements are equivalent*:
  (i) *J attracts compact sets*,
  (ii) *J is stable*,
  (iii) *J is a maximal compact invariant set*.

PROOF. Theorem 2.4.2 and the fact that J is compact and invariant imply that (ii) and (iii) are a consequence of (i).

We now prove (ii) implies (i). Suppose (ii) is satisfied and H is an arbitrary compact set of X. Since J attracts points, for any neighborhood U of J, there is an n_0 such that $T^{n_0} H \subset U$. Thus, we may suppose without loss of generality that $H \subset U$. As in the proof of Lemma 2.2.2, one shows that J attracts H. Thus, (ii) implies (i).

In order to prove that (iii) implies (ii), we need the following construction. Fix $\varsigma > 0$. We claim there is an $\varepsilon > 0$ such that $TN(J,\varepsilon) \subset N(J,\varsigma)$. To prove this, suppose $x \in J$, define $\eta(x) = \sup\{\delta : TN(x,\delta) \subset N(J,\varsigma)\}$ and let $\varepsilon = \inf_{x \in J} \eta(x)$. If $\varepsilon = 0$, then there is a sequence $\{x_i\} \subset J$ such that $\eta(x_i) \to 0$ as $i \to \infty$. Since J is compact, we may suppose that $x_i \to x_0$ as $i \to \infty$. If $\eta(x_0) > 0$, then, for all x_j such that $|x_j - x_0| < \eta(x_0)/2$, we have $\eta(x_i) > \eta(x_0)/2$ by the definition of $\eta(x)$. This contradicts the fact that $\eta(x_i) \to 0$ as $i \to \infty$. Thus $\varepsilon = \inf_{x \in J} \eta(x) > 0$ and the construction is completed.

Now, assume that (iii) is satisfied and J is not stable. Then there is an $\varepsilon > 0$, which we may choose so that $TN(J,\varepsilon) \subset N(J,\varsigma)$, a sequence of points $x_i \in N(J,\varepsilon/2^i)$ for which there exist integers $n_i = n_i(x_i)$ such that $T^j x_i \in N(J,\varepsilon)$ for $0 \leq j < n_i$, and $T^{n_i} x_i \notin N(J,\varepsilon)$. Since J is compact, we may also assume that $x_i \to x_0 \in J$ as $i \to \infty$. Let $A = \{T^j x_i, 0 \leq j < n_i, i > 0\}$, $B = \{x_i, i > 0\}$. Since $A \subset N(J,\varepsilon)$, we have $TA \subset N(J,\varsigma)$ and TA is bounded. Furthermore, since $\alpha(B) = 0$,

$$\alpha(A) \leq \alpha(B \cup (T(A) \cap A)) \leq \max(\alpha(B), \alpha(TA)) = \alpha(T(A)) < \alpha(A)$$

if A is not precompact. This is impossible so that A is precompact and we can assume without loss of generality that $T^{n_i} x_i \to y_0 \in \text{Cl } N(J,\varsigma)$ as $i \to \infty$. Since $x_i \to x_0 \in J$, J is invariant, and $y_0 \notin N(J,\varepsilon)$, it is clear that $n_i \to \infty$ as $i \to \infty$.

We now construct a negative orbit through y_0 which belongs to $N(J,\varsigma)$. Since $n_i \to \infty$ as $i \to \infty$, for any integer p, $n_i - p$ will be nonnegative for all but a finite number of i. Thus, for $p = 1$, $T^{n_i-1} x_i$ will be well defined for all but a finite number of terms. Thus, we can take a subsequence $\{n_{i,1}\} \subset \{n_i\}$, $\{x_{i,1}\} \subset \{x_i\}$ such that $T^{n_{i,1}} x_{i,1} \to y_1 \in \text{Cl } N(J,\varsigma)$ as $i \to \infty$. Continuity of T implies that $Ty_1 = y_0$.

This process can be continued for any integer p; that is, we can select sequences $\{n_{i,p}\} \subset \{n_{i,p-1}\}$, $\{x_{i,p}\} \subset \{x_{i,p-1}\}$ such that $T^{n_{i,p}-p} x_{i,p} \to y_p \in \text{Cl } N(J,\varsigma)$, and $Ty_{p+1} = y_p$. Since J attracts points, $\gamma^+(y_0)$ is bounded and so $K = \{y_j, j \geq 1\} \cup \gamma^+(y_0)$ is bounded. Clearly, $TK \subset K$ and T being α-condensing implies that

K is precompact. Thus, Cl K is compact and invariant and contains points which are not in J. But J is maximal, compact, and invariant. This contradiction proves that J is stable. This completes the proof of the theorem.

One of the important implications of Theorems 2.4.8 and 2.4.2 is the following corollary.

COROLLARY 2.4.9. *If T is α-condensing, point dissipative, and has a maximal compact, invariant set J, then J is uniformly asympotically stable, connected, and attracts neighborhoods of compact sets.*

2.5. Dependence on parameters.

Suppose Λ is a metric space, X is a complete metric space, and $T\colon \Lambda \times X \to X$ is continuous. For any $\lambda \in \Lambda$, let $T_\lambda = T(\lambda, \cdot) \colon X \to X$. If each T_λ has a maximal compact invariant set J_λ, we want to find additional conditions on T_λ to ensure that J_λ is *upper-semicontinuous* in λ at each $\lambda_0 \in \Lambda$; that is, $\delta(J_\lambda, J_{\lambda_0}) \to 0$ as $\lambda \to \lambda_0$ where the distance δ is defined in (2.1).

Another way to define upper semicontinuity of J_λ at $\lambda = \lambda_0$ is the following: Let $J^+ = \bigcap_{n \geq 1} \text{Cl} \bigcup_{|\lambda - \lambda_0| < 1/n} J_\lambda$. Then J_λ is upper semicontinuous at λ_0 if and only if $J^+ \subset \text{Cl}\, J_{\lambda_0}$. The set J^+ is similar to the ω-limit set of J_λ considering λ as the "time" parameter. This remark was pointed out by Yulin Cao.

Recall that T_λ is asymptotically smooth if, for any nonempty closed bounded set B such that $T_\lambda B \subset B$, there is a nonempty compact set $K_\lambda(B)$ in B such that $K_\lambda(B)$ attracts B under T_λ. This is equivalent to the following statement: for any nonempty closed bounded set B, there is a compact set $J_\lambda(B)$ such that $J_\lambda(B)$ attracts $L_\lambda(B) = \{x \in B\colon T_\lambda^n x \in B, n \geq 0\}$ under T_λ. The family of mappings $\{T_\lambda; \lambda \in \Lambda\}$ is *collectively asymptotically smooth* if each T_λ is asymptotically smooth and $\bigcup_{\lambda \in \Lambda} K_\lambda(B)$ (or equivalently $\bigcup_{\lambda \in \Lambda} J_\lambda(B)$) is precompact for each nonempty closed bounded set $B \subset X$. The family of mappings $\{T_\lambda, \lambda \in \Lambda\}$ is *collectively β-condensing* if each T_λ is β-condensing, and for any bounded set B, $\beta(B) > 0$, one has $\bigcup_{\lambda \in \Lambda} T_\lambda B$ bounded and $\beta(\bigcup_{\lambda \in \Lambda} T_\lambda B) < \beta(B)$.

LEMMA 2.5.1. *If the family $\{T_\lambda, \lambda \in \Lambda\}$ is collectively β-condensing, then it is collectively asymptotically smooth.*

PROOF. From Lemma 2.3.5, each T_λ is asymptotically smooth. Suppose B is a nonempty closed bounded set, $T_\lambda B \subset B$ for each $\lambda \in \Lambda$, and let K_λ be a compact set in B that attracts B under T_λ. Since B is a complete metric space, we can apply Theorem 2.4.6 to the space B and obtain a compact attractor in B which attracts B under T_λ. Thus, we can take K_λ to be this compact attractor. In particular, $T_\lambda K_\lambda = K_\lambda$. If $V = \bigcup_{\lambda \in \Lambda} K_\lambda$, then V is bounded. Also K_λ invariant under T_λ implies

$$\beta(V) = \beta\left(\bigcup_{\lambda \in \Lambda} T_\lambda K_\lambda\right) \leq \beta\left(\bigcup_{\lambda \in \Lambda} T_\lambda V\right).$$

Since the family $\{T_\lambda, \lambda \in \Lambda\}$ is collectively β-condensing, this implies that V is precompact, and thus, the family of maps is collectively asymptotically smooth. This completes the proof.

THEOREM 2.5.2. *Let X be a complete metric space, Λ a metric space, $T: \Lambda \times X \to X$ be continuous, $T_\lambda = T(\lambda, \cdot)$, and suppose there is a bounded set $B \subset X$ independent of $\lambda \in \Lambda$ such that B attracts compact sets of X under T_λ. If the family $\{T_\lambda, \lambda \in \Lambda\}$ is collectively asymptotically smooth, then the maximal compact inariant set A_λ of T_λ is upper semicontinuous in λ.*

PROOF. Without loss in generality suppose B is closed. Theorem 2.4.2 and Lemma 2.4.4 imply that each A_λ exists. The hypothesis that B attracts compact sets implies that $A_\lambda \subset B$ and $V = \bigcup_{\lambda \in \Lambda} A_\lambda$ is bounded. Also, the hypothesis of collectively asymptotically smooth implies that V is precompact.

To prove the upper semicontinuity of A_λ, we must show that if $\{\lambda_j, j \geq 1\} \subset \Lambda$, $\{x_j, j \geq 1\} \subset X$ are sequences such that $x_j \in A_{\lambda_j}$ for each j, $\lambda_j \to \lambda_0$, $x_j \to x_0$ as $j \to \infty$, then $x_0 \in A_{\lambda_0}$. Since A_{λ_j} is invariant under T_{λ_j}, there is a $y_j \in A_{\lambda_j}$ such that $x_j = T_{\lambda_j} y_j$. Since V is precompact and B is closed, we may assume without loss in generality that there is a $y_0 \in B$ such that $y_j \to y_0$ as $j \to \infty$. Since T is continuous, $T_{\lambda_0} y_0 = x_0$. In the same way, for any integer $n \geq 0$, there is a sequence $y_j^{(n)} \in A_{\lambda_j}$ and $y_0^{(n)} \in B$ such that $T^n y_j^{(n)} = x_j$, $y_j^{(n)} \to y_0^{(n)}$ as $j \to \infty$, and $T_{\lambda_0}^{(n)} y_0^{(n)} = x_0$. Since the set $\{y_j^{(n)}, j \geq 1, n \geq 0\}$ belongs to a compact set, we can use the triangularization procedure to obtain, for the map T_{λ_0}, a complete orbit $\gamma(x_0)$ through x_0, $\gamma(x_0) \subset B$. From Corollary 2.2.4, $\gamma(x_0)$ is precompact. Since $\gamma(x_0)$ is invariant under T_{λ_0} and A_{λ_0} is the maximal compact invariant set of T_{λ_0}, it follows that $\gamma(x_0) \subset A_{\lambda_0}$. In particular, $x_0 \in A_{\lambda_0}$. This proves the theorem.

THEOREM 2.5.3. *Let X be a complete metric space, Λ a metric space, $T: X \times \Lambda \to X$ be continuous, $T_\lambda = T(\lambda, \cdot)$, and suppose there is a bounded set B that attracts points of X under T_λ for each $\lambda \in \Lambda$ and, for any bounded set U, the set $V = \bigcup_{\lambda \in \Lambda} \bigcup_{n \geq 0} T_\lambda^n U$ is bounded. If the family $\{T_\lambda, \lambda \in \Lambda\}$ is collectively asymptotically smooth, then the global attractor A_λ for T_λ is upper semicontinuous.*

PROOF. Suppose H is an arbitrary compact set in X. Then A_λ attracts H under T_λ for each $\lambda \in \Lambda$ and so V attracts H under T_λ for each $\lambda \in \Lambda$. Theorem 2.5.2 implies the conclusion of the theorem.

Theorem 2.5.3 deals with the upper semicontinuity of the global attractor A_λ at λ_0. With appropriate modifications of the above arguments, one can prove the following result for local attractors.

THEOREM 2.5.4. *Let X be a complete metric space, Λ be a metric space, and $T: X \times \Lambda \to X$ be continuous and each $T_\lambda = T(\lambda, \cdot)$ be asymptotically smooth. Suppose one of the following conditions is satisfied.*

(i) A_λ is a local attractor for T_λ, there is a bounded set U in X such that $\bigcup_{\lambda \in \Lambda} A_\lambda \subset U$ and $T_\lambda \to T_{\lambda_0}$ pointwise as $\lambda \to \lambda_0$.

(ii) A_{λ_0} is a local attractor for T_{λ_0} and $T_\lambda y - T_{\lambda_0} y \to 0$ as $\lambda \to \lambda_0$ uniformly for y in a bounded neighborhood of A_0.

Then A_λ is upper semicontinuous at $\lambda = \lambda_0$.

REMARK 2.4.5. The above results on the upper semicontinuity of the attractor are easy to obtain because the mapping T_λ is continuous in λ. In applications, many of the parameters λ that occur are such that T_λ is continuous in λ. However, there are important cases where this is not true, for example, when the parameter λ corresponds to a singular perturbation problem in ordinary or partial differential equations or to a finite difference approximation to an evolutionary equation. Results on upper semicontinuity in cases of this type are much more difficult to obtain and require more detailed knowledge of the maps. Some results along this line will be presented in §4.10.

2.6. Fixed point theorems.

Our purpose in this section is to prove the following fixed point theorems.

THEOREM 2.6.1. *Suppose X is a Banach space. If $T: X \to X$ is α-condensing and compact dissipative, then T has a fixed point.*

From this result and Lemmas 2.3.5 and 2.4.5, we have

COROLLARY 2.6.2. *If $T: X \to X$ is α-condensing, point dissipative, and the orbit of any compact set is bounded, then T has a fixed point.*

Since a completely continuous map is α-condensing, Lemma 2.3.5, Theorem 2.6.1, and Theorem 2.4.7 imply

COROLLARY 2.6.3. *Suppose X is a Banach space. If $T: X \to X$ is completely continuous and point dissipative, then T has a fixed point.*

The proof of Theorem 2.6.1 makes use of the following result of Horn [1970] which is stated without proof.

LEMMA 2.6.4. *Let $S_0 \subset S_1 \subset S_2$ be convex subsets of a Banach space X with S_0, S_2 compact and S_1 open in S_2. Let $T: S_2 \to X$ be a continuous map such that, for some integer $m > 0$, $T^j(S_1) \subset S_2$, $0 \leq j \leq m-1$, $T^j(S_1) \subset S_0$, $m \leq j \leq 2m-1$. Then T has a fixed point.*

We need the following result.

LEMMA 2.6.5. *Suppose $K \subset B \subset S \subset X$ are convex subsets of a Banach space X with K compact, S closed and bounded, and B open in S. If $T: S \to X$ is continuous, $\gamma^+(B) \subset S$, and K attracts points of G, then there is a closed bounded subset J of S such that*

$$(6.1) \qquad J = \overline{co}\,\gamma^+(T(B \cap J)), \qquad J \cap K \neq \emptyset.$$

If, in addition, K attracts compact sets of B and J is compact, then T has a fixed point.

PROOF. Let F be the set of convex, closed, bounded subsets, of S such that $T^j(B \cap L) \subset L$ for $j \geq 1$ and $L \cap K \neq \emptyset$. The family F is not empty since $S \subset F$.

If $L \in F$, let $L_1 = \overline{co}\,\gamma^+(T(B \cap L))$. Since K attracts points of B and K is compact, there is a sequence $x_n \in L_1$ such that $x_n \to K$ as $n \to \infty$. Therefore, $(\text{Cl}\, L_1) \cap K \neq \emptyset$. But L_1 is closed so that $L_1 \cap K \neq \emptyset$. Also, L_1 is convex, closed, and $L_1 \subset S$. Since $L \subset F$, we have $L \supset L_1$ and $L_1 \supset T^j(B \cap L) \supset T^j(B \cap L_1)$ for all $j \geq 1$. Thus $L_1 \subset F$. It follows that a minimal element of F will satisfy the conditions of the lemma.

To prove such a minimal element exists, let $\{L_\alpha\}_{\alpha \in I}$ be a totally ordered family of sets in F. The set $L = \bigcap_{\alpha \in I} L_\alpha$ is closed, convex, and contained in S. Also, $T^j(B \cap L) \subset T^j(B \cap L_\alpha) \subset L_\alpha$ for all $\alpha \in I$ and $j \geq 1$. Thus, $T^j(B \cap L) \subset L$ for $j \geq 1$. If I^* is any finite subset of I, then by the same reasoning as for the set L_1 above, we have $K \cap (\bigcap_{\alpha \in I^*} L_\alpha) \neq \emptyset$. From compactness of K, it follows that $K \cap (\bigcap_{\alpha \in I} L_\alpha) \neq \emptyset$. Thus $L \in F$ and Zorn's lemma yields the existence of a minimal element J of F and it satisfies (6.1).

To prove the last part of the lemma, suppose that K attracts compact sets of B and J in (6.1) is compact. Without loss in generality, we can take $B = N(K, \varepsilon) \cap S$ for some $\varepsilon > 0$, since K is compact and convex. Here, $N(K, \varepsilon)$ is the ε-neighborhood of K. Let $S_0 = J \cap \text{Cl}\, N(K, \varepsilon/2)$, $S_1 = J \cap N(K, \varepsilon)$, and $S_2 = J$. Then $S_0 \subset S_1 \subset S_2$ are convex subsets of X with S_0, S_2 compact and S_1 open in S_2. Also, $T^j S_1 = T^j(J \cap N(K, \varepsilon)) \subset J$, $j \geq 1$, $\gamma^+(S_1) \subset J$. Therefore, $T^j S_1 \subset S_2$, $j \geq 1$. Also, the fact that K attracts compact sets of B implies there is an integer $n_1 = n_1(K, \varepsilon)$ such that $T^j S_1 \subset N(K, \varepsilon/2)$ for $j \geq n_1$. If $n_1 \geq 1$, then $T^j S_1 \subset J$ for $j \geq n_1$ and so $T^j S_1 \subset S_0$ for $j \geq n_1$. We can now apply Lemma 2.6.4 of Horn to obtain the existence of a fixed point of T. This completes the proof of the lemma.

LEMMA 2.6.6. *If T is α-condensing, then the set J in (6.1) is compact.*

PROOF. If $\tilde{J} = \gamma^+(T(B \cap J))$, then $\tilde{J} = T(B \cap J) \cup T(\tilde{J})$ and $\alpha(J) = \alpha(\tilde{J}) = \max[\alpha(T \cap J), \alpha(\tilde{J})]$. If $\alpha(\tilde{J}) > 0$, then T being α-condensing implies $\alpha(T\tilde{J}) < \alpha(\tilde{J})$. In this case, $\alpha(\tilde{J}) = \alpha(T(B \cap J))$. If $\alpha(T(B \cap J)) > 0$, then $\alpha(J) = \alpha(\tilde{J}) < \alpha(B \cap J) \leq \alpha(J)$ and this is a contradiction. Thus, $\alpha(\tilde{J}) = \alpha(J) = 0$. Since J is closed, this implies J is compact and the lemma is proved.

PROOF OF THEOREM 2.6.1. Let A be the maximal compact invariant set of Theorem 2.4.2 and let $K = \overline{co}\, A$. Part (ii) of Theorem 2.4.2 implies there is a convex neighborhood B of K such that $\gamma^+(B)$ is bounded and K attracts B. If $S = \overline{co}\,\gamma^+(B)$, then S, B, K satisfy the conditions of Lemma 2.6.5. Thus, J in (6.1) exists. Lemma 2.6.6 implies J is compact and Lemma 2.6.5 completes the proof of Theorem 2.6.1.

Question. If T is asymptotically smooth, is it possible to choose J in (6.1) so that it is compact? If so, Theorem 2.6.1 can be generalized.

In case T is a linear map satisfying some additional properties, one can prove a more general result than Theorem 2.6.1. To obtain such a result, we need the following Schauder-Tychonov theorem which is stated without proof.

THEOREM 2.6.7. *If Γ is a compact convex subset of a Banach space and $T: \Gamma \to \Gamma$ is continuous, then T has a fixed point in Γ.*

We can now prove the following result which will be useful for nonhomogeneous linear evolutionary equations.

THEOREM 2.6.8. *If X is a Banach space, $L: X \to X$ is linear and continuous, $y \in X$ is given, $T: X \to X$ is defined by $Tx = Lx + y$, and there is an $x_0 \in X$ such that $\bigcup_{n \geq 0} T^n x_0$ has compact closure, then T has a fixed point.*

PROOF. By hypothesis the set $\{T^k x_0, k \geq 0\}$ is bounded in X. Let $D = \mathrm{co}\{x_0, Tx_0, T^2 x_0, \ldots\}$. If $z \in D$, then $z = \sum_{i \in J} \alpha_i T^i x_0$ where J is a finite subset of the positive integers, $\alpha_i \geq 0$, $\sum_i \alpha_i = 1$. Furthermore,

$$Tz = Lz + y = \sum_{i \in J} \alpha_i L T^i x_0 + \sum_{i \in J} \alpha_i y = \sum_{i \in J} \alpha_i [L T^i x_0 + y] = \sum_{i \in J} \alpha_i T^{i+1} x_0 \in D$$

and $TD \subset D$. Since T is continuous, $T(\mathrm{Cl}\, D) \subset \mathrm{Cl}\, D$. Since $\mathrm{Cl}\, D$ is a compact convex set of X and T is continuous, there is a fixed point of T in $\mathrm{Cl}\, D$ by Theorem 2.6.7. This proves the result.

COROLLARY 2.6.9. *Suppose T is given as in Theorem 2.6.8 with L an α-condensing map. If there is an $x_0 \in X$ such that $\bigcup_{n \geq 0} T^n x_0$ is bounded, then T has a fixed point.*

PROOF. Since T is α-condensing, $\bigcup_{n \geq 0} T^n x_0$ has compact closure and one can apply Theorem 2.6.8.

2.7. Stability relative to the global attractor and Morse-Smale maps

The primary objective in the qualitative theory of discrete dynamical systems $T: X \to X$ is to study the manner in which the flow changes when T changes. Due to the infinite dimensionality of the space and the fact that T may not be one-to-one, a comparison of all orbits of two different maps is very difficult and is likely to lead to severe restrictions on the maps that are to be considered. For this reason, we restrict our discussion to mappings which have a global attractor and then make comparisons of orbits on the attractors.

Let $C^r(X, X)$, $r \geq 1$, be the space of C^r maps from X to X. Let $KC^r(X, X)$ be the subset of $C^r(X, X)$ with the property that

(i) $T \in KC^r(X, X)$ implies that T has a global attractor $A(T)$.

(ii) $A(T)$ is upper semicontinuous on $KC^r(X, X)$.

For $T, S \in KC^r(X, X)$, we say that T is *equivalent* to S, $T \sim S$, if there is a homeomorphism $h: A(T) \to A(S)$ such that $hT = Sh$ on $A(T)$. We say T is *A-stable* if there is a neighborhood V of T in $C^r(X, X)$ such that $T \sim S$ for every $S \subset V \cap KC^r(X, X)$.

The nonwandering set $\Omega(T)$ of $T \in KC^r(X, X)$ is the set of all $z \in A(T)$ such that, given a neighborhood V of z in $A(T)$ and an integer n_0, there is an $n > n_0$ such that $T^n(V) \cap V \neq \emptyset$. If $T \in KC^r(X, X)$, then $\Omega(T)$ is compact and, if T is one-to-one on $A(T)$, then $\Omega(T)$ is invariant. The proof of this latter fact is similar to the proof of Lemma 2.1.1.

A fixed point x of a map $T \in C^r(X, X)$ is *hyperbolic* if the spectrum of $DT(x)$ does not intersect the unit circle in $\mathbb{C}$ with center zero. For any hyperbolic fixed

point of T, let
$$W^s(x,T) = \{y \in X : T^n y \to x \text{ as } n \to \infty\},$$
$$W^u(x,T) = \{y \in X : T^{-n} y \text{ is defined for } n \geq 0 \text{ and } T^{-n} y \to x \text{ as } n \to \infty\}.$$

The sets $W^s(x,T)$, $W^u(x,T)$ are called respectively the stable and unstable sets of T. If x is a hyperbolic fixed point of T, then there is a neighborhood V of x such that
$$W_{\text{loc}}^s(x,T) \stackrel{\text{def}}{=} W^s(x,T,V) \stackrel{\text{def}}{=} \{y \in W^s(x,T) : T^n y \in V, n \geq 0\},$$
$$W_{\text{loc}}^u(x,t) \stackrel{\text{def}}{=} W^u(x,T,V) \stackrel{\text{def}}{=} \{y \in W^u(x,T) : T^{-n} y \in V, n \geq 0\}$$
are C^r-manifolds and will be referred to as the local stable and unstable manifolds. If the maps T and DT are one-to-one on X, then $W^s(x,T)$ and $W^u(x,T)$ are C^r-manifolds immersed in X (see the Appendix).

A point $x \in X$ is a *periodic point of period p of T* if $T^p x = x$, $T^j x \neq x$, $j = 1, 2, \ldots, p-1$. A periodic point of period p is hyperbolic if the spectrum of $DT^p(x)$ does not intersect the unit circle in $\mathbb{C}$ with center zero. The stable and unstable sets have the properties mentioned before.

A map $T \in KC^r(X,X)$ is *Morse-Smale* if

(i) T, DT are one-to-one on $A(T)$.

(ii) $\Omega(T)$ is finite and so consists of the periodic points of $T \stackrel{\text{def}}{=} \text{Per}(T)$.

(iii) All periodic points are hyperbolic with finite dimensional unstable manifolds.

(iv) $W^s(x,T)$ is transversal to $W^u(y,T)$ for all $x, y \in \text{Per}(T)$.

A basic result stated without proof is

THEOREM 2.7.1. *If $T \in KC^r(X,X)$ is Morse-Smale, then T is A-stable.*

For a Morse-Smale $T \in KC^r(X,X)$, the global attractor $A(T)$ can be written as
$$A(T) = \bigcup_{x \in \text{Per}(T)} W^u(x,T).$$

2.8. Dimension of the global attractor.

In this section, we record some results on the dimension of the attractor. The proofs can be obtained from the references.

Let K be a topological space. We say that K is *finite dimensional* if there exists an integer n such that, for every open covering A of K, there exists another open covering A' refining A such that every point of K belongs to at most $n+1$ sets of A'. In this case, the *dimension* of K, $\dim K$, is defined as the minimum n satisfying this property. Then $\dim R^n = n$ and, if K is a compact finite dimensional space, it is homeomorphic to a subset of R^n with $n = 2 \dim K + 1$. If K is a metric space, its *Hausdorff dimension* is defined as follows: for any $\alpha > 0$, $\varepsilon > 0$, let
$$\mu_\varepsilon^\alpha(K) = \inf \sum_i \varepsilon_i^\alpha$$

where the inf is taken over all coverings $B_{\varepsilon_i}(x_i)$, $i = 1, 2, \ldots$, of K with $\varepsilon_i < \varepsilon$ for all i, where $B_{\varepsilon_i}(x_i) = \{x: d(x, x_i) < \varepsilon_i\}$. Let $\mu^\alpha(K) = \lim_{\varepsilon \to 0} \mu^\alpha_\varepsilon(K)$. The function μ^α is called the *Hausdorff measure of dimension* α. For $\alpha = n$ and K a subset of R^n with $|x| = \sup |x_j|$, μ^n is the Lebesgue exterior measure. It is not difficult to show that, if $\mu^\alpha(K) < \infty$ for some α, then $\mu^{\alpha_1}(K) = 0$ if $\alpha_1 > \alpha$. Thus,

$$\inf\{\alpha: \mu^\alpha(K) = 0\} = \sup\{\alpha: \mu^\alpha(K) = \infty\}$$

and we define the Hausdorff dimension of K as

$$\dim_H(K) = \inf\{\alpha: \mu^\alpha(K) = 0\}.$$

It is known that $\dim(K) \leq \dim_H(K)$ and these numbers are equal when K is a submanifold of a Banach space. For general K, there is little that can be said relating these numbers.

To define another measure of the size of a metric space K, let $n(\varepsilon, K)$ be the minimum number of open balls of radius ε needed to cover K. Define the *limit capacity* $c(K)$ of K by

$$c(K) = \limsup_{\varepsilon \to 0} \frac{\log n(\varepsilon, K)}{\log(1/\varepsilon)}.$$

In other words, $c(K)$ is the minimum real number such that, for every $\sigma > 0$, there is a $\delta > 0$ such that

$$n(\varepsilon, K) \leq \left(\frac{1}{\varepsilon}\right)^{c(K)+\sigma} \quad \text{if } 0 < \varepsilon < \delta.$$

It is not difficult to show that $\dim_H(K) \leq c(K)$.

An important result in the applications is

THEOREM 2.8.1. *Suppose X is a Banach space, $T: X \to X$ is an α-contraction, point dissipative and orbits of bounded sets are bounded. Then the global attractor A has the following properties:*

(i) $c(A) < \infty$.

(ii) *If $d = 2\dim_H(A) + 1$ and S is any linear subspace of X with $\dim S \geq d$, then there is a residual set Π of the space of all continuous projections P of X onto S (taken with the uniform operator topology) such that $P \,|\, A$ is one-to-one for every $P \in \Pi$.*

The first part of this theorem says that the limit capacity $c(A)$ of A is finite which in turn implies that the Hausdorff dimension $\dim_H(A)$ is finite. The second part of the theorem says that the attractor can be "flattened" in a residual set of directions onto a finite dimensional subspace of dimension $2\dim_H(A) + 1$.

The proof of Theorem 2.8.1 will not be given. However, it is worthwhile pointing out that in proving that $c(A)$ is finite, explicit estimates are obtained. The estimates for the limit capacity of a compact attractor for a map T can be obtained by an application of general results for the capacity of compact subsets of a Banach space X with the property that $T(K) \supset K$ for some C^1

map $T: U \to E$, $U \supset K$, whose derivative can be decomposed as a sum of a compact map and a contraction—a special case of an α-contraction.

To describe the nature of these estimates, we need some notation. For $\lambda > 0$, the subspace of $\mathcal{L}(E)$ consisting of all maps $L = L_1 + L_2$ with L_1 compact and $\|L_2\| < \lambda$ is denoted by $\mathcal{L}_\lambda(E)$. Given a map $L \in \mathcal{L}_\lambda(E)$ we define $L_S = L$ restricted to S and

$$\nu_\lambda(L) = \min\{\dim S : S \text{ is a linear subspace of } E \text{ and } \|L_S\| < \lambda\}.$$

It is easy to prove that $\nu_\lambda(L)$ is finite for $L \in \mathcal{L}_{\lambda/2}(E)$.

The basic result for the estimate of $c(K)$ is contained in

THEOREM 2.8.2. *Let X be a Banach space, $U \subset E$ an open set, $T: U \to E$ a C^1 map, and $K \subset U$ a compact set such that $T(K) \supset K$. If the Fréchet derivative $D_xT \in \mathcal{L}_{1/4}(E)$ for all $x \in K$, then*

$$c(K) \leq \frac{\log\{\nu[2(\lambda(1+\sigma) + k^2)/\lambda\sigma]^\nu\}}{\log[1/2\lambda(1+\sigma)]}$$

where we have $k = \sup_{x \in K} \|D_xT\|$, $0 < \lambda < 1/2$, $0 < \sigma < (1/2\lambda) - 1$, $\nu = \sup_{x \in K} \nu_\lambda(D_xT^2)$. If $D_xT \in \mathcal{L}_1(E)$ for all $x \in K$, then $c(K) < \infty$.

2.9 Dissipativeness in two spaces.

In the applications, it often happens that the mapping T under consideration is defined on two Banach spaces with one compactly imbedded in the other. Under these circumstances, one sometimes can obtain much more information about the asymptotic behavior of iterates of T. This section is devoted to obtaining results in this direction.

Throughout this section X_1, X_2 will denote Banach spaces with norms respectively given by $|\cdot|_1, |\cdot|_2$. We also assume

(H_1) $i: X_1 \to X_2$ is a compact imbedding.

(H_2) $T = S + U$ where T, S, U are continuous operators mapping each X_j into itself $j = 1, 2$, $S(0) = 0$, and S is a contraction on X_1.

(H_3) If $B \subset X_1$, B and $U(B)$ are bounded in X_2, then $U(B)$ is bounded in X_1.

THEOREM 2.9.1. *Suppose $(H_1), (H_2)$, and (H_3) are satisfied. Then the following statements are equivalent:*

(i) *T is point dissipative in X_1,*
(ii) *T is bounded dissipative in X_1,*
(iii) *there is a bounded set in X_2 that attracts points in X_1.*

Using Theorem 2.9.1, Lemma 2.3.4, Theorem 2.4.6, Lemma 2.4.1, and Theorem 2.6.1, we obtain the following interesting corollary.

COROLLARY 2.9.2. *If Hypotheses* $(H_1), (H_2),$ *and* (H_3) *are satisfied,* (iii) *of Theorem* 2.9.1 *is satisfied, and* T *is an* α-*contraction in* X_1, *then there exists a connected global attractor* A *in* X_1. *Furthermore, there is a fixed point of* T *in* X_1.

Under a slightly stronger hypothesis on the map T, we can prove the following result.

THEOREM 2.9.3. *Suppose* $(H_1), (H_2),$ *and* (H_3) *are satisfied and* S *is also a contraction on* X_2. *If either*

(i) X_1 *is dense in* X_2

or

(ii) $U: X_2 \to X_1$ *is such that, if* B *and* $U(B)$ *are bounded in* X_2, *then* $U(B)$ *is bounded in* X_1,

then U *is conditionally completely continuous on* X_2 *and* T *is a conditional* α-*contraction on* X_2. *Also, if* J *is a compact invariant set in* X_2, *then* J *is the closure in* X_2 *of a bounded set in* X_1. *If, in addition,* T *is point dissipative in* X_2, *then there is a maximal, compact invariant set in* X_2 *which is u.a.s. and attracts a neighborhood of any compact set of* X_2. *In particular,* T *is compact dissipative.*

THEOREM 2.9.4. *Suppose* (H_1) *and* (H_2) *are satisfied and*

(i) S *is a linear contraction on both spaces.*

(ii) $U: X_2 \to X_1$ *is continuous and, if* $B, U(B)$ *are bounded in* X_2, *then* $U(B)$ *is bounded in* X_1.

(iii) U *is conditionally completely continuous on* X_1 *and* X_2.

Under these conditions, if J *is a closed bounded invariant set in* X_2, *then* $J \subset X_1$ *and* J *is a compact invariant set in* X_1.

Before proving these results, let us make a few remarks on the implications. Theorem 2.9.1 asserts that, under a very weak dissipative property in X_2 (namely, point dissipative with respect to points in X_1), one obtains that orbits of bounded sets in X_1 are bounded and, in addition, each such orbit enters a fixed bounded set and remains. Under the additional hypothesis that X_1 is dense in X_2, one is able to obtain the existence of a maximal compact invariant set which attracts neighborhoods of compact sets in X_2 only under the hypothesis that T is point dissipative in X_2. Theorem 2.9.4 is a type of regularity theorem asserting that closed bounded invariant sets in the big space X_2 must automatically be in the smaller space X_1 and be compact in X_1.

We state another consequence of the above theorems which will be especially useful in the discussion of partial differential equations.

COROLLARY 2.9.5. *Suppose*

(i) X_1 *is compactly embedded in* X_2, X_1 *dense in* X_2.

(ii) $T = S + U: X_j \to X_j$, $S: X_j \to X_j$ *is a linear contraction.*

(iii) $U: X_2 \to X_1$ *is continuous and, if* $B, U(B)$ *are bounded in* X_2, *then* $U(B)$ *is bounded in* X_1.

(iv) T is point dissipative in X_2.

Then there is a global attractor A in X_2 and $T(t)$ is bounded dissipative in X_1.

If, in addition,

(v) U is conditionally completely continuous on X_j, $j = 1, 2$, then $A \subset X_1$ and is a global attractor in X_1.

We now give the proofs of Theorems 2.9.1, 2.9.3, and 2.9.4. As notation, let $B_r^i = \{x \in X_i : |x|_i < r\}$. We need the following lemmas.

LEMMA 2.9.6. *If* $(H_1), (H_2), (H_3)$ *are satisfied, then, for any constant* $L > 0$ *and any* $A \subset X_1$ *such that* A *and* $U(A)$ *are bounded in* X_2, *there is a constant* $K(L, A)$ *such that, for any* n_0, $0 \leq n_0 < \infty$, *and for any* $B \subset B_L^1$, *if* $T^m(B) \subset A$ *for* $0 \leq m \leq n_0$, *then* $T^m(B) \subset B_{K(L,A)}^1$ *for* $0 \leq m \leq n_0$.

PROOF. From Hypothesis (H_3), there is an $M > 0$ such that $U(A) \subset B_M^1$. Suppose $T^m(B) \subset A$, $0 \leq m \leq n_0$, and let $\lambda \in [0, 1)$ be the contraction constant for S, $K = K(L, A) = (1 - \lambda)^{-1}(M + L)$. We use induction to show that $T^m(B) \subset B_K^1$ for $0 \leq m \leq n$ for each $n \leq n_0$. It is obviously true for $n = 0$. If it is assumed true for n, then, for any $x \in B$,

$$|T^{n+1}x|_1 = |(S+U)T^n x|_1 \leq |ST^n x|_1 + |UT^n x|_1$$
$$\leq \lambda |T^n x|_1 + M \leq \lambda(1-\lambda)^{-1}(M+L) + M$$
$$\leq \lambda(1-\lambda)^{-1}(M+L) + M + L \leq (1-\lambda)^{-1}(M+L) = K(L, A).$$

Thus, $T^m x \in B_K^1$ for $0 \leq m \leq n_0$ and the lemma is proved.

LEMMA 2.9.7. *Suppose* $(H_1), (H_2)$, *and* (H_3) *are satisfied. For any* $L > 0$ *and any* $A \subset X_1$ *for which* A, $U(A)$ *are bounded in* X_2, *there are constants* $n_1(L, A)$ *and* $Q(A)$ *such that, for any* n_0, $0 \leq n_0 < \infty$, *if* $T^m(B) \subset B_L^1 \cap A$, $0 \leq m \leq n_0$, *then* $T^m(B) \subset B_{Q(A)}^1$ *for* $n_1(L, A) \leq m \leq n_0$.

PROOF. Since (H_3) is satisfied, there is an $M > 0$ such that $U(A) \subset B_M^1$. Fix $\delta > 0$ and let $Q(A) = (1 - \lambda)^{-1}M + \delta$. If $B \subset X_i$ is bounded, let $|B|_i = \sup\{|x|_i : x \in B\}$. We also use the constant $K(L, A) = (1-\lambda)^{-1}(M+L)$ obtained in the proof of Lemma 2.9.6. If $L \leq K(0, A)$, let $n_1 = 0$. One can apply the same inductive argument as in the proof of Lemma 2.9.6 to obtain the conclusion of the lemma. In fact, if $L > K(0, A)$, choose n_1 so that $\lambda^{n_1}(L - K(0, A)) < \delta$. Then,

$$|T(B)|_1 = |(S+U)(B)|_1 \leq \lambda L + M = \lambda(L - K(0,A)) + \lambda(1-\lambda)^{-1}M + M$$
$$= \lambda(L - K(0,A)) + (1-\lambda)^{-1}M = \lambda(L - K(0,A)) + K(0,A).$$

By induction, for any n, $0 \leq n \leq n_0$, we obtain $|T^n(B)|_1 \leq \lambda^n(L - K(0,A)) + K(0, A)$. Hence, for $n_1 \leq n \leq n_0$, we have $|T^n B|_1 \leq K(0, A) + \delta = Q(A)$. This completes the proof of the lemma.

LEMMA 2.9.8. *Suppose* $(H_1)(H_2)$, *and* (H_3) *are satisfied. If there is a bounded set in* X_2 *which attracts points in* X_1, *then* T *is bounded dissipative in* X_1.

PROOF. We first show that T is point dissipative in X_1. Let B be a bounded set in X_2 which attracts points in X_1 and let $A = \{x \in B : T^n x \in B$ for $n \geq 0\}$. The set A is nonempty, attracts all points of X_1, and $T(A) \subset A$. If $x \in X_1$, there is an integer $n_1 = n_1(x, A)$ such that $T^n x \in A$ for $n \geq n_1$. Let $L_1 = |T^{n_1} x|_1$, $\tilde{A} = \bigcup_{n \geq n_1} T^n x$. If we apply Lemma 2.9.6 with $L = L_1$, $A = \tilde{A}$, and $B = \{T^{n_1} x\}$, then $|T^n x|_1 \leq K(L_1, \tilde{A})$ for $n \geq n_1$. Since $T(A) \subset A$, we can now apply Lemma 2.9.7 with L replaced by $K(L_1, \tilde{A})$ to obtain $|T^n x|_1 \leq Q(A)$ for $n \geq n_1(x, A)$; that is, $B^1_{Q(A)}$ attracts points of X_1.

Our objective is to show that $\gamma^+(B^1_{Q(A)})$ in X_1 attracts bounded sets of X_1. We first show that $\gamma^+(B^1_{Q(A)})$ is bounded in X_1. If $x \in B^1_{Q(A)}$, then $\gamma^+(x)$ is bounded in X_2. Let $C(x)$ be a neighborhood in X_2 of $\gamma^+(x)$ chosen so small that $U(C(x))$ is bounded in X_2. Since $T\gamma^+(x) \subset \gamma^+(x)$, we may apply Lemma 2.9.6 to obtain $\gamma^+(x) \subset B^1_{K(Q(A), C(x))}$. Since A attracts x, there is an $n_0(x)$ such that $T^n x \in A$ for $n \geq n_0(x)$. Let $n_1(x) = n_1(K(Q(A), C(x)), A)$ where $n_1(L, A)$ is defined in Lemma 2.9.7 and define $n^*(x) = n_0(x) + n_1(x)$. Since T is continuous, there is a $\delta(x)$ such that $T^m(B^2_\delta \cap B^1_{Q(A)}) \subset C(x)$ for $0 \leq m \leq n^*(x)$ and $T^m(B^2_\delta(x) \cap B^1_{Q(A)}) \subset A$ for $n_0(x) \leq m \leq n^*(x)$. Lemma 2.9.6 implies that $T^m(B^2_\delta(x) \cap B^1_{Q(A)}) \subset B^1_{K(Q(A), C(x))}$ and Lemma 2.9.7 implies that $T^{n^*(x)}(B^2_\delta(x) \cap B^1_{Q(A)}) \subset B^1_{Q(A)}$.

Since $B^1_{Q(A)}$ is a precompact set in X_2, the sets $\{B_\delta(x), x \in B^1_{Q(A)}\}$ form an open cover of $B^1_{Q(A)}$ in X_2 for which there is a finite subcover $\{B_\delta(x_i), i = 1, 2, \ldots, m\}$. If $N = \max\{n^*(x_i), 1 \leq i \leq m\}$, then it is clear that $\gamma^+(B^1_{Q(A)}) \subset \bigcup_{0 \leq m \leq N} T^m(B^1_{Q(A)})$ since any point in $B^1_{Q(A)}$ returns to $B^1_{Q(A)}$ by the Nth iteration of T.

A similar argument to the one above can be used to show that $\gamma^+(B^1_{Q(A)})$ attracts B^1_R for any $R > 0$. In fact, we form an open cover in X_2 of B^1_R by neighborhoods $\{B^2_\delta(x), x \in B^1_R\}$ such that there is an $n^*(x)$ with

$$T^{n^*(x)}(B^2_\delta(x) \cap B^1_R) \subset B^1_{Q(A)}.$$

Then there is a finite subcover $\{B^2_\delta(x_i), 1 \leq i \leq m\}$ and we can define $N = \max\{n^*(i), 1 \leq i \leq m\}$. One then shows that $n > N$ implies $T^n(B^1_R) \subset \gamma^+(B^1_{Q(A)})$. Hence, $\gamma^+(B^1_{Q(A)})$ attracts bounded sets in X_1. This completes the proof.

PROOF OF THEOREM 2.9.1. From Lemma 2.9.8, (iii)$\Rightarrow$(ii). But it is obvious that (ii)$\Rightarrow$(i)$\Rightarrow$(iii) and the theorem is proved.

PROOF OF THEOREM 2.9.3. Let us first show that T is a conditional α-contraction in X_2 by showing that U is conditionally completely continuous. In case (ii), this is obvious since X_1 is compactly imbedded in X_2. In case (i), if B and $U(B)$ are bounded in X_2, then there is a neighborhood N of B in X_2 such that N and $U(N)$ are bounded in X_2. But then (H3) implies that $U(N \cap X_1)$ is bounded in X_1 and, thus, precompact in X_2. Therefore, $U(B) \subset U(\text{Cl}(N \cap X_1)) \subset \text{Cl}\, U(N \cap X_1)$, which is compact. This completes the proof that T is a conditional α-contraction.

We next show that any compact invariant set in X_2 is the closure of a bounded set in X_1. Let k be the contraction constant for S in both spaces, $r = (1-k)^{-1}$. Suppose J is a compact invariant set in X_2. We first prove the result under hypothesis (ii). Since $TJ = J$ and J is bounded in X_2, it follows that $U(J)$ is bounded in X_1. Let ρ be chosen so that $rU(J) \subset B \stackrel{\text{def}}{=} B_\rho^1$. We show that $J \subset \text{Cl } B$ in X_2. Let $d_2(x, B) = \inf_{y \in B} |x - y|_2$. Let $\eta = \sup\{d_2(x, B), x \in J\}$. If we show that $\eta = 0$, then $J \subset \text{Cl } B$ in X_2. If $x \in J, y \in B, z = Sy + Ux$, then $z \in B$ since $|z|_1 \leq k|y|_1 + |Ux|_1 \leq k\rho + \rho r^{-1} = \rho$. Furthermore, $|Tx - z|_2 \leq k|x - y|_2$, which implies that

$$d_2(Tx, B) = \inf_{y \in B} |Tx - y|_2 \leq \inf_{y \in B} \{|Tx - z|_2, z = Sy + Ux\}$$
$$\leq k \inf_{y \in B} |x - y|_2 = k d_2(x, B).$$

Since $TJ = J$, this implies that

$$\eta = \sup_{x \in J} d_2(x, B) = \sup_{x \in J} d_2(Tx, B) \leq k \sup_{x \in J} d_2(x, B) = k\eta.$$

Thus, $\eta = 0$ and $J \subset \text{Cl } B$ in X_2.

Under Hypothesis (i), the proof is similar. Since J is compact and U is continuous, there is a bounded neighborhood N of J such that $U(N)$ is bounded in X_2. From (H$_3$), $U(N \cap X_1)$ is bounded in X_1 and there is a closed ball $B = B_\rho^1$ in X_1 such that $rU(N \cap X_1) \subset B$. Suppose $d_2(x, B)$ and η are defined as before. If $x \in J$, then X_1 dense in X_2 implies there is a sequence $\{x_n\} \subset N \cap X_1$ such that $x_n \to x$ in X_2 as $n \to \infty$. If $y \in B$, $z = Sy + Ux$, $z_n = Sy + Ux_n$, then the same proof as before shows that $|z_n|_1 \leq \rho$, $|z| \leq \rho$, and $z \in B$, $\{z_n\} \subset B$. Since

$$|Tx - z_n|_2 \leq |Tx - Tx_n|_2 + |Tx_n - z_n|_2$$
$$= |Tx - Tx_n|_2 + |Sx_n - Sy|_2 \leq |Tx - Tx_n|_2 + k|x_n - y|_2,$$

it follows that

$$\inf_{z \in B} |Tx - z|_2 \leq \inf_n |Tx - z_n|_2$$
$$= \inf_n |Sx - Sy + Ux - Ux_n| \leq k \inf_{y \in B} |x - y| = k d_2(x, y).$$

As in the previous case (ii), this implies $\eta \leq k\eta$ and $\eta = 0$. Thus, $J \subset \text{Cl } B$ in X_2. This completes the proof that any compact invariant set in X_2 is in the closure in X_2 of a bounded set in X_1.

We now prove that T being point dissipative in X_2 implies there is a maximal compact invariant set in X_2. From Theorem 2.9.1, there is a bounded set B in X_1 which attracts bounded sets of X_1. We can choose B closed so that, for any bounded set V in X_1, there is an n_0 such that $T^n V \subset B$ for $n \geq n_0$. Also, $\gamma^+(B)$ is bounded in X_1 and, thus, precompact in X_2. Therefore, its ω-limit set $\omega_2(B)$ in X_2 is nonempty, compact, and invariant in X_2 and $\omega_2(B)$ attracts B in X_2. We prove that $\omega_2(B)$ is maximal, compact invariant in X_2. In fact, if J is any compact invariant set in X_2, then there is a bounded set A in X_1 such that $J = \text{Cl } A$ in X_2. Since $T^n(A) \subset B$ for $n \geq n_0$, we have $\omega_2(A) \subset \omega_2(B)$. Since

T is continuous, $\omega_2(\operatorname{Cl} A) \subset \omega_2(B)$ and $J \subset \omega_2(B)$. Thus, $\omega_2(B)$ is a maximal, compact, invariant set in X_2. Since T is a conditional α-contraction on X_2, we now may apply Theorem 2.4.8 to see that $\omega_2(B)$ attracts compact sets of X_2. Theorem 2.4.2 implies the conclusion of the theorem.

PROOF OF THEOREM 2.9.4. With $U(n) = T^n - S^n$, we show first that $\bigcup_{n \geq 0} U(n)J$ is bounded in X_1. Since $J = T^n J$ for any $n \geq 0$ and S is linear,

$$J = TJ = SJ + UJ,$$
$$J = T^2 J = S(TJ) + U(TJ)$$
$$= S(SJ + UJ) + U(J) = S^2 J + SUJ + U(J).$$
$$\vdots$$
$$J = T^n J = S^n J + \sum_{m=0}^{n-1} S^m UJ.$$

Therefore, $U(n)J = \sum_{m=0}^{n-1} S^m UJ$. If k is the contraction for S in X_1, then $|U(n)J|_1 \leq |UJ|_1/(1-k)$. From Hypothesis (ii) of Theorem 2.9.4, $|UJ|_1$ is finite. Thus, $B \stackrel{\text{def}}{=} \bigcup_{n \geq 0} U(n)J$ is bounded in X_1. Since U is completely continuous, UB is precompact in X_1.

The next step is to show that J is a bounded invariant set in X_1. Since $TJ = J$ for all $n \geq 0$, it follows that

$$J = \lim_{n \to \infty} T^n J = \lim_{n \to \infty} [S^n J + U(n)J] = \lim_{n \to \infty} U(n)J$$

in X_2. We show that every limit point of $B = \bigcup_{n \geq 0} U(n)J$ is in X_1. Suppose n_p, $p \geq 0$, are integers, $n_p \to \infty$ as $p \to \infty$, $x_p \in U(n_p)J$, $x_p \to x_0$ in X_2 as $p \to \infty$. Since UB is precompact in X_1, the sequence $\{x_p\}$ is precompact in X_1. Thus, there is a subsequence $\{x_{p_j}\} \subset \{x_p\}$ such that $x_{p_j} \to \tilde{x}$ in X_1 as $j \to \infty$. Since X_1 is continuously imbedded in X_2, this implies $x_{p_j} \to \tilde{x}$ in X_2. Since $x_p \to x_0$ in X_2, it follows that $\tilde{x} = x_0$. Since this is true for every such subsequence of $\{x_p\}$, it follows that $x_p \to x_0$ in X_1 as $p \to \infty$. Thus, $J \subset X_1$. It is clear that $TJ = J$ and so J is invariant. Since J is bounded in X_1 and T is an α-contraction in X_1, it follows that J is precompact in X_1. This proves the theorem.

Notes and remarks.

The presentation of stability of invariant sets as presented in §2.2 is new although some partial results were obtained by Cooperman [1978] and Ize and dos Reis [1978]. Asymptotically smooth maps were introduced by Hale, LaSalle, and Slemrod [1972]. For a detailed discussion of α-contractions, see, for example, Sadovskii [1972], Martin [1976], and Deimling [1985]. The fact that α-contractions are asymptotically smooth was observed by Hale and Lopes [1973]. The fact that β-condensing maps are asymptotically smooth was proved by Massatt [1980]. For the case of α-condensing maps, the result is due to Cooperman [1978]. Lemma 2.3.6 is due to Lopes and Ceron [1984].

Lemma 2.4.1 is due to Massatt [1983a]. Theorem 2.4.1 (i)–(iv) is due to Hale, LaSalle, and Slemrod [1972]. Part (v) is due to Cooperman [1978]. Lemma 2.4.5 is due to Massatt [1983a]. Theorem 2.4.7 is due to Billotti and LaSalle [1971]. Theorem 2.4.8 is due to Cooperman [1978]. The results in §2.5 are motivated by Cooperman [1978]. Theorem 2.6.1 is due to Nussbaum [1972] and Hale and Lopes [1973] with the proof given following the latter. Lemma 2.6.5 is due to Hale and Lopes [1973]. The construction used in the proof of Lemma 2.6.5 was exploited by Chow and Hale [1974] to define a class of mappings called strongly limit compact, which includes the ones considered by Sadovskii [1972]. Corollary 2.6.9 and the method of proof in Theorem 2.6.8 are in Chow and Hale [1974]. Theorem 2.6.8 is a personal communication from Huang and Li (Hunan University). Theorem 2.7.1 is due to Oliva and a proof may be found in Hale, Magalhães, and Oliva [1984]. The one-to-oneness hypothesis can be dropped (see Quandt [1985]). The results as stated in §2.8 are due to Mañé [1981] and are extensions of Mallet-Paret [1976] (see also Hale, Magalhães, and Oliva [1984]). Other results on Hausdorff dimension have been given by A. Douady and J. Oesterle [1980]. §2.9 is based on Massatt [1983a], [1983b].

CHAPTER 3

Continuous Dynamical Systems

3.1. Limit sets.

Let X be a complete metric space, $R^+ = [0, \infty)$. A family of mappings $T(t)\colon X \to X, t \geq 0$, is said to be a C^r-semigroup, $r \geq 0$, provided that

(i) $T(0) = I$,

(ii) $T(t + s) = T(t)T(s), t \geq 0, s \geq 0$,

(iii) $T(t)x$ is continuous in t, x together with Fréchet derivatives in x up through order r for $(t, x) \in R^+ \times X$.

For any $x \in X$, the *positive orbit* $\gamma^+(x)$ through x is defined as $\gamma^+(x) = \{T(t)x, t \geq 0\}$. A *negative orbit* through x is a function $\phi\colon (-\infty, 0] \to X$ such that $\phi(0) = x$ and, for any $s \leq 0, T(t)\phi(s) = \phi(t + s)$ for $0 \leq t \leq -s$. A *complete orbit* through x is a function $\phi\colon R \to X$ such that $\phi(0) = x$ and, for any $s \in R, T(t)\phi(s) = \phi(t + s)$ for $t \geq 0$.

Since the range of $T(t)$ may not be all of X, to say there is a negative or complete orbit through x may impose restrictions on x. Also, since $T(t)$ need not be one-to-one it is not necessary for a negative orbit to be unique if it exists. Let *the negative orbit* through x be defined as the union of all negative orbits through x. Then

$$\gamma^-(x) = \bigcup_{t \geq 0} H(t, x),$$

$H(t, x) = \{y \in X\colon \text{there is a negative orbit through } x$
$\qquad \text{defined by } \phi\colon (-\infty, 0] \to X \text{ with } \phi(0) = x \text{ and } \phi(-t) = y\}.$

The *complete orbit* $\gamma(x)$ through x is defined as $\gamma(x) = \gamma^-(x) \cup \gamma^+(x)$. When a negative (or complete) orbit through x is defined, we sometimes write $T(t)x$ for an element on the orbit for $t < 0$ ($t \in R$). For any subset $B \subset X$, let $\gamma^+(B) = \bigcup_{x \in B} \gamma^+(x), \gamma^-(B) = \bigcup_{x \in B} \gamma^-(x), \gamma(B) = \bigcup_{x \in B} \gamma(x)$ be, respectively, the positive orbit, negative orbit, complete orbit through B if the latter exist.

For any set $B \subset X$, define the *ω-limit set* $\omega(B)$ *of B and the α-limit set* $\alpha(B)$ *of B* as

$$\omega(B) = \bigcap_{s \geq 0} \mathrm{Cl} \bigcup_{t \geq s} T(t)x, \qquad \alpha(B) = \bigcap_{s \geq 0} \mathrm{Cl} \bigcup_{t \geq s} H(t, x).$$

A set $B \subset X$ is said to *attract* a set $C \subset X$ under $T(t)$ if $\text{dist}(T(t)C, B) \to 0$ as $t \to \infty$. A set $S \subset X$ is said to be *invariant* if, for any $x \in S$, there is a complete orbit $\gamma(x)$ through x such that $\gamma(x) \subset S$. It is not difficult to show that S invariant is equivalent to $T(t)S = S$ for $t \geq 0$. Saying that a set S is invariant may impose restrictions upon S. An important class of invariant sets are the α-limit sets and ω-limit sets of orbits, as stated in

LEMMA 3.1.1. *For any set B in X for which $\omega(B)$ is compact and $\omega(B)$ attracts B, the set $\omega(B)$ is invariant. If, in addition, B is connected, then $\omega(B)$ is connected.*

If $\alpha(B)$ is compact, $\gamma^-(B) = \bigcup_{t \geq 0} H(t, B)$, and $\text{dist}(H(t, B), \alpha(B)) \to 0$ as $t \to \infty$, then $\alpha(B)$ is invariant. If, in addition, $H(t, B)$ is connected for each t then so is $\alpha(B)$.

PROOF. The proof of the first statement is omitted since it is so similar to the discrete case. Now suppose B is connected. If $\omega(B)$ is not connected, then $\omega(B)$ is the union of two disjoint compact subsets which are separated by a distance δ. Since $\omega(B)$ attracts B, this is clearly a contradiction and the lemma is proved.

LEMMA 3.1.2. *If B is a nonempty subset of X such that $\text{Cl}\,\gamma^+(B)$ is compact, then $\omega(B)$ is nonempty, compact, invariant, and $\omega(B)$ attracts B. If $\gamma^-(B)$ is nonempty and $\text{Cl}\,\gamma^-(B)$ is compact, then $\alpha(B)$ is nonempty, compact, and invariant.*

The proof is omitted since it is so similar to the discrete case.

3.2. Asymptotically smooth and α-contracting semigroups.

Let $T(t): X \to X$ be a C^r-semigroup for some $r \geq 0$. The semigroup $T(t)$ is *asymptotically smooth* if, for any nonempty, closed, bounded set $B \subset X$ for which $T(t)B \subset B$, there is a compact set $J \subset B$ such that J attracts B. As in the discrete case (Lemma 2.2.3), $T(t)$ is asymptotically smooth if and only if, for any nonempty, closed, bounded set $B \subset X$, there is a compact set $J = J(B) \subset X$ such that J attracts the set $L(B) = \{x \in B: T(t)x \in B \text{ for } t \geq 0\}$.

As in the discrete case, one can prove

LEMMA 3.2.1. *If $T(t)$ is asymptotically smooth and B is a nonempty set in X such that $\gamma^+(B)$ is bounded, then $\omega(B)$ is nonempty, compact, invariant, and $\omega(B)$ attracts B. If, in addition, B is connected, then $\omega(B)$ is connected. In particular, if, for some $x \in X, \gamma^+(x)$ is bounded, then $\text{Cl}\,\gamma^+(x)$ is compact and $\omega(x)$ is nonempty, compact, connected, and invariant.*

Let us now give some examples of asymptotically smooth maps.

A semigroup $T(t), t \geq 0$, is said to be *conditionally completely continuous* for $t \geq t_1$ if, for each $t \geq t_1$ and each bounded set B in X for which $\{T(s)B, 0 \leq s \leq t\}$ is bounded, we have $T(t)B$ precompact. A semigroup $T(t), t \geq 0$, is *completely continuous* if it is conditionally completely continuous and, for each $t \geq 0$, the set $\{T(s)B, 0 \leq s \leq t\}$ is bounded if B is bounded.

COROLLARY 3.2.2. *A semigroup $T(t), t \geq 0$, that is conditionally completely continuous for $t \geq t_0$ is asymptotically smooth.*

Using essentially the same proof as in Lemma 2.3.2 for the discrete case, one obtains the existence of a more interesting and less trivial example of an asymptotically smooth semigroup as stated in the following lemma.

LEMMA 3.2.3. *For each $t \geq 0$, suppose $T(t) = S(t) + U(t) \colon X \to X$ has the property that $U(t)$ is completely continuous and there is a continuous function $k \colon R^+ \times R^+ \to R^+$ such that $k(t, r) \to 0$ as $t \to \infty$ and $|S(t)x| \leq k(t, r)$ if $|x| \leq r$. Then $T(t), t \geq 0$, is asymptotically smooth.*

A semigroup $T(t), t \geq 0$, is said to be a *conditional β-contraction* with respect to the measure of noncompactness β if there is a continuous function $k \colon R^+ \to R^+$ such that $k(t) \to 0$ as $t \to \infty$, and, for each $t > 0$ and each bounded set $B \subset X$ for which $\{T(s)B, 0 \leq s \leq t\}$ is bounded, we have $\beta(T(t)B) \leq k(t)\beta(B)$. The semigroup $T(t), t \geq 0$, is said to be a *β-contraction* if it is a conditional β-contraction and, for each $t > 0$, the set $\{T(s)B, 0 \leq s \leq t\}$ is bounded if B is bounded. We refer to the function k as a *contracting function* of $T(t)$. A semigroup $T(t)$, $t \geq 0$, is said to be *β-condensing* if, for any bounded set B in X, the set $\{T(s)B, 0 \leq s \leq t\}$ is bounded for each $t > 0$ and $\beta(T(t)B) < \beta(B)$ if $\beta(B) > 0$.

LEMMA 3.2.4. *Suppose there is a nondecreasing function $M \colon R^+ \to R^+$ such that the semigroup $T(t), t \geq 0$, has the property that, for any $t > 0$ and any bounded set $B \subset X$, $\beta(T(t)B) \leq M(t)\beta(B)$. Then $T(t), t \geq 0$, is a β-contraction if and only if there is a $t_0 > 0$ such that the map $T(t_0)$ is a β-contraction.*

PROOF. Suppose $T(t), t \geq 0$, is a β-contraction and $k \colon R^+ \to R^+, k(t) \to 0$ as $t \to \infty$ is a contracting function for $T(t)$. Then there is a $t_0 > 0$ such that $k(t_0) < 1$ and, thus, $T(t_0)$ is a β-contraction.

Conversely, suppose $T(t_0), t > 0$, is a β-contraction with constant $\exp(-\alpha t_0)$, $\alpha > 0$. For any $t \in R^+$, there is an integer $n = n(t)$ such that $nt_0 \leq t < (n+1)t_0$. For any bounded set $B \subset X$, we have $T(t)B = [T(t_0)]^n T(s(t))B$, $s(t) = (n+1)t_0 - t$. Thus,

$$\beta(T(t)B) \leq e^{-n\alpha t_0}\beta(T(s(t))B) \leq e^{-n\alpha t_0} M(s(t))\beta(B) \leq M(t_0)e^{\alpha t_0}e^{-\alpha t}\beta(B).$$

Thus, $T(t)$ is a β-contraction with a contracting function $M(t_0)\exp{-\alpha(t - t_0)}$. This proves the lemma.

LEMMA 3.2.5. *Conditional β-contractions are asymptotically smooth.*

PROOF. Suppose B is a bounded set in X such that $T(t)B \subseteq B$. Then $\gamma^+(B)$ is bounded and

$$\beta(\operatorname{Cl}\gamma^+(T(t)B)) = \beta(\gamma^+(T(t)B)) = \beta(T(t)\gamma^+(B)) \leq k(t)\beta(\gamma^+(B)).$$

The definition of $\omega(B)$ and the fact that $k(t) \to 0$ implies that $\omega(B)$ is compact. To complete the proof, one shows that $\omega(B)$ attracts B as in the proof of Lemma 2.3.2.

The analogue of Lemma 2.3.2 for continuous semigroups is

LEMMA 3.2.6. *Suppose $T(t) = S(t) + U(t)$ where $U(t)$ is completely continuous for $t \geq 0$ and there is a continuous function $k\colon R^+ \times R^+ \to R^+$ such that $k(t,r) \to 0$ as $t \to \infty$ and $|S(t)x| \leq k(t,r)$ for $t \geq 0, |x| \leq r$. Then $T(t)$ is asymptotically smooth.*

3.3. Stability of invariant sets.

Suppose $T(t)\colon X \to X$ is a C^r-semigroup for some $r \geq 0$ and J is an invariant set. The set J is *stable* if, for any neighborhood V of J, there is a neighborhood U of J such that $T(t)U \subset V$ for all $t \geq 0$. As for the discrete case (Lemma 2.2.1), one shows that J is stable if and only if, for any neighborhood V of J, there is a neighborhood $V' \subseteq V$ of J such that $T(t)V' \subset V'$ for $t \geq 0$.

An invariant set J *attracts points locally* if there is a neighborhood W of J such that J attracts points of W. The set J is *asymptotically stable* (a.s.) if J is stable and attracts points locally. The set J is *uniformly asymptotically stable* (u.a.s.) if J is stable and attracts a neighborhood of J.

As for the discrete case, one can prove

LEMMA 3.3.1. *If J is a compact invariant set which is stable, then the following statements are equivalent:*

(i) *J attracts points locally.*

(ii) *There is a bounded neighborhood W of J such that $T(t)W \subset W$, $t \geq 0$, and J attracts compact sets of W.*

The proof of the following result is essentially the same as the proof of Theorem 2.2.5 for the discrete case.

THEOREM 3.3.2. *If $T(t)$ is asymptotically smooth and J is a compact invariant set that attracts points locally, then the following statements are equivalent:*

(i) *There is a bounded neighborhood W of J such that $T(t)W \subset W$, $t \geq 0$, and J attracts compact sets of W;*

(ii) *J is stable;*

(iii) *J is u.a.s.;*

and each condition implies that J is an isolated invariant set.

COROLLARY 3.3.4. *If $T(t), t \geq 0$, is asymptotically smooth, then a compact invariant set J is a.s. if and only if it is u.a.s.*

3.4. Dissipativeness and global attractors.

The semigroup $T(t)$ is said to be *point dissipative* (*compact dissipative*) (*locally compact dissipative*) (*bounded dissipative*) if there is a bounded set $B \subset X$ that attracts each point of X (each compact set of X) (a neighborhood of each compact set of X) (each bounded set of X) under $T(t)$.

For a C^r-semigroup $T(t), t \geq 0$, a compact invariant set A is said to be a *maximal compact invariant set* if every compact invariant set of the semigroup belongs to A. An invariant set A is said to be a *global attractor* if A is a maximal compact invariant set which attracts each bounded set $B \subset X$. In particular, $\omega(B)$ is compact and belongs to A. As for the discrete case (Lemma 2.4.1) (Theorem 2.4.2), one can prove the following results.

LEMMA 3.4.1. *If X is a Banach space and J is a maximal compact invariant set for a C^r-semigroup $T(t), t \geq 0$, which attracts compact sets, then J is connected.*

THEOREM 3.4.2. *If $T(t)\colon X \to X, t \geq 0$, is a C^r-semigroup and there is a nonempty compact set K that attracts compact sets of X and $A = \bigcap_{t \geq 0} T(t)K$, then*

(i) *A is independent of K and, if X is a Banach space, A is connected.*
(ii) *A is maximal, compact, and invariant.*
(iii) *A is stable and attracts compact sets of X.*

If, in addition, $T(t)$ is asymptotically smooth, then

(iv) *for any compact set H in X, there is a neighborhood H_1 of H such that $\gamma^+(H_1)$ is bounded and A attracts H_1. In particular, A is u.a.s.*
(v) *if C is any subset of X such that $\gamma^+(C)$ is bounded, then A attracts C.*

Finally, if $T(t)$ is one-to-one on A, then $T(t)$ is a C^r-group on A.

The only part of Theorem 3.4.2 that is not proved exactly as in the discrete case is the assertion about $T(t)|A$ being a group. But this is immediate from the one-to-oneness of $T(t)$ and the compactness of A.

COROLLARY 3.4.3. *If $T(t)\colon X \to X, t \geq 0$, is asymptotically smooth, then the following statements are equivalent:*

(i) *There is a compact set which attracts compact sets of X;*
(ii) *There is a compact set which attracts a neighborhood of each compact set of X.*

LEMMA 3.4.4. *If $T(t)\colon X \to X, t \geq 0$, is asymptotically smooth and $T(t)$ is compact dissipative, then there is a compact invariant set which attracts compact sets of X.*

LEMMA 3.4.5. *Let $T(t)\colon X \to X, t \geq 0$, be asymptotically smooth and point dissipative. If the orbit of any compact set is bounded, then $T(t)$ is locally compact dissipative. If the orbit of any bounded set is bounded, then $T(t)$ is bounded dissipative.*

THEOREM 3.4.6. *If $T(t)\colon X \to X, t \geq 0$, is asymptotically smooth, point dissipative, and orbits of bounded sets are bounded, then there exists a global attractor A. If $T(t)$ is also one-to-one on A, then $T(t)|A$ is a C^r-group. If, in addition, X is a Banach space, then A is connected.*

An *equilibrium point* for a C^r-semigroup $T(t), t \geq 0$, is a point $x \in X$ such that $T(t)x = x$ for $t \geq 0$. For the case of an α-contraction, we can prove there is an equilibrium point if the conditions of Theorem 3.4.6 are satisfied.

THEOREM 3.4.7. *If X is a Banach space, $T(t)\colon X \to X, t \geq 0$, is an α-contraction, point dissipative, and orbits of bounded sets are bounded, then there exists a connected compact attractor A. If, in addition, $k(t) \in [0,1)$ is a contracting function for $T(t)$, there is an equilibrium point for $T(t)$. If, in addition, $T(t)$ is one-to-one on A, then $T(t)|A$ is a C^r-group.*

PROOF. Since an α-contraction is asymptotically smooth, it is only necessary to prove the existence of an equilibrium point for $T(t)$. For any $s_1 > 0$, consider the mapping $T(s_1)$ on X. Since $k(t) \in [0,1)$ for each $t > 0$, the map $T(s_1)$ is an α-contraction which is compact dissipative. Thus, from Theorem 2.6.1, there is a fixed point x_1 of $T(s_1)$. Thus, there is a complete orbit through x_1 given by $\phi(ns_1 + s) = T(s)x_1, 0 \leq s \leq s_1, n = 0, \pm 1, \pm 2, \ldots$. This orbit is periodic of period s_1, compact, and invariant and, therefore, must belong to the compact attractor A. Thus, $x_1 \in A$. Choose a sequence $s_m \to 0$ as $m \to \infty$. By the same reasoning as above, the map $T(s_m)$ has a fixed point $x_m \in A$ corresponding to a periodic orbit of $T(t)$ of period s_m for all m. Since $\{x_m, m \geq 1\} \subset A$ and A is compact, there must be a subsequence (which we label the same) which converges to a point x^* in A. For any t and any integer m, there is an integer $k_m = k_m(t)$ so that $k_m s_m \leq t < (k_m + 1)s_m$. Then, for any fixed t,

$$|T(t)x^* - x^*| \leq |T(t)x^* - T(t)x_m| + |T(t)x_m - x_m| + |x_m - x^*|$$
$$= |T(t)x^* - T(t)x_m| + |T(t - k_m s_m)x_m - x_m| + |x_m - x^*|,$$

and the right hand side of this expression $\to 0$ as $m \to \infty$ since $x_m \to x^*$, $t - k_m s_m \to 0$ as $m \to \infty$, and $T(t)x$ is continuous in t, x. This proves the theorem.

Theorems 3.4.6 and 3.4.7 are probably the most easily applied in specific problems. If there is a $t_1 > 0$ such that $T(t)$ is completely continuous for $t > t_1$, one can improve Theorem 3.4.6 as in the discrete case (Theorem 2.4.7) and prove the existence of an equilibrium point as in the proof of Theorem 3.4.7 when $t_1 = 0$. The precise statement is

THEOREM 3.4.8. *If there is a $t_1 \geq 0$ such that the C^r-semigroup $T(t)\colon X \to X, t \geq 0$, is completely continuous for $t > t_1$ and point dissipative, then there is a global attractor A. If X is a Banach space, then A is connected and, if $t_1 = 0$, there is an equilibrium point of $T(t)$. If, in addition, $T(t)$ is one-to-one on A, then $T(t)|A$ is a C^r-group.*

The statement about the existence of an equilibrium point in Theorems 3.4.7 and 3.4.8 will be seen to be effective in determining the existence of solutions of boundary value problems for partial differential equations.

3.5. Dependence on parameters.

Suppose Λ is a metric space, X is a complete metric space and, for each $\lambda \in \Lambda, T(t,\lambda)\colon X \to X$ is a C^r-semigroup with $T(t,\lambda)x$ continuous in t, λ, x. If, for each $\lambda \in \Lambda$, $\{T(t,\lambda), t \geq 0\}$ is asymptotically smooth, then for any closed bounded set $B \subset X$ for which $T(t,\lambda)B \subset B, t \geq 0$, there is a compact set

$K_\lambda(B) \subset B$ such that $K_\lambda(B)$ attracts B under $T(t,\lambda)$. We say the family of semigroups $\{T(t,\lambda), t \geq 0\}, \lambda \in \Lambda$, is *collectively asymptotically smooth* if $\bigcup_{\lambda \in \Lambda} K_\lambda(B)$ is compact. The same family of semigroups is *collectively β-contracting* if $\{T(t,\lambda), t \geq 0\}$ is a β-contraction for each $\lambda \in \Lambda$, and there is a $t_0 > 0, k \in [0,1)$ such that, for any bounded set $B, \beta(B) > 0, V_{t_0}(B) \stackrel{\text{def}}{=} \bigcup_{\lambda \in \Lambda}\{T(s,\lambda)B, 0 \leq s \leq t_0\}$ is bounded for each t and $\beta(V_{t_0}(B)) \leq k\beta(B)$.

As in the discrete case one proves

LEMMA 3.5.1. *The family of semigroups $\{T(t,\lambda), t \geq 0\}, \lambda \in \Lambda$, is collectively asymptotically smooth if it is collectively β-contracting.*

PROOF. From Lemma 3.2.5, for each $\lambda \in \Lambda, T(t,\lambda)$ is asymptotically smooth. Suppose B is a nonempty, closed, bounded set, $T(t,\lambda)B \subset B$ for each $\lambda \in \Lambda$ and $t \geq 0$, and let $K_\lambda \subset B$ be a compact set that attracts B under $T(\cdot,\lambda)$. Since B is a complete metric space, we may apply Theorem 3.4.6 to the space B and thus there is a compact attractor in B which attracts B under $T(\cdot,\lambda)$. Then we can take K_λ to be this compact attractor. In particular, $T(t,\lambda)K_\lambda = K_\lambda$ for $t \geq 0$. If $V_0 \stackrel{\text{def}}{=} \bigcup_{\lambda \in \Lambda} K_\lambda, V_0$ is bounded. Also, K_λ invariant under $T(\cdot,\lambda)$ implies

$$\beta(V_0) = \beta\left(\bigcup_{\lambda \in \Lambda}\{T(s,\lambda)K_\lambda, 0 \leq s \leq t_0\}\right)$$
$$\leq \beta\left(\bigcup_{\lambda \in \Lambda}\{T(s,\lambda)V_0, 0 \leq s \leq t_0\}\right)$$
$$= \beta(V_{t_0}(V_0)) \leq k\beta(V_0) < \beta(V_0).$$

Thus, V_0 is precompact and the lemma is proved.

The following results also are proved in a manner completely analogous to the ones for the discrete case.

THEOREM 3.5.2. *Suppose $T(t,\lambda): X \to X, t \geq 0$, is a C^r-semigroup for each $\lambda \in \Lambda$ and suppose there is a bounded set $B \subset X$ independent of λ such that B attracts compact sets of X under $T(t,\lambda)$. If the family of semigroups is collectively asymptotically smooth, then the maximal compact invariant set A_λ of T_λ is upper semicontinuous in λ.*

THEOREM 3.5.3. *Suppose $T(t,\lambda): X \to X, t \geq 0$, is a C^r-semigroup for each $\lambda \in \Lambda$ and there is a bounded set B that attracts points of X under $T(t,\lambda)$ for each $\lambda \in \Lambda$ and, for any bounded set U, the set $V \stackrel{\text{def}}{=} \bigcup_{\lambda \in \Lambda} \bigcup_{t \geq 0} T(t,\lambda)U$ is bounded. If the family of semigroups is collectively asymptotically smooth, then the global attractor A_λ is upper semicontinuous in λ.*

3.6. Periodic processes.

Suppose X is a complete metric space, $R^+ = [0,\infty), u: R^+ \times R \times X \to X$ is a given mapping and define $U(t,\sigma): X \to X$ for $t \in R^+, \sigma \in R$, by $U(t,\sigma)x = u(t,\sigma,x)$. A *process on X* is a mapping $u: R^+ \times R \times X \to X$ satisfying the

following properties:
  (i) u is continuous;
  (ii) $U(0,\sigma) = I$, the identity;
  (iii) $U(t, \sigma + s)U(s,\sigma) = U(s+t,\sigma)$.

A process u is said to be an ω-*periodic process* if there is an $\omega > 0$ such that $U(t, \sigma + \omega) = U(t,\sigma)$ for $t \in R^+, \sigma \in R$. A process is an *autonomous process* if $U(t,\sigma)$ is independent of σ, that is, if $T(t) = U(t,0), t \geq 0$, then $T(t), t \geq 0$, is a C^0-semigroup on X. A *trajectory* through (σ, x) for a process is the set $\{(t,y): y = u(t,\sigma,x), t \geq 0\}$.

In a process, $u(t,\sigma,x)$ can be considered as the state of a system at time $\sigma + t$ if the state at time σ is x.

We say the family of maps $\{U(t,\sigma), t \geq 0, \sigma \in R\}$ is *asymptotically smooth* if, for any $\sigma \in R$ and any nonempty closed bounded set B in X such that $U(t,\sigma)B \subset B, t \geq 0$, there is a nonempty compact set $K \subset B$ such that K attracts B under $U(t,\sigma), t \geq 0$. The family of maps $\{U(t,\sigma), t \geq 0, \sigma \in R\}$ is said to be *completely continuous* for $t \geq t_1$, if, for each $t \geq t_1, \sigma \in R$, and each bounded set B in X, the set $\{U(s,\sigma)B, 0 \leq s \leq t\}$ is bounded and $U(t,\sigma)B$ is precompact. The family of maps $\{U(t,\sigma), t \geq 0, \sigma \in R\}$ is said to be α-*condensing* if, for each $t > 0$, $\sigma \in R$, and any bounded set B in X, the set $\{U(s,\sigma)B, 0 \leq s \leq t\}$ is bounded and $\alpha(U(t,\sigma)B) < \alpha(B)$ if $\alpha(B) > 0$. We say $\{U(t,\sigma), t \geq 0, \sigma \in R\}$ is *point dissipative* if there is a bounded set B such that, for each $\sigma \in R$, the set B attracts points of X under $U(t,\sigma), t \geq 0$.

A set $M \subset R \times X$ is said to be *invariant* under the process u if $(\sigma_0, x_0) \in M$ implies that $u(t,\sigma_0,x_0)$ can be defined for all $t \in R$ and $(t, u(t,\sigma_0,x_0)) \in M$ for all $t \in R$. An invariant set $M \subset R \times X$ is said to be a *global attractor* if, for any $\sigma \in R$ and any bounded set B in X, $\mathrm{dist}(U(t,\sigma)B, M_t) \to 0$ as $t \to \infty$ where $M_t = \{x: (t,x) \in M\}$ is the cross-section of M at t.

We can now state

THEOREM 3.6.1. *Suppose u is an ω-periodic process which is asymptotically smooth, point dissipative, and orbits of bounded sets under $U(t,\sigma)$ for each $\sigma \in R$ are bounded. If $T_\sigma = U(\omega,\sigma)$, then T_σ is asymptotically smooth, and there is a global attractor A_σ for T_σ, connected if X is a Banach space, and $A_\sigma = U(\sigma,0)A_0$ for each $\sigma \in R$. The set $M = \{(t,x): x = u(t,0,x_0), x_0 \in A_0, t \in R\} \subset R \times X$ is invariant and a global attractor.*

PROOF. It is clear that T_σ is asymptotically smooth and the global attractor A_σ for T_σ exists by Theorem 2.4.6. Since the process is ω-periodic, the remaining assertions will follow if we show that $A_\sigma = U(\sigma,0)A_0$ for $0 \leq \sigma \leq \omega$. Since $U(\omega,0)A_0 = A_0$ and $U(\sigma,\omega) = U(\sigma,0)$, we have

$$U(\omega,\sigma)U(\sigma,0)A_0 = U(\sigma,\omega)U(\omega - \sigma,\sigma)U(\sigma,0)A_0 = U(\sigma,\omega)U(\omega,0)A_0$$
$$= U(\sigma,\omega)A_0 = U(\sigma,0)A_0$$

and $U(\sigma,0)A_0$ is invariant under T_σ But it is clear that $U(\sigma,0)A_0$ attracts bounded sets under T_σ since A_0 attracts bounded sets under T_0. Since $U(\sigma,0)A_0$

is compact and invariant under T_σ and attracts the attractor A_σ of T_σ, it follows that $A_\sigma = U(\sigma,0)A_0$ and the theorem is proved.

COROLLARY 3.6.2. *If X is a Banach space, the ω-periodic process u is α-condensing, point dissipative, and orbits of bounded sets under $U(t,\sigma)$ for each $\sigma \in R$ are bounded, then the conclusions of Theorem 3.6.1 are satisfied and, in addition, there is a fixed point of $U(\omega,0)$, that is, an ω-periodic trajectory of the process.*

We can also prove the following result following the ideas of the proof of Theorem 3.4.8.

THEOREM 3.6.3. *If X is a Banach space, the ω-periodic process is completely continuous for $t \geq t_1 > 0$ and is point dissipative, then the conclusions of Theorem 3.6.1 are satisfied and, in addition, there is a fixed point of $U(\omega,0)$, that is, an ω-periodic trajectory of the process.*

A process u is affine if $u(t,\sigma,x) = L(t,\sigma,x) + y_\sigma$ where $L(t,\sigma,x)$ is linear in x. For affine processes, one obtains the following result as a consequence of Corollary 2.6.9.

THEOREM 3.6.4. *If X is a Banach space, the ω-periodic process is α-condensing and affine, and there is a bounded trajectory of the process, then there is an ω-periodic trajectory.*

3.7. Skew product flows.

In the previous section, we have discussed the ω-periodic process which includes certain types of evolutionary equations for which the vector field is ω-periodic in the independent variable. There is another way to treat this problem which allows one to generalize some of the results to other types of nonautonomous process.

Let us begin by considering an ordinary differential equation in R^n. Suppose $f: R \times R^n \to R^n$ is continuous and consider the equation

$$\dot{u} = f(t,u). \tag{7.1}$$

Suppose f is smooth enough to ensure that the initial value problem has a unique solution which depends continuously upon the initial data. Also, suppose each solution $\phi(t,\sigma,x,f), \phi(\sigma,\sigma,x,f) = x$, exists for $t \geq \sigma$. For $t \in [0,\infty]$, let $u(t,\sigma,x,f) = \phi(t+\sigma,\sigma,x,f)$, and, for any $t \in [0,\infty)$, let $\sigma(t)f: R \times R^n \to R^n$ be defined by $(\sigma(t)f)(s,x) = f(t+s,x)$, the *translation of f*. We suppose that $\mathcal{F}$ is a complete metric space of functions with the property that $f \in \mathcal{F}$ implies the translation $\sigma(t)f \in \mathcal{F}$. Then $\sigma(0) = I, \sigma(t+s)f = \sigma(t)\sigma(s)f$, and $\sigma(t), t \geq 0$, is a C^0-semigroup if $\sigma(t)f$ is continuous in (t,f).

For any $t \geq 0$, define

$$\pi(t): R^n \times \mathcal{F} \to R^n \times \mathcal{F}, \qquad \pi(t)(x,f) = (u(t,0,x,f), \sigma(t)f). \tag{7.2}$$

We have $\pi(0) = I$,

$$\pi(t+s)(x,f) = (u(t+s,0,x,f), \sigma(t+s)f) = (u(t,s,u(s,0,x,f),f), \sigma(t)\sigma(s)f)$$
$$= (u(t,0,u(s,0,x,f), \sigma(s)f), \sigma(t)\sigma(s)f) = \pi(t)(u(s,0,x,f), \sigma(s)f)$$
$$= \pi(t)\pi(s)(x,f)$$

and the semigroup property is satisfied. If the family $\mathcal{F}$ is so that $\pi(t)(x,f)$ is continuous in (t,x,f), then $\pi(t), t \geq 0$, is a C^0-semigroup on $R^n \times \mathcal{F}$. Finding conditions on $\mathcal{F}$ which ensure this latter property is an interesting but nontrival problem which has received considerable attention in the literature (see notes and remarks at the end of the chapter for references). Our objective is to extend this idea to the more general processes defined in §3.6.

Let us recall that a process on a complete metric space X is a continuous function $u: R^+ \times R \times X \to X$ such that, for all $t \geq 0, \tau \geq 0, s \in R, x \in X$,

$$u(0,s,x) = x, \qquad u(t+\tau, s, x) = u(t, \tau+s, u(\tau, s, x)).$$

Let W denote the set of all processes endowed with some metric, and for $u \in W$, define the translation $\sigma(\tau), \tau \in R$, of the process as

(7.3) $\qquad (\sigma(\tau)u)(t,s,x) = u(t, \tau+s, x).$

Then $\sigma(0) = I, \sigma(t+\tau) = \sigma(t)\sigma(\tau)$ for all $t, \tau \in R$, and $\sigma(\tau), \tau \in R$, is a semigroup. Also, let $\alpha: R^+ \times R \times W \to X$ be defined by

(7.4) $\qquad \alpha(t, x, u) = u(t, 0, x).$

With α and σ as in (7.3) and (7.4), define

(7.5) $\qquad \pi(t): X \times W \to X \times W, \qquad \pi(t)(x, u) = (\alpha(t, x, u), \sigma(t)u).$

One verifies that $\pi(0) = I$,

$$\pi(t+\tau)(x, u)$$
$$= (\alpha(t+\tau, x, u), \sigma(t+\tau)u) = (u(t+\tau, 0, x), \sigma(t)\sigma(\tau)u)$$
$$= (u(t, \tau, u(\tau, 0, x)), \sigma(t)\sigma(\tau)u) = ((\sigma(\tau)u)(t, 0, u(\tau, 0, x)), \sigma(t)\sigma(\tau)u)$$
$$= (\alpha(t, u(\tau, 0, x), \sigma(\tau)u), \sigma(t)\sigma(\tau)u) = \pi(t)(u(\tau, 0, x), \sigma(\tau)u)$$
$$= \pi(t)(\alpha(\tau, x, u), \sigma(\tau)u) = \pi(t)\pi(\tau)(x, u)$$

and the semigroup property is satisfied. Consequently, the family of mappings $\pi(t), t \geq 0$, would be a C^0-semigroup if $\pi(t)(x, u)$ were continuous in t, x, u. One cannot expect this to be true in the space of all processes W. However, we can hope that the continuity properties hold on some subset $V \subset W$ of the processes in some appropriate topology. We will not address this question in detail at this time and simply make the following hypotheses:

(7.6) There is a subset $V \subset W$ which, with the metric of W, forms a complete metric space.

(7.7) $\qquad \pi(t): X \times V \to X \times V$ is a C° − semigroup.

When (7.7) is satisfied, we refer to $\pi(t), t \geq 0$, as the *skew product flow of the processes u*.

We note that (7.7) implies that $\sigma(t)\colon V \to V$ for $t \geq 0$ and is a C^0-semigroup. With this notation, the following lemma is obvious.

LEMMA 3.7.1. *If* (7.6), (7.7) *are satisfied and the skew product flow* $\pi(t), t \geq 0$, *is asymptotically smooth, then the translation semigroup* $\sigma(t)$ *is asymptotically smooth.*

The following results give the existence and structure of the global attractor for skew product flows.

THEOREM 3.7.2. *Suppose* (7.6), (7.7) *are satisfied and the skew product flow* $\pi(t)\colon X \times V \to X \times V$ *is point dissipative. If either*

(i) $\pi(t), t \geq 0$, *is asymptotically smooth and orbits of bounded sets are bounded*

or

(ii) $\pi(t), t \geq 0$, *is completely continuous for* $t > t_1$,

then there is a global attractor $A = A(V)$ *for* $\pi(t)$ *and a global attractor* B *for the translation semigroup* $\sigma(t)$. *Furthermore, if*

$$K = \{y \in X \colon (y, u) \in A, u \in B\},$$

then K *is compact and the set* K *is invariant in the following sense: for any* $y \in K$, *there is a* $v \in B$ *such that* $v(t, 0, y) \in K$ *for all* $t \in R$.

PROOF. Theorems 3.4.6 and 3.4.8 imply the existence of the global attractor A. From Lemma 3.7.1, $\sigma(t)\colon V \to V, t \geq 0$, is asymptotically smooth. It is also point dissipative and satisfies either (i) or (ii). Thus, there is a global attractor $B \subset V$ of $\sigma(t)$. If we define K as above, then the last statement of the theorem is just a restatement of invariance of A. This proves the theorem.

Suppose V is a complete metric space of processes on X and $\sigma(t)\colon V \to V, t \geq 0$, is a C^0-semigroup. For any $u \in V$, we let $H(u) = \operatorname{Cl} \gamma_\sigma^+(u)$, where $\gamma_\sigma^+(u) = \bigcup_{t \geq 0} \sigma(t) u$. The set $H(u)$ is called the *hull* of u. In the same way, one defines the *hull* $H(S)$ *of a subset* S of V. We let $\omega_\sigma(u)$ denote the ω-limit set of u with respect to the translation semigroup $\sigma(t)$.

Let us now discuss some special cases of Theorem 3.7.2.

(i) *Autonomous case*. Recall that an autonomous process u is one for which $u(t, s, x)$ is independent of s; that is, $u(t, 0, x) = u(t, s, x)$ for all $s \in R$. Therefore, $H(u) = \omega(u) = \{u\}$ under $\sigma(t)$. If V in Theorem 3.7.2 contains only autonomous processes, this implies that the discussion of the attractor A is reduced to the discussion of the attractor separately for each $u \in V$. If u is a given autonomous process, $V = \{u\}$, then Theorem 3.7.2 gives a global attractor $A = A(u)$ under $\pi(t), A = K \times \{u\}$, and K is invariant under $u(t, 0, \cdot)$. This is just a restatement of Theorem 3.4.6 or 3.4.8.

Now suppose Λ is a complete metric space, $V(\Lambda) = \{u_\lambda, \lambda \in \Lambda\}$, where each u_λ is an autonomous process continuous in λ and suppose $V(\Lambda)$ is bounded. Suppose $\pi(t)\colon X \times V(\Lambda) \to X \times V(\Lambda)$ satisfies the conditions of Theorem 3.7.2.

Let A be the global attractor. For any $\lambda \in \Lambda$, let $T_\lambda(t) \colon X \to X$ be defined by $T_\lambda(t)x = u_\lambda(t,0,x), x \in X$. Since the processes are autonomous, $T_\lambda(t)$ satisfies the conditions of Theorem 3.5.3 and there is a global attractor A_λ under $T_\lambda(t)$ which is upper semicontinuous in λ. The set $A = \bigcup_{\lambda \in \Lambda}(A_\lambda \times \{u_\lambda\})$.

(ii) *Asymptotically autonomous processes.* Suppose V is a complete metric space of processes for which $\sigma(t) \colon V \to V$ is asymptotically smooth and $\omega(V) = \{v\}$ where v is an autonomous process. We refer to this case as *asymptotically autonomous*. If $\pi(t) \colon X \times V \to X \times V$ satisfies the conditions of Theorem 3.7.2, then the global attractor A exists and $A - K \times \{v\}$. Also, from Theorem 3.7.2, the set K is invariant under the semigroup $v(t,0,\cdot)$ generated by the autonomous process v. Also, since A attracts bounded sets of $X \times V$, this implies that, for any bounded set $C \subset X$ and any $u \in V$, the ω-limit set of the set $\{u(t,0,C), t \geq 0\}$ in X belongs to K and is, therefore, the union of invariant sets of the semigroup $v(t,0,\cdot)$.

(iii) *Periodic processes.* Recall that a process u is ω-periodic if $u(t, s+\omega, x) = u(t,s,x)$ for all $(t,s,x) \in R^+ \times R \times X$. For any periodic process u, $\sigma(t)u$ is also a periodic process, $H(u) = \{\sigma(t)u, 0 \leq t \leq \omega\}$ and $\omega_\sigma(u) = H(u)$. The set $H(u)$ is homeomorphic to the circle S^1. As in the autonomous case, if $\pi(t) \colon X \times V \to X \times V$ is a semiflow for which V consists only of periodic processes, then the discussion of the attractor A can be restricted to the case where V consists of $H(u)$ for a periodic process u.

Assume u is a periodic process, $H(u)$ is the hull of u, and $\pi(t) \colon X \times H(u) \to X \times H(u)$ is a skew product flow that satisfies the conditions of Theorem 3.7.2. Let $A = A(u)$ be the attractor. Let $K = \{y \in X \colon (y,v) \in A, v \in H(u)\}$. If $y \in K$, then invariance of A implies there is a $\tau \in [0,\omega]$ such that $(\sigma(\tau)u)(t,0,y) \in K$ for all $t \in R$. Also, from the invariance of A, we can reproduce the results on periodic processes obtained in §3.6. In fact, if $M = \{(t, u(t,0,y)) \colon t \in R, y \in K, (u(t,0,y), \sigma(t)u) \in A\}$ then M is the same as the set M in Theorem 3.6.1.

(iv) *Quasiperiodic processes.* A process $u \colon R^+ \times R \times X \to X$ is said to be *quasiperiodic* if there is an integer $p \geq 1$ and a continuous function $v \colon R^+ \times R^p \times X \to X$ such that $v(t, \theta_1, \ldots, \theta_p, x)$ is periodic in θ_j of some period $\omega_j, j = 1, 2, \ldots, p$, $v(t, \theta, x)$ is continuous in θ uniformly for t, x in bounded sets, and $u(t,s,x) = v(t,s,s,\ldots,s,x)$. We can assume also that the numbers $\{\nu_j = 2\pi/\omega_j\}$ are rationally independent. If u is quasiperiodic, then $H(u) = \{v(\cdot, \cdot + \theta_1, \ldots, \cdot + \theta_p, \cdot), \theta_j \in R, j = 1, 2, \ldots, p\}$; that is, $H(u)$ is homeomorphic to the torus T^p and $\omega(u) = H(u)$.

Let u be a quasiperiodic process, $U = H(u)$, and suppose $\pi(t) \colon X \times H(u) \to X \times H(u)$ is a skew product flow satisfying the conditions of Theorem 3.7.2. Let $A = K \times H(u)$ be the attractor for $\pi(t), K = \{y \in X \colon (y,v) \in A, v \in H(u)\}$.

Question. Let $N = \{(t, u(t,0,y)) \colon t \in R, y \in K, (u(t,0,y), \sigma(t)u) \in A\}$ and define $N_s = \{x \colon (s,x) \in N\}$. Is the set N_s quasiperiodic in s?

The set N is invariant under the map $U(t,s) \stackrel{\text{def}}{=} u(t,s,\cdot)$ in the sense that if $(s,x) \in N$, then $(t, U(t,s)x) \in N$ for all $t \in R$. Furthermore, N attracts bounded sets under $U(t,s)$. If the answer to the above question is yes, then we obtain some consequences which are of some interest. If N_s is a singleton for each s, then there is a quasiperiodic trajectory of the process which attracts bounded sets.

(v) *Almost periodic processes.* A process $u\colon R^+ \times R \times X \to X$ is said to be *almost periodic* if $u(t,s,x)$ is almost periodic in s uniformly with respect to t,x in bounded sets. For an almost periodic process u, Bochner's theorem implies that $\omega_\sigma(u) = H(u)$ is a minimal set of $\sigma(t)$. If one takes $U = H(u)$ and $\pi(t)\colon X \times H(u) \to X \times H(u)$ is a skew product flow satisfying the conditions of Theorem 3.7.2, then there is a global attractor.

Our next objective is to obtain some relations about the stability of a trajectory of a process and the stability of the corresponding skew product flow. To do this, we need some terminology. Suppose u is a process on X, $0 \in X$. If $u(t,s,0) = 0$ for all $(t,s) \in R^+ \times R$, we say that 0 is an *equilibrium point* for the process u. The equilibrium point zero is *uniformly stable* (u.s.) with respect to the process u if, for any neighborhood V of zero in X, there is a neighborhood U of zero in X, such that $u(t,s,U) \subset V$ for $(t,s) \in R^+ \times R$. The equilibrium point zero is *uniformly asymptotically stable* (u.a.s.) with respect to the process u if it is u.s. and there is a neighborhood W of zero in X such that, for any neighborhood V of zero in X, there is a $T > 0$ such that $u(t,s,W) \subset V$ for $t \geq T, s \in R$.

LEMMA 3.7.3. *Suppose u is a process on X for which $H(u)$ is compact and let $\omega_\sigma(u)$ be the ω-limit set of u with respect to $\sigma(t)$. Suppose $\pi(t)\colon X \times H(u) \to X \times H(u)$ is defined by (7.5) and is a skew product flow. If zero is an equilibrium point for the process u, then zero is u.s. (u.a.s.) with respect to the process u if and only if $\{0\} \times \omega_\sigma(u)$ is u.s. (u.a.s.) with respect to skew product flow $T(t)$.*

PROOF. If zero is u.s. with respect to the process u and $\gamma_\sigma^+(u) = \bigcup_{t \geq 0} \sigma(t)u$, then, for any neighborhood V of zero in X, there is a neighborhood U of zero in X such that $u(t,s,U) \subset V$ for $(t,s) \in R^+ \times R$; that is, $(\sigma(s)u)(t,0,U) \subset V$ for $(t,s) \in R^+ \times R$; that is, $\pi(t)(U \times \gamma_\sigma^+(u)) \subset V \times \gamma_\sigma^+(u)$ for $t \geq 0$. Since $\gamma_\sigma^+(u)$ is dense in $H(u)$ and $\pi(t)$ is continuous, we have $\pi(t)(U \times H(u)) \subset V \times H(u)$ for $t \geq 0$ and $\{0\} \times H(u)$ is u.s. Since $\omega_\sigma(u)$ attracts $H(u)$ under $\sigma(t)$, it follows that $\{0\} \times \omega_\sigma(u)$ is u.s. for $\pi(t)$.

Conversely, if $\{0\} \times \omega_\sigma(u)$ is u.s. for $\pi(t)$, then $\{0\} \times H(u)$ is u.s. for $\pi(t)$. This clearly implies that zero is u.s. for the process u.

Now suppose that zero is u.a.s. for the process u. Then $\{0\} \times \omega_\sigma(u)$ is u.s. for $\pi(t)$. Also, there is a neighborhood W of zero in X such that, for any neighborhood V of zero in X, there is a $T > 0$ such that $u(t,s,W) \subset V$ for $t \geq T$, $s \in R$; that is, $\pi(t)(W \times \gamma_\sigma^+(u)) \subset V \times \gamma_\sigma^+(u)$ for $t \geq T$. Consequently, $\pi(t)(W \times H(u)) \subset V \times H(u)$ for $t \geq T$ and $\{0\} \times H(u)$ attracts $W \times H(u)$. Since $\omega_\sigma(u)$ attracts $H(u)$ under $\sigma(t)$, this implies that $\{0\} \times \omega_\sigma(u)$ attracts $W \times H(u)$ and $\{0\} \times \omega_\sigma(u)$ is u.a.s. The converse is obvious and the theorem is proved.

THEOREM 3.7.4. *Suppose u is a process on X for which $H(u)$ is compact and let $\omega_\sigma(u)$ be the ω-limit set of u with respect to $\sigma(t)$. Also, suppose $\pi(t)\colon X\times H(u) \to X\times H(u)$ defined by (7.5) is a skew product flow which is asymptotically smooth. If zero is an equilibrium point for the skew product flow, then zero is u.a.s. for the process u if and only if zero is u.s. for the process u and there is a neighborhood V of zero in X such that, for every $x \in V, v \in \omega_\sigma(u), v(t,0,x) \to 0$ as $t \to \infty$.*

PROOF. Suppose the latter statement of the theorem is satisfied. Since zero is u.s. for the process u, Lemma 3.7.3 implies $\{0\}\times\omega_\sigma(u)$ is u.s. for $\pi(t)$. We may choose U so that $Z = \mathrm{Cl}\bigcup_{t\geq 0} \pi(t)(U\times H(u))$ is bounded and $\{0\}\times\omega_\sigma(u)$ attracts points of Z. Since $\pi(t)$ is asymptotically smooth and $\{0\}\times\omega_\sigma(u)$ is the maximal compact invariant set in Z, we have from Theorem 3.4.6 that $\{0\}\times\omega_\sigma(u)$ attracts Z and $\{0\}\times\omega_\sigma(u)$ is u.a.s. Lemma 3.7.3 implies zero is u.a.s. for the process u.

Conversely, if zero is u.a.s. for the process u, then $\{0\}\times\omega_\sigma(u)$ is u.a.s. for $\pi(t)$ from Lemma 3.7.3. This clearly implies the last statement of the theorem and the proof is complete.

To express the implications of Theorem 3.7.4 in another way, let us make another definition. We say an equilibrium point zero of a process u is *asymptotically stable* (a.s.) *for the process u* if zero is uniformly stable and there is a neighborhood $V(u)$ of zero in X such that $v(t,\sigma,x) \to 0$ as $t \to \infty$ for every $\sigma \in R, x \in V(u)$. The conclusion of Theorem 3.7.4 states that zero is a.s. for each process $v \in \omega_\sigma(u)$ with the neighborhood V of zero in X independent of v if and only if zero is u.a.s. for the process u.

An interesting consequence of Theorem 3.7.4 relates to linear skew product flows. Suppose $\pi(t)\colon X \times U \to X \times U$ is a skew product flow for the complete metric space of processes U. The flow $\pi(t)$ is said to be *linear* if each process u is linear; that is, $u(t,s,x)$ is linear in x.

COROLLARY 3.7.5. *Suppose u is a linear process in a Banach X with $H(u)$ compact. Suppose the skew product flow $\pi(t)\colon X\times H(u) \to X\times H(u)$ is asymptotically smooth and let $\omega_\sigma(u)$ be the ω-limit set of $H(u)$ with respect to $\sigma(t)$. Then zero is u.a.s. for u if and only if $v(t,0,x)\to 0$ as $t\to\infty$ for every $x\in X$ and every $v \in \omega_\sigma(u)$.*

PROOF. Since the process u is linear, every $v \in \omega_\sigma(u)$ is linear. If $v(t,0,x) \to 0$ as $t \to \infty$ for every $x \in X$ and every $v \in \omega_\sigma(u)$, it follows that there is a constant $k_s(v)$ such that $|v(t,s,x)| \leq k_s(v)|x|$ for $t \geq 0, x \in X$. The principle of uniform boundedness implies that we can choose $k_s(v)$ independent of s. Therefore, zero is u.s. for every process $v \in \omega_\sigma(u)$. An application of Theorem 3.7.4 completes the proof of the result.

We end this section with some remarks about the relationship between the definition of a skew product flow for processes and the definition for evolutionary equations. Suppose $\dot{x} = f(t,x)$ is an abstract evolutionary equation for which we know that there is a unique solution $\phi(t,s,x,f)$ through any $(s,x) \in R \times X$ where X is a complete metric space, $f \in \mathcal{F}$ a complete metric space which contains $\tilde{\sigma}(t)f$ if it contains f, $\tilde{\sigma}(t)f(s,x) = f(t+s,x)$. Let $\gamma_{\tilde{\sigma}}^+(f) = \bigcup_{t\geq 0} \tilde{\sigma}(t)f$, $H(f) = \text{Cl}\,\gamma_{\tilde{\sigma}}^+(f)$. Now, suppose that the solution $\phi(t,s,x,f)$ exists for all $t \geq s$. If we define $u(t,s,x,f) = \phi(t+s,s,x,f)$ then $u(\cdot,\cdot,\cdot,f)$ is a process on X.

Let $\gamma_\sigma^+(u), H(u)$ be defined as before for the process u. If $v \in \gamma_\sigma^+(u)$, then $v = \sigma(\tau)u$ for some τ, and

$$v(t,s,x,f) = u(t, s+\tau, x, f) = \phi(t+s+\tau, s+\tau, x, f) = \phi(t, 0, x, \tilde{\sigma}(s+\tau)f).$$

If appropriate continuity conditions are satisfied, then

$$H(u) = \phi(\cdot, 0, \cdot, H(f)).$$

We have already defined the skew product flow on $H(u)$ by

$$\pi(t)\colon X \times H(u) \to X \times H(u),$$
$$\pi(t)(x,v) = (v(t,0,x,f), \sigma(t)v) = (\phi(t,0,x,f), \phi(\cdot,0,\cdot,\sigma(t)g))$$

where $v(t,s,x,g) = \phi(t+s,s,x,g)$ for some $g \in H(f)$.

We can also define a skew product flow on $H(f)$ by

$$\tilde{\pi}(t)\colon X \times H(f) \to X \times H(f), \qquad \tilde{\pi}(t)(x,g) = (\phi(t,0,x,g), \sigma(t)g).$$

Under appropriate continuity conditions, the skew product flow $\tilde{\pi}$ on $H(u)$ and the skew product flow π on $H(f)$ should be equivalent. In particular situations, it is not difficult to verify that one can work with $\tilde{\pi}(t)$ on $H(f)$ or $\pi(t)$ on $H(u)$, whichever is more convenient.

3.8. Gradient systems.

In this section, we consider a special class of systems, called gradient systems, for which the structure of the flow on the global attractor can be described in some detail.

Let $T(t), t \geq 0$, be a C^r-semigroup on a complete metric space X and let E be the set of equilibrium points of $T(t)$; that is, $x \in E$ if and only if $T(t)x = x$ for $t \geq 0$. Recall that an equilibrium point is hyperbolic if the spectrum $\sigma(DT(t)(x))$ does not intersect the unit circle with center zero in $\mathbf{C}$.

If $x \in E$, the unstable set of x is $W^u(x) = \{y \in X\colon T(-t)y \text{ is defined for } t \geq 0 \text{ and } T(-t)y \to x \text{ as } t \to \infty\}$. If x is hyperbolic, then there is a neighborhood U of x such that $W^u_{\text{loc}}(x) \stackrel{\text{def}}{=} \{y \in W^u(x)\colon T(-t)y \in U, t \geq 0\}$ is a submanifold of X. The stable set $W^s(x)$ is the set $\{y \in X\colon T(t)y \to x \text{ as } t \to \infty\}$. If x is hyperbolic, there is a neighborhood U of x such that $W^s_{\text{loc}}(x) \stackrel{\text{def}}{=} \{y \in W^s(x)\colon T(t)x \in U, t \geq 0\}$ is a submanifold of X. If $T(t)$ is one-to-one and $DT(t)(y), t \geq 0, y \in X$, is an isomorphism, then $W^u(x)$ is an immersed submanifold of X (see the Appendix).

DEFINITION 3.8.1. A strongly continuous C^r-semigroup $T(t)\colon X \to X, t \geq 0, r \geq 1$, is said to be a *gradient system* if

(i) Each bounded positive orbit is precompact.

(ii) There exists a Lyapunov function for $T(t)$; that is, there is a continuous function $V\colon X \to R$ with the property that

(ii$_1$) $V(x)$ is bounded below,

(ii$_2$) $V(x) \to \infty$ as $|x| \to \infty$,

(ii$_3$) $V(T(t)x)$ is nonincreasing in t for each x in X,

(ii$_4$) if x is such that $T(t)x$ is defined for $t \in R$ and $V(T(t)x) = V(x)$ for $t \in R$, then x is an equilibrium point.

For gradient systems, one can prove the following result.

LEMMA 3.8.2. *If $T(t)$ is a gradient system, then the ω-limit set $\omega(x)$ for each $x \in X$ belongs to E. If $\gamma^-(x)$ is a precompact orbit through x, then the α-limit set $\alpha(x)$ of x belongs to E.*

PROOF. Since the positive orbit $\gamma^+(x)$ through x is precompact, the set $\{V(T(t)x), t \geq 0\}$ is bounded below and then (ii$_3$) implies $V(T(t)x) \to c$, a constant, as $t \to \infty$. Since $\gamma^+(x)$ is precompact, $\omega(x)$ is compact and invariant. The fact that V is continuous implies $V(T(t)y) = c$ for $t \in R$ and all $y \in \omega(x)$. Hypothesis (ii$_4$) implies $y \in E$.

Suppose $\gamma^-(x)$ is a precompact orbit, $x \notin E$. Then $\alpha(x)$ is compact. If $y \in \alpha(x)$, then there exists a sequence $t_n \to -\infty$ as $n \to \infty$ such that $T(t_n)x \to y$ as $n \to \infty$. Choose the t_n so that $t_n - t_{n-1} \geq 1$, for all n. Then, for any $t \in (0,1)$, condition (ii$_3$) implies that $V(T(t_{n-1})x) \leq V(T(t_n + t)x) \leq V(T(t_n)x)$ for all n and so $V(T(t_n + t)x) \to y$ as $n \to \infty$. Since $V(T(t_n + t)x)$ also converges to $V(T(t)y)$ as $n \to \infty$, it follows that $V(T(t)y) = V(y)$ for all $t \in (0,1)$ and, thus, for all $t \in R$. Property (ii$_4$) implies y is an equilibrium point.

Lemma 3.8.2 implies that $T(t), t \geq 0$, is point dissipative if and only if E is bounded.

THEOREM 3.8.3. *Suppose $T(t)$ is a gradient system on X for which $V(T(t)x) < V(x)$ for $t > 0$ and $x \notin E$. If $x_0 \in E$ is hyperbolic with $\dim W^u(x_0) < \infty$, $T(t)$ is one-to-one on $W^u(x_0)$, and $DT(t)(x)$ is an isomorphism for each $x \in W^u(x_0)$, then $W^u(x_0)$ is an embedded submanifold of X.*

For the proof of this theorem, we need the following

LEMMA 3.8.4. *If $T(t)$ is a gradient system for which $V(T(t)x) < V(x)$ for $t > 0$ and $x_0 \in E$ is hyperbolic, then there is a neighborhood U_0 of x_0 such that, if $x \in U_0 \setminus W^s_{\mathrm{loc}}(x_0)$, then there is a $t_0 > 0$ such that $T(t)x \notin U_0$ for $t \geq t_0$. Thus, $T(t)x$ eventually leaves U_0 never to return.*

PROOF. It is clear that we need only consider the case where $\dim W^u(x_0) \geq 1$. Choose $\delta_1 > 0$ so that $T(t)x \in B_{\delta_1}(x_0) = \{x \colon |x - x_0| < \delta_1\}$ for all $t \leq 0$ implies that $x \in W^u_{\mathrm{loc}}(x_0)$. There are constants $K, \alpha > 0$ such that $\mathrm{dist}(T(t)x, W^u_{\mathrm{loc}}(x_0)) \leq K \exp(-\alpha t)$ as long as $T(t)x$ remains inside $B_{\delta_1}(x_0)$. For any δ_0, $0 < \delta_0 < \delta_1$, there is a $t_2 > 0$ such that $T(t_2)B_{\delta_0}(x_0) \subset B_{\delta_1}(x_0)$. Let

$$W_\eta = \{x \colon \delta_0 \leq |x - x_0| \leq \delta_1, \mathrm{dist}(x, W^u_{\mathrm{loc}}(x_0)) < \eta\}$$

and choose η so that $\sup\{V(x), x \in W_\eta\} < V(x_0)$. This is always possible since $V(T(t)x) < V(x)$ if $x \notin E$ and $\dim W^u_{\mathrm{loc}}(x_0) \geq 1$.

Choose $t_1 > 0$ so that $K\exp(-\alpha t_1) < \eta$ and choose a neighborhood U_0 of x_0 such that $T(t)U_0 \subset B_{\delta_0}(x_0)$ for $0 \leq t \leq t_1$ and $\sup\{V(x), x \in W\} < \inf\{V(y), y \in U_0\}$. If $x \in U_0 \backslash W_{\mathrm{loc}}^s(x_0)$, then $T(t)x$ eventually leaves $B_{\delta_1}(x_0)$ and there is a $t_0 > 0$ and $\varepsilon_0 > 0$ such that $T(t)x \in \mathrm{Cl}\, B_{\delta_1}(x_0), 0 \leq t \leq t_0$, and $T(t_0 + \varepsilon)x \notin \mathrm{Cl}\, B_{\delta_1}(x_0)$. Since $t_0 \geq t_1$, we have $T(t_0)x \in W$ and $V(T(t)x) \leq V(T(t_0)x) < \inf\{V(y), y \in U_0\}$ for $t \geq t_0$. Thus $T(t)x \notin U_0$ for $t \geq t_0$ and the lemma is proved.

PROOF OF THEOREM 3.8.3. Since $T(t)$ is one-to-one, $DT(t)$ is an isomorphism, and $W_{\mathrm{loc}}^u(x_0)$ is a submanifold of X, it follows that $W^u(x_0)$ can be immersed into X. To prove that it can be embedded, we must prove that it is isomorphic to a submanifold of X. We do this by showing that, if $x \in W^u(x_0) \backslash W_{\mathrm{loc}}^u(x_0)$, there is a neighborhood U_1 of x and a $t_0 > 0$ such that $U_1 \cap W^u(x_0) \subset T(t_0)W_{\mathrm{loc}}^u(x_0)$. Using the set W_η and constant δ_1 in the proof of Lemma 3.8.4, there are $t_1 \geq 0, t_0 > 0$, such that $T(-t)x \to x_0$ as $t \to \infty, |T(-t_1)x - x_0| > \delta_1$ and $T(-t_0)x \in W_1 = W_\eta \cap W_{\mathrm{loc}}^u(x_0)$. Since $\dim W_{\mathrm{loc}}^u(x_0) < \infty$, the set W_1 is compact and there is a $t_2 > 0$ such that $\sup\{y, y \in T(t_2)W_1\} < V(x)$ and there is neighborhood U_1 of x such that $\sup\{V(y), y \in T(t_2)W_1\} < \inf\{V(z), z \in U_1\}$. If $y \in U_1 \cap W^u(x_0)$, then there is a $t_3 \geq 0$ such that $T(-t_3)y \in W$. Thus, $t_3 < t_2$ and $U_1 \cap W^u(x_0) \subset T(t_2)W_{\mathrm{loc}}^u(x_0)$. This proves the theorem.

THEOREM 3.8.5. *If $T(t), t \geq 0$, is a gradient system, asymptotically smooth, and E is bounded, then there is a global attractor A for $T(t)$ and $A = W^u(E) = \{y \in X \colon T(-t)y$ is defined for $t \geq 0$ and $T(-t)y \to E$ as $t \to \infty\}$. If X is a Banach space, then A is connected. If, in addition, each element of E is hyperbolic, then E is a finite set and*

$$A = \bigcup_{x \in E} W^u(x).$$

PROOF. Since $T(t)$ is a gradient system, Lemma 3.8.2 and the fact that E is bounded imply that $T(t)$ is point dissipative. The conditions on V imply that orbits of bounded sets are bounded. Theorem 3.4.6 implies that the global attractor A exists and is connected if X is a Banach space.

Since every element in A has a globally defined orbit through it and A is compact, it follows from Lemma 3.8.2 that $\alpha(x) \subset E$ for each $x \in A$; that is, $A = W^u(E)$.

If each element of E is hyperbolic, then $E \subset A$ and A compact imply that E is a finite set. Thus, $W^u(E) = \bigcup_{y \in E} W^u(y)$. The theorem is proved.

For completely continuous gradient systems, one can state

THEOREM 3.8.6. *If $T(t), t \geq 0$, is a gradient system and there is a $t_1 \geq 0$ such that $T(t)$ is completely continuous for $t > t_1$ and E is bounded, then the conclusions of Theorem 3.8.3 are valid.*

Suppose the conditions of either Theorem 3.8.5 or 3.8.6 are satisfied so that there is a global attractor A. Let $E = (x_1, \ldots, x_p)$ where each x_j is hyperbolic and let $v^1 > v^2 > \cdots > v^q$ be the distinct points of the set $\{V(x_1),\ldots,V(x_p)\}$. Define

(8.1) $$E^j = \{x \in E\colon V(x) = v^j\}, \qquad j = 1, 2, \ldots, q.$$

From property (ii)$_3$ and (ii)$_4$ in Definition 3.8.1, the sets $\{E^1,\ldots,E^q\}$ form a Morse decomposition of A; that is, each E^j is compact, invariant, and, for any $x \in A\backslash\bigcup_{j=1}^{q} E^j$, there are integers k and l with $k < l$ such that $\alpha(x) \in E^k$, $\omega(x) \in E^l$. With this notation, define, for $1 \leq k \leq q$,

(8.2) $$A^k = \bigcup\{W^u(x)\colon x \in \bigcup_{j=k}^{q} E^j\}, \qquad U^k = \{x \in X\colon V(x) < v^k\}.$$

We say that A^k is the *compact attractor for $T(t)$ restricted to the set U^{k-1}* if, for any closed bounded set W in U^{k-1} such that $\sup\{V(x)\colon x \in W\} < v^{k-1}$, A^k attracts W under $T(t)$.

We can now prove

THEOREM 3.8.7. *Suppose the conditions of either Theorem 3.8.5 or 3.8.6 are satisfied and the sets A^k, U^k are defined in (8.1) and (8.2). Then $A^1 = A$, the global attractor for $T(t)$, and for $k > 1$, A^k is the compact attractor for $T(t)$ restricted to the set U^{k-1}. In particular, for any compact set in $A^{k-1}\backslash E^{k-1}$, $\mathrm{dist}(T(t)K, A^k) \to 0$ as $t \to \infty$.*

PROOF. It is clear that $T_0(t)U^{k-1} \subset U^{k-1}$ for $t \geq 0$. Also, for any $x \in U^{k-1}$, $\omega(x) \in \bigcup_{j=k}^{q} E^j$, which is bounded. Furthermore, for any constant $c < v^{k-1}$, the set $\{x\colon V(x) \leq c\}$ is positively invariant under $T(t)$ since $T(t)$ is a gradient system. This set also is a complete metric space. Since $T(t)$ is asymptotically smooth, Theorem 3.4.6 implies that $T(t)$ restricted to U^{k-1} has a compact attractor. The same arguments as used in the proof of Theorem 3.8.5 imply that the attractor is A^k.

Let us now consider a family of gradient semigroups $T_\lambda(t), t \geq 0$, on X, where λ is a parameter in a subset of a Banach space, and $T_\lambda(t)x$ is continuous in (t,x,λ) and C^r in x for some $r \geq 1$. Let E_λ be the set of equilibrium points of $T_\lambda(t)$ and let A_λ be the global attractor of $T_\lambda(t)$. Let us recall that a family of subsets B_λ, $\lambda \in \Lambda$, of X is said to be *upper semicontinuous* at λ_0 [*lower semicontinuous* at λ_0] if $\delta(B_\lambda, B_{\lambda_0})$ [$\delta(B_{\lambda_0}, B_\lambda)$] $\to 0$ as $\lambda \to \lambda_0$. It is said to be *continuous* at λ_0 if it is both upper and lower semicontinuous at λ_0. The distance $\delta(B, A)$ from B to A is defined in Chapter 2, formula (2.1).

THEOREM 3.8.8. *Suppose Λ is a subset of a Banach space and, for each $\lambda \in \Lambda, T_\lambda(t), t \geq 0$, is a C^r-gradient semigroup on $X, r \geq 1$, and the family $\{T_\lambda(t), \lambda \in \Lambda\}$ of semigroups is collectively asymptotically smooth. If $\bigcup\{E_\lambda, \lambda \in \Lambda\}$ is bounded, then the global attractor A_λ of $T_\lambda(t)$ exists and A_λ is upper semicontinuous at λ_0.*

PROOF. Since $\bigcup\{E_\lambda, \lambda \in \Lambda\}$ is bounded and $T_\lambda(t)$ is gradient, it follows that there is a bounded set $B \subset X$ such that B attracts points of X under $T_\lambda(t)$ for $\lambda \in \Lambda$. The conclusion in the theorem is now a consequence of Theorem 3.5.3.

Theorem 3.8.8 says that the "size" of the global attractor A_λ cannot increase much for λ near λ_0 if the hypotheses of the theorem are satisfied. It may, however, become smaller unless additional conditions are imposed on the flow on A_{λ_0}. The following result is a special case of a more general fact that will be mentioned in §4.10.4.

THEOREM 3.8.9. *Assume the conditions of Theorem 3.8.8 are satisfied. If the equilibrium points of E_λ are hyperbolic for each λ in a neighborhood of λ_0, then the global attractor A_λ is continuous at λ_0.*

Theorem 3.8.9 gives good information about the size of the global attractor A_λ near λ_0. Under the local condition of hyperbolicity of the equilibria, the size of the global attractor does not change. However, the flows may not stay the same for each λ. To state a general result in this direction, we need more notation. If $x \in E_\lambda$, let $W_\lambda^s(x)$ be the stable set of x, and let $W_\lambda^u(x)$ be the unstable set of x. If x is hyperbolic, then $W_{\lambda,\text{loc}}^s(x)$ is a submanifold of codimension equal to $\dim W_{\lambda,\text{loc}}^u(x)$ at x.

As in §2.7, we say T_λ is *equivalent* to $T_{\lambda_0}(t)$, that is, $T_\lambda(t) \sim T_{\lambda_0}(t)$, if there is a homeomorphism $h\colon A_\lambda \to A_{\lambda_0}$ which preserves orbits and the sense of orientation in t. We say $T_\lambda(t)$ is *A-stable* at λ_0 if there is a neighborhood V of λ_0 such that $T_\lambda(t) \sim T_{\lambda_0}(t)$ for all $\lambda \in V$.

We can now prove the following result.

THEOREM 3.8.10. *Suppose $T_\lambda(t)$ satisfies the conditions of Theorem 3.8.8 and, as $\lambda \to \lambda_0, T_\lambda(1) \to T_{\lambda_0}(1)$ in $C^r(U,X), r \geq 1$, where U is a neighborhood of A_{λ_0}. In addition, suppose*

(i) *T_λ is one-to-one on A_λ and the derivative $DT_\lambda(t)\psi$ is an isomorphism for each $\lambda \in \Lambda, \psi \in A_\lambda$.*

(ii) *E_{λ_0} is hyperbolic and $W_{\lambda_0}^u(x) \stackrel{+}{\pitchfork} W_{\lambda_0,\text{loc}}^s(y)$ for each $x, y \in E_{\lambda_0}$ ($\stackrel{+}{\pitchfork}$ means transversal).*

Then, $T_{\lambda_0}(t)$ is A-stable.

PROOF. Theorem 3.8.8 implies that A_λ is upper semicontinuous at λ_0. The hypotheses (i), (ii) imply that $T_{\lambda_0}(t)$ is Morse-Smale. We obtain the result from §2.7 using the family of mappings $\{T_\lambda(1), \lambda \in \Lambda\}$.

For gradient systems, it is frequently true that the stable manifolds for hyperbolic equilibrium points are embedded submanifolds. A result in this direction is the following

THEOREM 3.8.11. *Suppose $T(t)$ is a gradient system for which $V(T(t)x) < V(x)$ for $t > 0$ if $x \notin E$. If x_0 is hyperbolic, $\dim W^u(x_0) < \infty$, and $W^s(x_0)$ is an invariant injectively immersed submanifold of X, then $W^s(x_0)$ is an embedded submanifold and there is an open neighborhood Q of $W^s(x_0)$ such that, if $x \in Q \setminus W^s(x_0)$, there exists a $t_0 > 0$ such that $T(t)x$ is defined for $t \geq 0$ and $T(t)x \notin Q$ for $t \geq t_0$.*

PROOF. If $x \in W^s(x_0)$, there is a $t_1 > 0$ such that $T(t_1)x \in W^s(x_0) \cap U_0$ where U_0 is the neighborhood provided by Lemma 3.8.4. There is a neighborhood

U_1 of x with $T(t_1)U_1 \subset U_0$. With t_0 as in Lemma 3.8.4, $T(t)(U_1 \cap W^s(x_0)) \subset W^s_{\text{loc}}(x_0)$ for $t \geq t_1 + t_0$, which proves $W^s(x_0)$ is embedded. To complete the proof, let $Q = \{y \in X \colon \text{there is a } t \geq 0 \text{ such that } T(t)y \in U_0\}$. The set Q is a neighborhood of $W^s(x_0)$ and satisfies the conditions stated in the theorem.

3.9. Dissipativeness in two spaces.

In this section, we state the analogues for semigroups of the results of §2.9 for mappings. The proofs are omitted since they follow along the same lines as the ones for mappings.

We now suppose that X_1, X_2 are Banach spaces with norms respectively given by $|\cdot|_1, |\cdot|_2$. We also assume

(H_1) $i\colon X_1 \to X_2$ is a compact embedding.

(H_2) $T(t) = S(t) + U(t)$ where $T(t)x, S(t)x, U(t)x$ are continuous in t, x for $t \geq 0$ and $x \in X_j, j = 1, 2$, and map X_j into $X_j, j = 1, 2$. Furthermore, $S(t)0 = 0$, and $S(t)$ is a contraction on X_1 with contraction constant $k(t) \to 0$ as $t \to \infty$.

(H_3) For any $t \geq 0$, if $B \subset X_1$, B and $\{U(t)B, 0 \leq t \leq \tau\}$ are bounded in X_2, then $\{U(t)B, 0 \leq t \leq \tau\}$ is bounded in X_1.

THEOREM 3.9.1. *If* (H_1), (H_2), (H_3) *are satisfied, then the following statements are equivalent*:

(i) $T(t)$ *is point dissipative in* X_1,

(ii) $T(t)$ *is bounded dissipative in* X_1,

(iii) *there is a bounded set in* X_2 *that attracts points in* X_1.

COROLLARY 3.9.2. *If* (H_1), (H_2), (H_3) *are satisfied*, (iii) *of Theorem 3.9.1 is satisfied, and* $T(t), t \geq 0$, *is asymptotically smooth in* X_1, *then there exists a connected global attractor in* X_1. *Furthermore, if* $T(t)$ *is* α-*condensing in* X_1, *then there is a fixed point in* X_1.

THEOREM 3.9.3. *Suppose* (H_1), (H_2), (H_3) *are satisfied and* $S(t), t > 0$, *is also a contraction on* X_2 *with contraction constant* $k(t) \to 0$ *as* $t \to \infty$. *If either*

(i) X_1 *is dense in* X_2;

(ii) $U(t)\colon X_2 \to X_1$ *is such that, for any* $\tau \geq 0$, *if* B *and* $\{U(t)B, 0 \leq t \leq \tau\}$ *are bounded in* X_2, *then* $\{U(t)B, 0 \leq t \leq \tau\}$ *is bounded in* X_1. *Then* $T(t)$ *is a conditional* α-*contraction on* X_2. *Also, if* J *is a compact invariant set in* X_2, *then* J *is the closure in* X_2 *of a bounded set in* X_1. *If, in addition,* $T(t)$ *is point dissipative in* X_2, *then there is a maximal, compact invariant set in* X_2 *which is u.a.s. and attracts a neighborhood of any compact set of* X_2.

THEOREM 3.9.4. *Suppose* (H_1), (H_2) *are satisfied and*

(i) $S(t)$ *is a linear semigroup on both spaces with norm* $\leq K(t)$ *and* $K(t) \to 0$ *as* $t \to \infty$,

(ii) $U(t)\colon X_2 \to X_1$ *is continuous and, for any* $\tau \geq 0$, *if* $B, \{U(t)B, 0 \leq t \leq \tau\}$ *are bounded in* X_2, *then* $U(t)(B)$ *is bounded in* X_1,

(iii) $U(t)$ *is conditionally completely continuous on* X_1 *and* X_2; *that is, for* $j = 1$ *or* 2, *if* $B, \{T(s)B, 0 \leq s \leq t\}$ *are bounded in* X_j, *then* $U(t)B$ *is precompact in* X_j.

Under these assumptions, if J is a closed bounded invariant set in X_2, then $J \subset X_1$ and J is a compact invariant set in X_1.

A consequence of the above theorems which is useful in differential equations is

COROLLARY 3.9.5. *Suppose*
(i) X_1 *is compactly embedded in* X_2, X_1 *dense in* X_2.
(ii) $T(t) = S(t) + U(t)\colon X_j \to X_j$, $S(t)\colon X_j \to X_j$ *is a linear semigroup with norm* $\leq K(t)$ *and* $K(t) \to 0$ *as* $t \to \infty$, $j = 1, 2$.
(iii) $U(t)\colon X_2 \to X_1$ *is continuous and, for any* $\tau \geq 0$, *if* $B, \{U(t)B, 0 \leq t \leq \tau\}$ *are bounded in* X_2, *then* $\{U(t)B, 0 \leq t \leq \tau\}$ *is bounded in* X_1.
(iv) $T(t)$ *is point dissipative in* X_2.
Then there is a global attractor A for $T(t)$ in X_2 and $T(t)$ is bounded dissipative in X_1.

If, in addition,
(v) $U(t)$ *is conditionally completely continuous in* $X_j, j = 1, 2$,
then $A \subset X_1$ and is a global attractor for $T(t)$ in X_1.

Theorem 3.9.4 is a result on regularity; that is, one assumes $J \subset X_2$, the large space, and concludes $J \subset X_1$, the small space. For evolutionary equations, this has important implications. For example, consider an evolutionary equation

$$(9.1) \qquad du/dt = Au + f(u),$$

where A is the infinitesimal generator of a linear C^0-semigroup $S(t)$ on two Banach spaces X_1, X_2 with X_1 compactly embedded in X_2 and dense in X_2. Also, suppose there are constants $k > 0, \delta > 0$ such that

$$(9.2) \qquad |S(t)|_{X_j} \leq ke^{-\delta t}, \qquad t \geq 0.$$

Suppose also that (9.1) generates a C^0-semigroup $T(t), t \geq 0$, on X_1 and X_2 with the variation of constants formula being valid:

$$T(t)u_0 = S(t)u_0 + \int_0^t S(t-\tau)f(u(\tau))d\tau.$$

Define

$$U(t)u_0 = \int_0^t S(t-\tau)f(u(\tau))d\tau.$$

Let us now suppose that X_1 is a Banach space obtained by using the graph norm of domain $D(A + f)$ of $A + f$ in X_2. If the conditions of Theorem 3.9.4 are satisfied, then any compact invariant set J in X_2 must belong to X_1 and must be a compact invariant set in X_1. In particular, $J \subset D(A + f)$ and $T(t)|J$ must be a C^1-function of t. In particular, if J corresponds to a periodic orbit Γ, then $T(t)|\Gamma$ is a C^1-function of t and we have: *every periodic orbit Γ is a C^1-manifold*. This remark is important for obtaining the behavior of solutions of the evolutionary equation near Γ.

3.10. Properties of the flow on the global attractor.

If $T(t), t \geq 0$, is a C^0-semigroup on a complete metric space X which has a global attractor A, then the semigroup restricted to A gives a semigroup $T_A(t): A \to A, T_A(t)x = T(t)x, x \in A, t \geq 0$. The semigroup $T_A(t)$ can have interesting properties that are not shared by $T(t)$ and are a consequence of the existence of a global attractor.

We have already observed the following result, but restate it for emphasis.

THEOREM 3.10.1. *If $T_A(t): A \to A, t \geq 0$, is one-to-one, then $T_A(t)$ is a group.*

It should be emphasized that the hypothesis in the theorem says that $T(t)$ is one-to-one on A. The semigroup $T(t)$ can have this property and not be one-to-one on the whole space X. Examples are easy to construct. Cases of this type arise in evolutionary equations—especially delay differential equations (see §4.1).

It is certainly much nicer to have $T_A(t)$ be a group. In each particular application, this leads to the natural problem of discussing the one-to-oneness of the semigroup $T(t)$ restricted to compact invariant sets. As applications are considered, the reader will find it instructive to always consider this point.

For semigroups $T(t)$ defined by evolutionary equations in infinite dimensional spaces, it is not to be generally expected that, for each $x \in X$, $T(t)x$ is differentiable in t. However, for many applications, the function $T(t)x$ will be differentiable in t if x belongs to the attractor. Results of this type have been obtained by Hale and Scheurle [1985]. We now summarize some special cases of these results with an indication of the proofs.

Consider an abstract evolutionary equation

$$(10.1) \qquad \dot{u} = -Bu + f(u)$$

with $u(0) = \phi \in X$, a Banach space. Suppose that $B: D(B) \subset X \to X$ is a linear operator which is the infinitesimal generator of a C^0-semigroup $e^{-Bt}, t \geq 0$, which has the property that there is a positive constant δ such that the radius $r_e(e^{-Bt})$ of the essential spectrum satisfies

$$(10.2) \qquad r_e(e^{-Bt}) \leq e^{-\delta t}, \qquad t \geq 0.$$

Suppose (10.1) generates a semigroup $T(t)$ on X. One can prove the following result.

THEOREM 3.10.2. *Suppose B satisfies (10.2), $f \in C^k(X, X), 0 \leq k \leq \infty$ [or f is analytic on X], and there is a global attractor A for (10.1). Then there is a constant $r > 0$ such that $T_A(t) = T(t)|A$ has the property that, for each $\phi \in A$, the mapping $t \mapsto T_A(t)\phi$ is C^k [or analytic] if $|f'(u)| < r$ for u in a neighborhood of A.*

This is a rather surprising result because it implies that the semigroup $T(t)$ restricted to A is as smooth in t as the vector field f. This is a consequence of the fact that each orbit through a point $\phi \in A$ can be extended to $-\infty$ and remains in a compact set. The theorem could be stated for orbits of this type with the assumption on the existence of the global attractor being unnecessary. The bound r on the derivative of f is related to the bound on the semigroup e^{Bt} and the decay rate δ. In applications that are considered in Chapter 4, there has been no example given to show that any restriction of f is necessary.

It is certainly an interesting problem to determine those general classes of functions f for which the regularity properties of $T_A(t)\phi$ in t in Theorem 3.10.2 hold.

Let us pose the problem in a more precise way. Suppose e^{-Bt} has the exponential decay as before and suppose that $T(t)\phi$ is defined for all t in R and is uniformly bounded. Is $T(t)\phi$ smooth in t? Without some restrictions on the functions f, this is obviously not true. In fact, the operator B could be of the form $B = C + \delta I$ where e^{-Ct} is a group with $|e^{-Ct}| = 1$ for all t and $e^{-Ct}\phi$ is not differentiable in t for every ϕ. With $f(u) = \delta u$, we have $T(t) = e^{-Ct}$ and the answer to the above question is negative. In many applications, the function f is compact. Let us therefore pose the question for f compact. A rather surprising result which was communicated to the author by Shmuel Kaniel is

PROPOSITION 3.10.3. *There exists an equation* (10.1) *satisfying* (10.2) *with f compact and continuously differentiable and a $\phi \in X$ such that $T(t)\phi$ is defined and uniformly bounded for t in R and $T(t)\phi$ is not differentiable in t.*

PROOF. Suppose there exists a sequence of real numbers $\lambda_k \to \infty$ as $k \to \infty$ such that $-i\lambda_k$ are eigenvalues of the operator C acting on a Hilbert space X with a complete set of eigenvectors ϕ_k. For simplicity, take $\lambda_k = k$. Also, let $a_k = 1/k$. The function

$$u(t) = \sum_{k=1}^{\infty} a_k e^{i\lambda_k t} \phi_k$$

is a solution of the equation $\partial u/\partial t = -Cu$. It is also nondifferentiable since $\sum |a_k|^2 \lambda_k^2 = \infty$.

Let $B = C + \delta I$. We construct an f which is compact and C^1 such that $f = \delta I$ on the set S:

$$S = \left\{ b\colon b = \sum b_k \phi_k,\ |b_k| \leq 1/k,\ k = 1, 2, \ldots \right\}.$$

This will prove the proposition since $u(t) \in S$ for all t in R and is a solution of (10.1) which is not C^1 in t.

To construct such an f, let $\varsigma(h)$ be a C^∞ real-valued function on $[0, \infty)$ such that

$$\varsigma(r) = r, \quad 0 \leq h \leq 1,$$
$$\varsigma'(r) \text{ monotone decreasing to zero on } [1, 2],$$
$$\varsigma(r) = p, \quad 2 \leq r, \quad \text{where } p = \varsigma(2).$$

Let $h(z)$ be a function defined on the complex plane $z = x + iy$ such that $h(z) = \varsigma(r)e^{i\theta}$ if $z = re^{i\theta}$. Then h is C^∞ and there is a constant M such that h satisfies

$$h(z) = z \quad \text{for } |z| \leq 1,$$
$$|h(z)| \leq p \quad \text{for all } z,$$
$$|\partial h(z)/\partial x| \leq M, |\partial h(z)/\partial y| \leq M \quad \text{for all } z.$$

With this notation, define $f: X \to X$ as follows:

$$f(u) = \sum b_k \phi_k, \quad b_k = \frac{\delta}{k} h(ka_k) \quad \text{if } u = \sum a_k \phi_k.$$

On the set S, $k|a_k| \leq 1$ and so $h(ka_k) = ka_k$ and $f(u) = \delta u$. The function is compact since its range is in the set δpS and S is compact.

Let us compute the derivative of f at u in the direction of v. If $v = \sum c_k \phi_k = \sum (\xi_k + i\eta_k)\phi_k$, then

$$Df(u)(v) = \sum \left[\xi_k \frac{\partial h}{\partial x}(ka_k) + i\eta_k \frac{\partial h}{\partial y}(ka_k) \right] \phi_k$$

and $|Df(u)(v)| \leq M|v|$. Thus, f has a bounded Fréchet derivative.

To show that $Df(u)(v)$ is continuous in u, let $u_j = \sum a_k^{(j)} \phi_k, j = 1, 2$, and suppose that $\sum |c_k| \leq N$. Fix $\varepsilon > 0$ and take m so that

$$M \left(\sum_{k \geq m} |c_k|^2 \right)^{1/2} \leq \varepsilon/4.$$

Now choose $\eta_1 > 0$ so that, if $|a_k^{(1)} - a_k^{(2)}| \leq \eta_1$ for $k \leq m$, then

$$\left| \frac{\partial h}{\partial x}(ka_k^{(1)}) - \frac{\partial h}{\partial x}(ka_k^{(2)}) \right| < \frac{\varepsilon}{4mN}, \quad \left| \frac{\partial h}{\partial y}(ka_k^{(1)}) - \frac{\partial h}{\partial y}(ka_k^{(2)}) \right| \leq \frac{\varepsilon}{4mN}.$$

With these estimates, it is now a straightforward computation to show that there is an $\eta_2 > 0$ such that $|u_1 - u_2| < \eta_2$ implies $|Df(u_1)(v) - Df(u_2)(v)| < \varepsilon$. This completes the proof of the proposition.

In this example given in Proposition 3.10.3, the continuity of $Df(u)(v)$ in u is not uniform in v. Moreover, the second derivative of f does not exist. For this and after several unsuccessful efforts to prove otherwise, we propose the following.

CONJECTURE. *If f is compact and C^2 and $T(t)\phi$ is defined and uniformly bounded for t in R, then $T(t)\phi$ is C^1 in t.*

In the study of the structure of the dynamics of $T(t)$, Theorem 3.10.2 will play a major role. In fact, one cannot begin the detailed structure of a flow without the concept of normal hyperbolicity of compact invariant sets. The definition of this concept uses in a significant way the "tangential" direction of the flow on the invariant set.

We do not want to dwell on this abstract point since it will not be used in these notes. However, it is easy to see the problems involved by considering the following situation. Suppose (10.1) has a periodic orbit γ described by a solution $p(t)$ of period ω. Then γ is a closed curve. In any qualitative theory of differential equations, one must use the concept of hyperbolicity of a periodic orbit. One way to introduce this concept is to use the *Poincaré map* defined in the following way. Let S be a transversal to γ (note that this requires γ to be a C^1-manifold). The fact that γ is periodic allows one to define a mapping π, the Poincaré map, of a subset of S into itself such that γ corresponds to a fixed point p_0 of π. The orbit γ is said to be hyperbolic if the spectrum $\sigma(D\pi(p_0))$ has the property that $|\lambda| \neq 1$ for $\lambda \in \sigma(D\pi\omega(p_0))$ (note again that π must be C^1).

There are two major problems here. The first one is that γ must be a C^1-manifold in order to define the Poincaré map. This problem is overcome if the conditions of Theorem 3.10.2 are satisfied. The second difficulty which is much more serious concerns the fact that π must be a C^1-map. For a general transversal S, one cannot expect that π is C^1 because $\pi(p) = T(t_p)p$ where t_p depends on p. Differentiability of π will require some regularity of $T(t)p$ in t, which in general cannot be expected to be valid.

Another way to define hyperbolicity is to use the *linear variational equation* about $p(t)$:

$$(10.3) \qquad \dot{v} = -Bv + f'(p(t))v.$$

This is a linear equation with periodic coefficients and one can define the characteristic multipliers for (10.3). If $U(t, \sigma)$ is the solution operator of (10.3), then the characteristic multipliers of (10.3) are the elements of the point spectrum of $U(\sigma + \omega, \sigma)$. Let us assume that the radius of the essential spectrum of this operator is < 1 which will be true if f is compact and (10.2) is satisfied. If γ is a C^1-manifold (which is true if the conditions of Theorem 3.10.2 are satisfied), then one is always a multiplier since $\dot{p}(t)$ satisfies (10.3). If one is a simple multiplier of (10.3) and no other multiplier has modulus one, we say γ is hyperbolic.

Let us now indicate why Theorem 3.10.2 is true for the case in which

$$(10.4) \qquad |e^{-Bt}| \leq ke^{-\delta t}, \qquad t \geq 0.$$

If $u(t)$ is a solution of (10.1), defined and bounded for $t \leq 0$, then u is a solution of the equation

$$(10.5) \qquad \dot{v} = -Bv + f(u(t)).$$

Since B satisfies (10.4), the solution of (10.5), which is bounded for $t \leq 0$, is unique, must be $u(t)$, and must satisfy the integral equation

$$(10.6) \quad u = KFu, \qquad (Kg)(t) = \int_{-\infty}^{t} e^{-B(t-s)} g(s)\, ds, \qquad (Fu)(t) = f(u(t)).$$

The main part of the proof is the analysis of the behavior of the operator K on C^k functions on $(-\infty, 0]$ (or functions analytic in a neighborhood of $(-\infty, 0)$). In particular, one must show that K maps these classes of functions into themselves. The next step is to apply the contraction mapping principle. The restriction on the derivative of f in the statement of the theorem arises in the verification of these two steps.

As a final remark, we emphasize that there are more general results as well as applications in Hale and Scheurle [1985].

Notes and remarks.

Skew product flows for processes in §3.7 follow Dafermos [1975] and LaSalle [1976] which in turn are extensions of skew product flows associated with ordinary differential equations, integral equations, and delay equations which were considered extensively by Miller and Sell. The problem of showing when such problems have enough smoothness properties to define skew product flows has been completely solved in some particular cases (see Sell [1971], Miller and Sell [1976], Artstein [1977a,b], Kakakostas [1982], and the references therein). Theorem 3.7.4 is essentially contained in Hale and Kato [1978] (see also Cooperman [1978]). It generalizes a result of Artstein [1978] for the case where $X = R^n$. For functional differential equations, see Palmer [1978] and Murakami [1985].

For linear skew product flows generated by processes u on a finite dimensional space with $H(u)$ compact, many interesting results are known concerning exponential dichotomies, spectral theory, etc. (see Sell [1975]). Possible extensions of these results to a Banach space with the corresponding skew product flow asymptotically smooth should be considered.

Definition 3.8.1 can be found in Hale [1985] and is a weakened version of a definition given by Babin and Vishik [1983]. It is a special case of Conley [1972]. Theorem 3.8.3 is due to Henry [1981, p. 156] and Lemma 3.8.4 to Chafee [1977]. Theorem 3.8.7 is essentially due to Babin and Vishik [1983]. The other results are based on Hale [1985] except for Theorem 3.8.9 which was stated in Babin and Vishik [1986]. A complete proof under more general hypotheses is contained in Hale and Raugel [1987] (see §4.10.4).

If the semigroup is not asymptotically smooth and the base space X is reflexive, then it is possible to discuss the existence of global attractors in the weak topology (see Hale and Stavrakakis [1987]).

CHAPTER 4

Applications

4.1. Retarded functional differential equations (RFDE's).

4.1.1. *Properties of the semigroup.* For a given $\delta > 0$, let $C = C([-\delta, 0], R^n)$. Suppose $f: C \to R^n$ is a C^r-function, $r \geq 1$, and that f is a bounded map, that is, takes bounded sets into bounded sets. For any function $x: [-\delta, \alpha] \to R^n$, $\alpha \geq 0$, let x_t for $t \in [0, \alpha]$ be the function from $[-\delta, 0]$ into R^n defined by $x_t(\theta) = x(t+\theta)$. An RFDE is a relation

$$\dot{x}(t) = f(x_t) \tag{1.1}$$

where $\dot{x}(t)$ is the right hand derivative of $x(t)$ at t.

For a given $\phi \in C$, there is a unique solution $x(t, \phi)$ of (1.1) defined on $[-\delta, \alpha_\phi)$, $\alpha_\phi > 0$, with $x_t(\cdot, \phi) \in C$, $t \in [0, \alpha_\phi)$, $x_0(\cdot, \phi) = \phi$. Furthermore, if $\alpha_\phi < \infty$, then $|x_t(\cdot, \phi)| \to \infty$ as $t \to \alpha_\phi^-$ (see Hale [1977]). Let $T(t)\phi = x_t(\cdot, \phi)$ for $t \in [0, \alpha_\phi)$.

THEOREM 4.1.1. *If $T(t)$ is defined for $t \geq 0$, then $T(t): C \to C$, $t \geq 0$, is a C^r-semigroup which is conditionally completely continuous for $t \geq \delta$. If $T(t)$ is a bounded map for each $t \in [0, \delta]$, then $T(t)$ is completely continuous for $t \geq \delta$.*

More generally, if $T(t)$ is a bounded continuous map and $S(t): C \to C$, $t \geq 0$, is the linear semigroup given by

$$S(t)\phi(\theta) = \begin{cases} \phi(t+\theta) - \phi(0) & \text{for } t + \theta < 0, \\ 0 & \text{for } t + \theta \geq 0, \end{cases}$$

then $T(t) = S(t) + U(t)$ where $U(t)$ is compact for $t \geq 0$. Furthermore, for any $\alpha > 0$, there is an equivalent norm in C so that $|S(t)| \leq e^{-\alpha t}$, $t \geq 0$, and $T(t)$ is an α-contraction in this norm.

PROOF. This result can be found in Hale [1977]. We only indicate the proof for the smoothing properties of $T(t)$ mentioned in the theorem. If B is a bounded set in C and $\{T(s)B, 0 \leq s \leq \delta\}$ is bounded, then there is a constant K such that $|f(T(s)B)| \leq K$, $0 \leq s \leq \delta$. If $\phi \in B$ and $x(t) = T(t)\phi(0)$, then $|\dot{x}(s)| \leq K$ for $s \in [0, \delta]$. By the Arzela-Ascoli theorem, this implies that $T(\delta)$ is conditionally completely continuous. For $t \geq \delta$, $T(t) = T(t-\delta)T(\delta)$ and so $T(t)$ is conditionally completely continuous. If $T(t)$ is a bounded map for each $t \geq 0$, then one sees that $T(t)$ is completely continuous for $t \geq \delta$.

Now suppose that $S(t)$ is defined as in the theorem. It is easy to show that $S(t)$ is a semigroup on C. Since $S(t) = 0$ for $t \geq \delta$, for any $\alpha > 0$, there is an M_α such that $|S(t)| \leq M_\alpha \exp(-\alpha t)$, $t \geq 0$. For any $\phi \in C$, let $|\phi|^* = \sup_{t \geq 0} |S(t)\phi| \exp \alpha t$. Then

$$|S(t)\phi|^* = \sup_{\tau \geq 0} |S(\tau)S(t)\phi|e^{\alpha \tau} \leq e^{-\alpha t}|\phi|^*$$

and $|\phi| \leq |\phi|^* \leq M_\alpha |\phi|$. This shows the existence of the equivalent norm as stated in the theorem.

With $U(t)\phi = T(t)\phi - S(t)\phi$, we have

$$U(t)\phi(\theta) = \begin{cases} \phi(0) & \text{for } t + \theta < 0, \\ \phi(0) + \int_0^{t+\theta} f(T(s)\phi)\,ds & \text{for } t + \theta \geq 0. \end{cases}$$

For any bounded set $B \subset C$, the set $\{T(s)B,\ 0 \leq s \leq t\}$ is bounded by hypothesis. Since f is a bounded map, there is a constant M such that $|f(T(s)B)| \leq M$ for $0 \leq s \leq t$. This shows that $U(t)B$ consists of bounded equicontinuous functions and $U(t)$ is completely continuous. This completes the proof of the theorem.

The decomposition of $T(t)$ as $S(t) + U(t)$ is very natural. In fact, let us consider equation (1.1) as a perturbation of the zero vector field

$$\dot{x}(t) = 0 \cdot x_t + f(x_t).$$

Also let $\tilde{S}(t)\colon C \to C$ be the semigroup defined by the equation $\dot{x}(t) = 0 \cdot x_t$,

$$\tilde{S}(t)\phi(\theta) = \begin{cases} \phi(t+\theta) & \text{for } t + \theta < 0, \\ \phi(0) & \text{for } t + \theta > 0, \end{cases}$$

and let $\tilde{U}(t)\colon C \to C$ be defined by

$$\tilde{U}(t)\phi(\theta) = \begin{cases} 0 & \text{for } t + \theta < 0, \\ \int_0^{t+\theta} f(T(s)\phi)\,ds & \text{for } t + \theta \geq 0; \end{cases}$$

then $T(t)\phi = \tilde{S}(t)\phi + \tilde{U}(t)\phi$. The semigroup $\tilde{S}(t)$ has 1 as an eigenvalue for every $t \geq 0$ since it leaves the constant functions invariant. If we write C as $C = C_0 \oplus P$ where $C_0 = \{\phi \in C\colon \phi(0) = 0\}$ and P is the subspace of C given by the constant functions, then $\tilde{S}(t)|C_0 = 0$ for $t \geq \delta$. The semigroup $S(t)$ in Theorem 4.1.1 is precisely $\tilde{S}(t)|C_0$.

The decomposition $T(t)\phi = \tilde{S}(t)\phi + \tilde{U}(t)\phi$ also can be considered as the variation of constants formula for (1.1) treating $f(x_t)$ as the nonhomogeneous term for the linear system $\dot{x}(t) = 0$. In fact, if $X_0(\theta)$, $\theta \in [-\delta, 0]$, is an $n \times n$ matrix with $X_0(\theta) = 0$ for $\theta < 0$, $X_0(0) = I$, then, for all $t \geq 0$, $\theta \in [-\delta, 0]$,

(1.2)
$$T(t)\phi(\theta) = \tilde{S}(t)\phi(\theta) + \int_0^{t+\theta} \tilde{S}(t-s)X_0(\theta)f(T(s)\phi)\,ds$$
$$= \tilde{S}(t)\phi(\theta) + \int_0^t \tilde{S}(t-s)X_0(\theta)f(T(s)\phi)\,ds$$

since $\tilde{S}(t-s)X_0(\theta) = X_0(t-s+\theta) = 0$ for $s > t+\theta$. Formally, this can be written as the variation of constants formula

$$T(t)\phi = \tilde{S}(t)\phi + \int_0^t \tilde{S}(t-s)X_0 f(T(s)\phi)\,ds, \qquad t \geq 0.$$

This is not a function space integral, but must be considered as an ordinary integral (1.2) in R^n for each $\theta \in [-\delta, 0]$. If this were a function space integral, then we could write it as an abstract evolutionary equation

$$(d/dt)u(t) = Au(t) + X_0 f(u(t))$$

where $u(t) = T(t)\phi$ and A is the infinitesimal generator of the semigroup $\tilde{S}(t)$. One can easily show that the domain $D(A)$ of A is $\{\phi \in C : d\phi/d\theta \in C$ and $\dot\phi(0) = 0\}$ and $A\phi = d\phi/d\theta$ if $\phi \in D(A)$. Since the range of A and the function $X_0 f$ lie in different spaces, it is not easy to interpret this as a differential equation. Chow and Mallet-Paret [1977] enlarged the space to $C \oplus \langle X_0 \rangle$ (which also was used by Hale [1977]) and modified the domain and range of A appropriately so as to be able to do averaging for Hopf bifurcation. Using the dual semigroup theory of Hille and Phillips, Diekmann and his colleagues have shown how the above equation can be interpreted as an abstract evolutionary equation (see Diekmann [1987] and references therein).

It is possible also to consider (1.2) in other spaces so that the above variation of constants formula becomes an integral in function space. The one most frequently used is $L^p([-\delta, 0], R^n) \times R^n$ (see, for example, Banks and Burns [1975]).

4.1.2. *Global attractor.* Using the results of Chapter 3 and Theorem 4.1.1, we have the following result

THEOREM 4.1.2. *Suppose $T(t)$, $t \geq 0$, for (1.1) is defined and a bounded map for each $t \geq 0$. If $T(t)$ is point dissipative, then*

(i) *there is a global connected attractor A for (1.1).*

(ii) *There is an equilibrium point of (1.1), that is, a constant solution.*

(iii) *If $f \in C^k(C, R^n)$, $0 \leq k \leq \infty$ [or analytic], then $\phi \in A$ implies ϕ is a C^{k+1} function [or an analytic function].*

(iv) *If f is analytic, then $T(t)$ is one-to-one on A.*

PROOF. Theorem 4.1.1 and Theorem 3.4.8 with $t_1 = \delta$ imply the existence of the compact attractor. Using the representation of $T(t) = S(t) + U(t)$ and Theorem 3.4.7, one obtains the existence of an equilibrium point. The group property in (iv) follows from Theorem 3.4.8. The C^{k+1}-smoothness is an easy consequence of the fact that the solution $x(t)$ of (1.1) with $x_0 \in A$ satisfies $x_t(\theta) = x(t+\theta)$ and is defined and bounded for $t \leq 0$. The proof of analyticity is more difficult and will not be given (see Nussbaum [1973]). Part (iv) is immediate from properties of analytic functions.

4.1.3. *An example.* As an example, consider the equation

(1.3) $$\dot{x}(t) = -\alpha x(t-1)[1 + x(t)]$$

where $\alpha > 0$ is a constant. This is a special case of (1.1) with $\delta = 1$, $f(\phi) = -\alpha\phi(-1)[1 + \phi(0)]$ and is one of the simplest nonlinear equations with a delay. This equation arises in number theory and has been used in population modeling.

We want to show that (1.3) has a global attractor in some subset of C. Let $C_{-1} = \{\phi \in C \colon \phi(0) > -1\}$ and let $T_\alpha(t)\phi$ correspond to the solution of (1.3) through ϕ.

LEMMA 4.1.3. $T_\alpha(t) \colon C_{-1} \to C_{-1}$ for $t \geq 0$, $T_\alpha(t)$ is a bounded map for each t and is point dissipative.

PROOF. Notice first that a solution $x(t)$ of (1.3) satisfies, for any $t \geq t_0 \geq 0$,

$$(1.4) \qquad 1 + x(t) = [1 + x(t_0)] \exp\left\{-\alpha \int_{t_0-1}^{t-1} x(\xi)\, d\xi\right\}.$$

Therefore, if $\phi \in C_{-1}$, then $x_t \in C_{-1}$ as long as $x(t)$ is defined. But this formula also shows that x cannot become unbounded on a finite interval. Thus, $T_\alpha(t)\phi = x_t(\cdot, \phi)$ is defined for $t \geq 0$. From this formula, it is also clear that $T_\alpha(t)$ is a bounded map for each $t \geq 0$.

To show $T_\alpha(t)$ is point dissipative, we consider first the case where the zeros of $x(t)$ are bounded, $x_0 \in C_{-1}$. Then there is a $t_1 > 0$ such that $x(t)$ has constant sign for $t \geq t_1 - 1$. Since $x(t) > -1$ for all $t \geq 0$, we have $\dot{x}(t)x(t-1) < 0$ for $t \geq t_1$ and $x(t)$ is bounded and, thus, $\dot{x}(t) \to 0$ as $t \to \infty$. Thus, $x(t) \to 0$ as $t \to \infty$ (since -1 is excluded).

If the zeros of $x(t)$ are unbounded, then there is a sequence of nonoverlapping intervals I_k of $[0, \infty)$ such that x is zero at the end points of each I_k and has constant sign on I_k. Thus, there is a $t_k \in I_k$ such that $\dot{x}(t_k) = 0$ and $x(t_k-1) = 0$. Using (1.4) for $t_0 = t_k - 1$, $t = t_k$, we have

$$\ln(1 + x(t_k)) = -\alpha \int_{t_k-2}^{t_k-1} x(\xi)\, d\xi < \alpha$$

since $x(t) > -1$ for $t \geq 0$. Thus, $x(t_k) \leq e^\alpha - 1$ for all t_k. Since the t_k are the extreme values of $x(t)$, this implies that $|x_t| \leq e^\alpha - 1$ for t sufficiently large. This shows that $T(t)$ is point dissipative and completes the proof of the lemma.

Lemma 4.1.3 and Theorem 4.1.2 imply that (1.2) has a global attractor $A_\alpha \subset C_{-1}$ for each $\alpha > 0$.

LEMMA 4.1.4. A_α is upper semicontinuous in α.

PROOF. In the proof of the previous lemma, we have shown that the global attractor A_α lies in the interior of the closed ball B_α with center zero and radius $e^\alpha - 1$. Fix $\alpha_0 > 0$ and consider $\alpha_1 \in (0, \alpha_0)$. The set A_{α_1} attracts the closed ball B_{α_0}. Since the semigroup $T_\alpha(t)$ depends continuously on α, this is enough to show that A_α is upper semicontinuous at α_1.

LEMMA 4.1.5 (Walther [1975]). If $S = \{\alpha \in (0, \pi/2] \colon A_\alpha = \{0\}\}$, then S is nonempty and open.

PROOF. Without proving it, we use a very difficult theorem of Wright [1955] which states that every solution of (1.3) approaches zero if $\alpha < e^{-1}$. For $\alpha < e^{-1}$, the roots of the characteristic equation $\lambda = -\alpha \exp(-\lambda)$ for the linear variational equation around zero have negative real parts. Thus, the origin is uniformly asymptotically stable in this range of α. From Theorem 2.4.8 with $T = T(1)$, one concludes that $\{0\}$ attracts compact sets of C_{-1}. Since $T(1)B$ is compact for any bounded set B in C_{-1}, it follows that $\{0\}$ attracts B. Thus $A_\alpha = \{0\}$ for $0 < \alpha < e^{-1}$. Since A_α is upper semicontinuous in α, the set S is open and the lemma is proved.

CONJECTURE. *S is closed.*

For $\alpha = \pi/2$, there is a supercritical Hopf bifurcation (see Hale [1977], Chow and Mallet-Paret [1977]). Thus, for $\alpha > \pi/2$, but very close to $\pi/2$, the set A_α must contain at least a disk.

4.1.4. *A gradient system.* Our next example is a gradient system.

Suppose $b \in C^2([-1, 0], R)$, $b(-1) = 0$, $b(\theta) > 0$, $b'(\theta) \geq 0$, $b''(\theta) \geq 0$ for $\theta \in (-1, 0)$ and

(1.5) $\quad\quad\quad\quad\quad\quad \exists \theta \in [-1, 0]$ such that $b''(\theta_0) > 0$.

Let $g: R \to R$ be a C^1-function such that

(1.6) $$G(x) = \int_0^x g(s)\,ds \to \infty \quad \text{as } |x| \to \infty$$

and consider the equation

(1.7) $$\dot{x}(t) = -\int_{-1}^0 b(\theta) g(x(t+\theta))\,d\theta.$$

Equation (1.7) is a special case of (1.1). If $\phi \in C = C([-1, 0], R)$, then there is a unique solution $x(t, \phi)$ of (1.7) through ϕ. Let $T(t)\phi = x_t(\cdot, \phi)$.

THEOREM 4.1.6. *The semigroup $T(t)$, $t \geq 0$, defined by (1.7) is a gradient system. If the set of equilibrium points E (the zeros of g) is bounded, then there is a global attractor $A_{b,g}$ for (1.7). Furthermore, if each element of E is hyperbolic, then*

$$A_{b,g} = \bigcup_{\phi \in E} W^u(\phi).$$

PROOF. This theorem is essentially contained in Hale [1977, p. 120ff]. We give an indication of the proof. First observe that any solution of (1.7) satisfies

(1.8) $$\ddot{x}(t) + b(0)(g(x(t))) = \int_{t-1}^t \dot{b}(u-t) g(x(u))\,du$$
$$= \dot{b}(-1) \int_{-1}^0 g(x(t+\theta))\,d\theta - \int_{-1}^0 \ddot{b}(\theta) \left[\int_\theta^0 g(x(t+u))\,du\right] d\theta.$$

If, for $\phi \in C$,

$$V(\phi) = G(\phi(0)) + \frac{1}{2} \int_{-1}^0 \dot{b}(\theta) \left[\int_\theta^0 g(\phi(s))\,ds\right]^2 d\theta,$$

then we can show that

(1.9)
$$\frac{d}{dt}V(T(t)\phi) = -\frac{1}{2}\dot{b}(-1)\left[\int_{-1}^{0} g((T(t)\phi)(\theta))\,d\theta\right]^2$$
$$-\frac{1}{2}\int_{-1}^{0}\dot{b}(\theta)\left[\int_{\theta}^{0} g((T(t)\phi)(s))\,ds\right]^2 d\theta.$$

The hypotheses on b, g imply that $V(\phi) \to \infty$ as $|\phi| \to \infty$ and $dV(T(t)\phi)/dt \le 0$. This proves that all solutions are bounded and $T(t)$ is a C^1-semigroup on C.

Relations (1.5) and (1.6) imply that $V: C \to R$ satisfies the conditions (ii)$_1$–(ii)$_3$ of Definition 3.8.1 of a gradient system. To prove (ii)$_4$, suppose ϕ is such that $T(t)\phi$ is defined for $t \in R$, and $V(T(t)\phi) = V(\phi)$ for $t \in R$. Then $dV(T(t)\phi)/dt = 0$ for all $t \in R$. From (1.8) and (1.9), this implies that ϕ must correspond to the initial value of a bounded solution on R of the equation

$$\ddot{x} + b(0)g(x) = 0$$

satisfying $\int_{-s}^{0} g(x_t(\theta))\,d\theta = 0$ for s in some interval I containing θ_0. This implies that $\dot{x}(t)$ is periodic of period s for each s in I. Therefore, $\dot{x}(t)$ is constant. Boundedness of $x(t)$ implies $x(t)$ is a constant. This proves that $T(t)$ is a gradient system. Theorem 4.1.1, E being bounded, and Theorem 3.4.8 imply there is a global attractor $A_{b,g}$. The proof of the theorem is completed by applying Theorem 3.8.5.

The equilibrium set E consists of constant functions corresponding to the zeros of g. If $g(x_0) = 0$, then the linear variational equation about x_0 is

$$\dot{x}(t) = -\int_{-1}^{0} b(\theta)g'(x_0)x(t+\theta)\,d\theta$$

and the characteristic equation is

(1.10)
$$\lambda = -\int_{-1}^{0} b(\theta)g'(x_0)e^{\lambda\theta}\,d\theta.$$

The equilibrium point x_0 is hyperbolic if and only if $\text{Re}\,\lambda \ne 0$ for every λ satisfying (1.10). One can show that hyperbolicity of x_0 is equivalent to $g'(x_0) \ne 0$. Furthermore, x_0 is stable if $g'(x_0) > 0$ and unstable with $\dim W^u(x_0) = 1$ if $g'(x_0) < 0$. From Theorem 4.1.6, this implies that the global attractor $A_{b,g}$ is a one dimensional set.

The next question of interest is to determine the flow on the attractor $A_{b,g}$. More specifically, suppose b is a fixed function and consider the attractor $A_{b,g}$ as a function of g. For each zero x_0 of g with $g'(x_0) < 0$, we know that x_0 is unstable with $\dim W^u(x_0) = 1$. Thus, there are two orbits leaving x_0; that is, there are two distinct solutions $\phi(t), \psi(t)$ of (1.7), defined for $t \le 0$, which approach x_0 as $t \to -\infty$. The basic problem is to determine the ω-limit sets of $\phi(t), \psi(t)$ as $t \to \infty$. We know that these ω-limit sets must be equilibrium points, but will one of these be less than x_0 and the other be greater than x_0 for each unstable point x_0? If $E = (x_1, x_2, x_3)$, with $x_1 < x_2 < x_3$, this is obviously

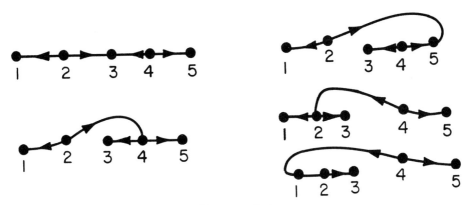

FIGURE 1.1
Possible attractors for (1.7) for g with five zeros $x_1 < x_2 < x_3 < x_4 < x_5$ (only subscripts labeled).

true since $A_{b,g}$ is connected. The flow on $A_{b,g}$ preserves the natural order of the points x_1, x_2, x_3 on the real line.

To describe the situation where g has five simple zeros, it is convenient to systematize the notation. Let us use the symbol $j[k, l]$ to mean that the unstable point x_j is connected to x_k, x_l by an orbit. If g has five simple zeros $x_1 < x_2 < x_3 < x_4 < x_5$, then x_2, x_4 are unstable with one dimensional unstable manifolds and x_1, x_3, x_5 are stable. The flow on $A_{b,g}$ is determined by how x_2, x_4 are connected to the other points by orbits. The proof of the following rather surprising result can be found in Hale and Rybakowski [1982] (see Figure 1.1).

THEOREM 4.1.7. *For a given b, one can realize each of the following flows on $A_{b,g}$ by choosing an appropriate g with five simple zeros*
  (i) $2[1, 3]$, $4[3, 5]$,
  (ii) $2[1, 4]$, $4[3, 5]$,
  (iii) $2[1, 5]$, $4[3, 5]$,
  (iv) $2[1, 3]$, $4[2, 5]$,
  (v) $2[1, 3]$, $4[1, 5]$.

The only flow which preserves the natural order $x_1 < x_2 < x_3 < x_4 < x_5$ is (i). Case (ii) has a nontransverse intersection between $W^u(x_2)$ and $W^s(x_4)$ and case (iv) between $W^s(x_2)$ and $W^u(x_4)$. For cases (iii) and (v), the natural order of the equilibrium points on R is not preserved on the attractor.

4.1.5. *Equations with negative feedback.* Our next example is concerned with the properties of the flow on the global attractor for equations with negative feedback. The results are due to Mallet-Paret [1988] and any details of proofs that are not given below can be found in that paper.

Consider the equation

(1.11) $$\dot{x}(t) = -\beta x(t) - g(x(t-1))$$

where $\beta \geq 0$. The hypotheses assumed are

(1.12) g has negative feedback; that is, $xg(x) > 0$ for $x \neq 0$ and $g'(0) > 0$,

(1.13) there is a constant k such that $g(x) \geq -k$ for all x,

(1.14) the zero solution of (1.11) is hyperbolic.

The results hold in more general situations. The right-hand side of (1.11) can be replaced by a function $f(x(t), x(t-1))$ with a modified definition of negative feedback. Hypothesis (1.13) can be replaced by the hypothesis that the semigroup generated by (1.11) is point dissipative. Hypothesis (1.14) is not necessary, but to eliminate it requires more complicated definitions.

Equation (1.3) can be transformed into an equation with negative feedback that satisfies (1.13) provided we consider $\phi \in C$ with $\phi(\theta) > -1$. In fact, if $1+x \longmapsto e^x$ then (1.2) is a special case of (1.11) with $f(x) = \alpha(e^x - 1)$. Hypothesis (1.14) excludes the discussion of a discrete set of α where Hopf bifurcations occur.

LEMMA 4.1.8. *Under hypotheses* (1.12), (1.13), *the semigroup $T(t)$ generated by* (1.11) *is a bounded map and is point dissipative. Thus, there is a global attractor A.*

PROOF. By integrating (1.11) over $[0,1]$, one obtains $T(1)$ is a bounded map. Thus, $T(t)$ is a bounded map for any $t > 0$.

To show point dissipative, observe that

$$(d/dt)(e^{\beta t} x(t)) \leq e^{\beta t} k$$

for all $t \geq 0$. Therefore

$$x(t) \leq x(0) e^{-\beta t} + k(1 - e^{-\beta t})/\beta$$

and $\limsup_{t \to \infty} x(t) \leq 2k/\beta$. Since $x(t)$ is bounded above, it follows that $-g(x(t-1))$ is bounded below by a constant K_1. Therefore, arguing as above, one obtains that $\liminf_{t \to \infty} x(t) \geq -2K_1/\beta$. Theorem 4.1.2 completes the proof of the lemma.

One of the basic properties of the flow of (1.11) on the global attractor is that there is a Morse decomposition.

A *Morse decomposition* of the global attractor A is a finite ordered collection $A_1 < A_2 < \cdots < A_M$ of disjoint compact invariant subsets of A (called *Morse sets*) such that, for any $\phi \in A$, there are integers $N \geq K$ such that $\alpha(\phi) \subset S_N$ and $\omega(\phi) \subset S_K$ and $N = K$ implies that $\phi \in S_N$. In the case $N = K$, $T(t)\phi \in S_N$ for $t \in R$.

The Morse sets together with the connecting orbits $C_K^N = \{\phi \in A : \alpha(\phi) \subset S_N$ and $\omega(\phi) \subset S_K\}$ for $N > K$ give the global attractor A.

To obtain a Morse decomposition of A, it is convenient to think of the flow on A in the following way. For any $\phi \in A$, $T(t)\phi \in A$ for all $t \in (-\infty, \infty)$. Since $(T(t)\phi)(\theta) = (T(t+\theta)\phi)(0)$ for $\theta \in [-1, 0]$, the orbit $T(t)\phi$, $t \in (-\infty, \infty)$, can be identified with the function $x(t, \phi) = T(t)\phi(0)$, $t \in R$. This observation is important for the following definition. For any $\phi \in A$, $\phi \neq 0$, let $\sigma \geq t$ be the first zero of $x(t,\phi)$ in $[t,\infty)$ if it exists. We define $V(T(t)\phi)$ as the number of zeros of x (counting multiplicity) in the half-open interval $(\sigma - 1, \sigma]$. If σ does not exist, define $V(T(t)\phi) = 1$. Thus, $V(T(t)\phi)$ is either a positive integer or ∞.

THEOREM 4.1.9.
(i) $V(T(t)\phi)$ *is nonincreasing in t for each $\phi \in A$, $\phi \neq 0$.*
(ii) $V(T(t)\phi)$ *is an odd integer for each $\phi \in A$, $t \in R$, $\phi \neq 0$.*
(iii) *There is a constant K such that $V(T(t)\phi) \leq K$ for all $t \in R$, $\phi \in A$, $\phi \neq 0$.*

PROOF. We only give some intuitive ideas of why Theorem 4.1.9 is true. The details can be found in Mallet-Paret [1986]. To see why $V(T(t)\phi)$ should be nonincreasing in t, suppose that the zeros of $x(t,\phi)$ are simple. Let $\sigma_0 < \sigma_1$ be consecutive zeros with $x(t,\phi) > 0$ in between. Then $\dot{x}(\sigma_0,\phi) > 0$ and $\dot{x}(\sigma_1,\phi) < 0$. From the negative feedback condition on (1.11), it follows that $x(\sigma_0 - 1, \phi) < 0$ and $x(\sigma_1 - 1, \phi) > 0$. Thus, $x(t,\phi) = 0$ at some point in $(\sigma_0 - 1, \sigma_1 - 1)$; that is, $x(t,\phi)$ can have no more zeros in $(\sigma_1 - 1, \sigma_1]$ than it does in $(\sigma_0 - 1, \sigma_0]$. This shows that $V(T(t)\phi)$ is nonincreasing. Again, assuming that the zeros of $x(t,\phi)$ are simple, then $x(\sigma,\phi) = 0$, $\dot{x}(\sigma,\phi) > 0$, implies $x(\sigma - 1, \phi) < 0$. This implies immediately that the number of zeros in $(\sigma - 1, \sigma]$ is odd. The proof of (i) and (ii) in the general case requires some technical estimates which are omitted.

The proof of property (iii) in the theorem is technical and difficult. The difficulty arises in the determination of the behavior of the flow on A near zero and upper bounds on the value of $V(T(t)\phi)$ for ϕ near zero. One must show that the eigenfunctions of the unstable manifold of zero yield an upper bound for $V(T(t)\phi)$. As a corollary of the proof, one obtains the result that there is no solution on A which is superexponential; that is, $e^{\alpha t}x(t,\phi) \to 0$ as $t \to \infty$ for every $\alpha > 0$.

We are now prepared to define the Morse sets for A. For any odd integer N, define
$$S_N = \{\phi \in A, \phi \neq 0 \colon V(T(t)\phi) = N \text{ for } t \in R \text{ and } 0 \notin \alpha(\phi) \cup \omega(\phi)\}.$$
For $N^* = \dim W^u(0)$, define $S_{N^*} = \{0\}$.

THEOREM 4.1.10. *The sets S_N, $N \in \{N^*, 1, 3, 5, \dots\}$, form a Morse decomposition of A with the ordering $S_K < S_N$ if and only if $K < N$.*

The main part of the proof here is to show that the sets S_N are closed sets. Once this is known, the fact that it is a Morse decomposition follows from the fact that $V(T(t)\phi)$ is nonincreasing in t.

Further properties are also known about the sets S_N. In particular, for $N \neq N^*$, if $\phi \in S_N$, then the zeros of $x(t,\phi)$ are simple. This allows one to prove that each S_N for $N < N^*$ is not empty and contains a periodic orbit $x_N(t)$ with least period τ satisfying $2/N < \tau < 2/(N-1)$. Furthermore, $x_N(t)$ has exactly two zeros in $[0,\tau)$.

Since $S_N \neq \varnothing$ for $N < N^*$, it follows that the connecting orbits $C_K^N \neq \varnothing$ for various N and K. An eventual goal is to understand the topological structure of these sets and how they fit together to give the attractor A. Fiedler and Mallet-Paret [1987] have shown that $C_N^{N^*} \neq \emptyset$ for $N < N^*$.

As we have seen, the existence of the above Morse decomposition gives a much better picture of the flow on A. This does not mean that the flow is simple. In fact, numerical studies suggest that, in many cases, the sets S_N have a very complicated interval structure involving dynamics which appears to be chaotic (see, for example, Mackey and Glass [1977], Farmer [1982], Chow and Green [1986], and Hale and Sternberg [1987]). Some theoretical results relating this to orbits which are homoclinic to periodic orbits may be found in Walther [1981], an der Heiden and Walther [1983], and Hale and Lin [1986].

4.1.6. Periodic equations. Let us now consider a nonautonomous equation

$$\dot{x}(t) = f(t, x_t) \tag{1.15}$$

with $f: R \times C \to R^n$ a continuous bounded map, and $f(t, \phi)$ a C^r-function in ϕ. For any $(\sigma, \phi) \in R \times C$, there is a solution $x(t, \sigma, \phi)$ of (1.15) defined on an interval $[\sigma - r, \sigma + \alpha_{\sigma,\phi})$, $\alpha_{\sigma,\phi} > 0$ and, if $\alpha_{\sigma,\phi} < \infty$, then $|x_t(\cdot, \sigma, \phi)| \to \infty$ as $t \to \sigma + \alpha_{\sigma,\phi}$. Let $[T(t,\sigma)\phi](\theta) = x(t+\sigma+\theta, \sigma, \phi)$, $-\delta \leq \theta \leq 0$, $0 \leq t < \alpha_{\sigma,\phi}$, that is, the solution in C through (σ, ϕ) at time $t+\sigma$. If we assume that each solution is defined on $[\sigma - \delta, \infty)$, then equation (1.15) defines a process on $R^+ \times R \times C$ and $T(t, \sigma)\phi$ is C^r in ϕ for all $t \geq 0$, $\sigma \in R$, $\phi \in C$.

If there is an $\omega > 0$ such that $f(t + \omega, \phi) = f(t, \phi)$, then the process is ω-periodic, $T(t, \sigma + \omega) = T(t, \sigma)$ for $t \geq 0$, $\sigma \in R$. As in the proof of Theorem 4.1.1, one shows that

$$T(t, \sigma) = S(t) + U(t, \sigma)$$

where $S(t)$ is the same operator as in Theorem 4.1.1 and

$$U(t,\sigma)\phi(\theta) = \begin{cases} \phi(0) + \int_0^{t+\theta} f(s + \sigma, T(s,\sigma)\phi)\,ds, & t + \theta \geq 0, \\ \phi(0), & t + \theta < 0. \end{cases}$$

As before, $U(t, \sigma)$ is completely continuous if $T(t, \sigma)$ is a bounded map for each $t \geq 0$. In an appropriate norm, $\{T(t, \sigma), t \geq 0\}$ is an α-contraction for each $\sigma \in R$. We can now state the following consequence of Theorems 3.6.1, 3.6.3, and 3.6.4.

THEOREM 4.1.11.

(i) *Suppose $T(t, \sigma)$ as defined above for (1.15) exists for $t \geq 0$, is a bounded map for each $t \geq 0$, and there is an $\omega > 0$ such that $f(t + \omega, \phi) = f(t, \phi)$ for all $t \in R$, $\phi \in C$. If $T(t, \sigma)$ is point dissipative, then there is a connected global attractor A_σ for the map $T(\omega, \sigma)$ and the set*

$$M = \{(t, \psi): \psi = T(t, 0)\phi, \phi \in A_0, t \in R\}$$

is invariant and a global attractor for (1.15). Finally, there is an ω-periodic solution of (1.15).

(ii) *If $f(t,\phi)$ is linear in ϕ and there is a bounded solution of (1.15), then there is an ω-periodic solution of (1.15).*

It should be emphasized that the period of the periodic solution given in Theorem 4.1.11 is the same as the period of the vector field $f(t,\phi)$. In particular, this period need not be related to the length δ of the delay interval. If $\omega \geq \delta$, then one can obtain the existence of ω-periodic solutions by using asymptotic fixed point theorems for compact operators and the fact that $T(\omega,0)$ is compact if $\omega \geq \delta$ (see, for example, Yoshizawa [1966], [1975]; Fennel and Waltman [1969]). If $\omega < \delta$, then one needs the information that $T(\omega,0)$ is an α-contraction and the fixed point theorems in §2.6.

4.2. Sectorial evolutionary equations.

A linear operator A on a Banach space X is sectorial if A is a closed, densely defined operator such that there exist constants $\phi \in (0,\pi/2)$, $M \geq 0$, such that

$$S_{b,\phi} \stackrel{\text{def}}{=} \{\lambda \in \mathbf{C}: \phi \leq |\arg(\lambda - b)| \leq \pi, \lambda \neq b\} \subseteq \rho(A),$$

where $\rho(A)$ is the resolvent set of A, and

$$|(\lambda I - A)^{-1}| \leq M/|\lambda - b| \quad \text{for all } \lambda \in S_{b,\phi}.$$

It is known (see, for example, Henry [1981, pp. 20–23]) that A is sectorial if and only if the semigroup e^{-At} generated by A is an analytic semigroup; that is, for each $x \in X$, $e^{-At}x$ is real analytic on $0 < t < \infty$.

If A is a sectorial operator on X with $\operatorname{Re}\sigma(A) > 0$, then, for any $\alpha > 0$, one can define

$$A^{-\alpha} = \frac{1}{\Gamma(\alpha)}\int_0^\infty t^{\alpha-1}e^{-At}\,dt,$$

where $\Gamma(\alpha)$ is the gamma function evaluated at α. The operator $A^{-\alpha}$ is a bounded linear operator on X which is one-to-one and satisfies $A^{-\alpha}A^{-\beta} = A^{-(\alpha+\beta)}$ for $\alpha > 0$, $\beta > 0$ (see Henry [1981, p. 25]). One can now define A^α for $\alpha > 0$ as the inverse of $A^{-\alpha}$. The operator A^α, $\alpha > 0$, is closed and densely defined. Define $A^0 = I$. Also, if $\operatorname{Re}\sigma(A) > \delta > 0$, then for any $\alpha \geq 0$, there exists a constant $C_\alpha < \infty$ such that

$$|A^\alpha e^{-At}| \leq C_\alpha t^{-\alpha}e^{-\delta t}, \qquad t > 0,$$

and, if $0 < \alpha < 1$, $x \in D(A^\alpha)$, then

$$|(e^{-At} - I)x| \leq \frac{1}{\alpha}C_{1-\alpha}t^\alpha|A^\alpha x|, \qquad t \geq 0.$$

The constant C_α is bounded as $\alpha \to 0+$ (see Henry [1981, pp. 26–27]).

If A is a sectorial operator on X, then there is an $a \in R$ such that $A_1 = A + aI$ has $\operatorname{Re}\sigma(A_1) > 0$. If we define $X^\alpha = D(A_1^\alpha)$, $\alpha \geq 0$, with the graph norm $|x|_\alpha = |A_1^\alpha x|$, $x \in X^\alpha$, then X^α is a Banach space with norm $|\cdot|_\alpha$. Furthermore, if X^α is defined by using a different a with the above property, then the norms are equivalent and so the dependence on a is suppressed. Furthermore, $X^0 = X$

and X^α is a dense subspace of X^β for any $\alpha \geq \beta \geq 0$. If A has compact resolvent, the inclusion $X^\alpha \subset X^\beta$ is compact when $\alpha > \beta \geq 0$.

As an example, $(A\phi)(x) = -\partial^2\phi(x)/\partial x^2$, $0 < x < 1$, where ϕ is a C^2-function on $[0,1]$ with compact support in $(0,1)$. Considering this set of functions as a subset of $L^2(0,1)$, we have

$$(A\phi, \psi) = -\int_0^1 \phi''(x)\psi(x)\, dx = -\int_0^1 \phi(x)\psi''(x) = (\phi, A\psi),$$

$$(A\phi, \phi) = -\int_0^1 \phi''(x)\phi(x)\, dx = \int_0^1 [\phi'(x)]^2\, dx \geq 0.$$

Using Friedrich's extension theorem, the operator A can be extended to a selfadjoint densely defined linear operator on $X = L^2(0,1)$. Then

$$D(A) = \{\phi \in L^2(0,1) \colon A\phi \in L^2(0,1)\} = H_0^1(0,1) \cap H^2(0,1).$$

Furthermore, $\sigma(A) = \{n^2\pi^2,\ n = 1, 2, \ldots\}$ and $\lambda_n = n^2\pi^2$ is a simple eigenvalue with eigenfunction $\phi_n(x) = 2^{1/2}\sin n\pi x$. If one defines the projection operator E_n onto the span of ϕ_n as $E_n\psi = (\psi, \phi_n)\phi_n$, then one can show that

$$(\lambda I - A)^{-1}\psi = \sum_{n=1}^\infty (\lambda - \lambda_n)^{-1} E_n\psi \quad \text{if } \lambda \neq \lambda_n.$$

This formula allows one to show that A is sectorial. It also shows that the resolvent operator is compact. Furthermore,

$$D(A^\alpha) = \left\{\psi \in L^2(0,1) \colon \sum_{n=1}^\infty \lambda_n^{2\alpha}(\phi_n, \psi)^2 < \infty\right\},$$

$$A^\alpha \psi = \sum_{n=1}^\infty \lambda_n^\alpha(\phi_n, \psi)\phi_n.$$

More generally, if Ω is a bounded domain in R^n with smooth boundary and Δ is the Laplacian operator, define $A\phi = -\Delta\phi$ on the space C_0^∞ of C^∞-functions with compact support in Ω. Then one can extend A to a selfadjoint positive definite operator in $X = L^2(\Omega)$ with $D(A) = H_0^1(\Omega) \cap H^2(\Omega)$. The fractional power spaces X^α can then be defined. One can also show as in the one dimensional case that the resolvent operator of A is compact.

To give another illustration of the theory of Chapter 3, let us consider an abstract evolutionary equation

(2.1) $\qquad \dot{u} + Au = f(u), \qquad t > 0, \qquad u(0) = u_0,$

where A is a sectorial operator on X and there is an $\alpha \in [0, 1)$ such that $f\colon X^\alpha \to X$ is locally Lipschitz continuous; i.e., f is continuous and, for any bounded set U in X^α, there is a constant k_U such that

$$|f(u) - f(v)| \leq k_U |u - v|_\alpha \quad \text{for } u, v \in U.$$

A *solution* of (2.1) on $[0, \tau)$ is a continuous function $u\colon [0, \tau) \to X^\alpha$, $u(0) = u_0$, such that $f(u(\cdot))\colon [0, \tau) \to X$ is continuous, $u(t) \in D(A)$, and u satisfies (2.1) on

$(0, \tau)$. This definition is due to M. Miklavčič [1985] and differs slightly from the one in Henry [1981, p. 53]. The above definition requires that $f(u(\cdot))$ be continuous from the semiclosed interval $[0, \tau)$ to X whereas Henry requires $f(u(\cdot))$ to be Hölder continuous from the open interval $(0, \tau)$ to X and $\int_0^\rho |f(u(s))|\, ds < \infty$ for some $\rho > 0$. Miklavčič [1985] gives an example showing that solutions through u_0 may not be unique if one uses Henry's definition whereas they are with his definition. The modification of the definition from the one in Henry [1981] does not alter in any essential way the results in Henry [1981] or their proofs.

Miklavčič [1985] shows that the solutions of (2.1) coincide with those solutions of the integral equation

$$(2.2) \qquad u(t) = e^{-At}u_0 + \int_0^t e^{-A(t-s)} f(u(s))\, ds, \qquad 0 \leq t < \tau,$$

for which $u\colon [0, \tau) \to X^\alpha$ is continuous and $f(u(\cdot))\colon [0, \tau) \to X$ is continuous. One can then prove (see Henry [1981, pp. 54–57, 62–65]) the following result.

THEOREM 4.2.1. *Under the above hypotheses on A, f, there is a unique solution of (2.1) on a maximal interval of existence $[0, \tau_{u_0})$. If $\tau_{u_0} < \infty$, then there is a sequence $t_n \to \tau_{u_0}^-$ such that $|u(t_n)|_\alpha \to \infty$. If, in addition, f is a C^r-function in u, then the solution $u(t, u_0)$ is a C^r-function in (t, u_0) in the domain of definition of the function.*

If we make the additional hypothesis that all solutions are defined for $t \geq 0$, then we can let $T(t)\colon X^\alpha \to X^\alpha$, $t \geq 0$, be defined by $T(t)u_0 = u(t, u_0)$ and obtain a strongly continuous semigroup of operators. One can now prove the following theorem.

THEOREM 4.2.2. *If A, f satisfy the conditions imposed above, and in addition, the resolvent of A is compact and $T(t)$ takes bounded sets into bounded sets for each $t > 0$, then $T(t)$ is compact on X^α for $t > 0$.*

For the proof, we need the following lemma.

LEMMA 4.2.3. *If A is a sectorial operator with compact resolvent in X, then e^{-At} is compact on X^α, $0 \leq \alpha < 1$, for $t > 0$.*

PROOF. For any $t > 0$, $e^{-At}x \in D(A) = X^1$ for any $x \in X$. Suppose B is a bounded set in X^α, $0 \leq \alpha < 1$. Then, for $x \in B$,

$$|e^{-At}x|_1 = |Ae^{-At}x| = |A^{1-\alpha}e^{-At}A^\alpha x|$$
$$\leq |A^{1-\alpha}e^{-At}| \cdot |A^\alpha x| \leq \frac{k}{t^{1-\alpha}}|A^\alpha x|$$

and so $e^{-At}B$ is bounded in X^1. Since $X^1 \subset X^\alpha$ is a compact embedding for $0 \leq \alpha < 1$, this proves the lemma.

PROOF OF THEOREM 4.2.2. We know that $T(t)$ satisfies the equation

$$T(t)x = e^{-At}x + \int_0^t e^{-A(t-\tau)} f(T(\tau)x)\, d\tau, \qquad t \geq 0, \qquad x \in X^\alpha.$$

Suppose M is bounded in X^α. Let $M_1 = \{\int_0^t e^{-A\tau} f(T(t-\tau)x)\,d\tau : x \in M\}$. If $t > 0$ is fixed and $N = \{T(t-\tau)x,\ x \in M,\ 0 \leq \tau \leq t\}$ is bounded, choose $0 < \varepsilon < t$ and write

$$\int_0^t e^{-A\tau} f(T(t-\tau)x)\,d\tau = \int_0^\varepsilon e^{-A\tau} f(T(t-\tau)x)\,d\tau$$
$$+ e^{-A\varepsilon} \int_\varepsilon^t e^{-A(\tau-\varepsilon)} f(T(t-\tau)x)\,d\tau$$

and note that

$$\left| \int_0^\varepsilon A^\alpha e^{-A\tau} f(T(t-\tau)x)\,d\tau \right| \leq c \int_0^\varepsilon \frac{1}{\tau^\alpha}\,d\tau \leq c \frac{\varepsilon^{1-\alpha}}{1-\alpha},$$
$$\left| \int_\varepsilon^t A^\alpha e^{-A(\tau-\varepsilon)} f(T(t-\tau)x)\,d\tau \right| \leq c \int_\varepsilon^t \frac{1}{(\tau-\varepsilon)^\alpha}\,d\tau \leq \frac{c}{1-\alpha} t^{1-\alpha},$$

where $c = c(N)$. Since $e^{-A\varepsilon}$ is compact by the above lemma, this implies the α-measure of noncompactness of M_1, $\alpha(M_1)$ satisfies $\alpha(M_1) \leq [c/(1-\alpha)]\varepsilon^{1-\alpha}$. Since ε is arbitrary, this implies $\alpha(M_1) = 0$ and M_1 is compact. This proves the theorem.

Under the hypotheses of Theorem 4.2.2, the operator $T(t)$ satisfies the smoothness properties of Theorem 3.4.8. Therefore, if $T(t)$ is point dissipative, there is a global attractor A. Furthermore, there is an equilibrium point of $T(t)$, that is, a solution of

(2.3) $$Au = f(u).$$

This is summarized in

THEOREM 4.2.4. *If $T(t)$ satisfies the conditions of Theorem 4.2.2 and is point dissipative, then there exists a connected global attractor $A = A_f$. Furthermore, the equation (2.3) has at least one solution.*

It is also possible to prove

THEOREM 4.2.5. *If the conditions of Theorem 4.2.4 are satisfied, then, for any $u_0 \in A_f$, the map $t \longmapsto T(t)u_0$ is as smooth as the vector field $f(u)$.*

Since the hypotheses imply that e^{At} generates an analytic semigroup that is compact for $t > 0$, the evolutionary equation (2.1) falls into the class considered by Hale and Scheurle [1985]. In this paper, there were limitations on the bound of the derivative of f. However, these restrictions on f can be eliminated if one uses the fact that e^{At} is an analytic semigroup.

Let us now consider a nonautonomous equation

(2.4) $$\dot{u} + Au = f(t, u),$$

where A is sectorial on X with compact resolvent and there is an $\alpha \in [0, 1)$ such that $f \colon R \times X^\alpha \to X$ is locally Hölder continuous in t and locally Lipschitz in u. More precisely, if $(t_1, u_1) \in R \times X^\alpha$, then there is a neighborhood U of (t_1, u_1) and constants $c > 0$, $0 < \nu \leq 1$, such that, for all $(t, u), (s, v) \in U$,

(2.5) $$|f(t, u) - f(s, v)|_X \leq c[|t - s|^\nu + |u - v|_{X^\alpha}].$$

Using the same definition of a solution as before, one can show that (2.4) under hypothesis (2.5) has a unique solution $u(t, \sigma, \phi)$ through the point $(\sigma, \phi) \in R \times X^\alpha$ existing on an interval $[\sigma, \sigma + \beta_{\sigma,\phi})$ and, if $\beta_{\sigma,\phi} < \infty$, then $|u(t, \sigma, \phi)|_{X^\alpha} \to \infty$ as $t \to \sigma + \beta_{\sigma,\phi}$.

Let us suppose that each solution $u(t, \sigma, \phi)$ exists for $t \geq \sigma$ and define $T(t, \sigma)\phi = u(t+\sigma, \sigma, \phi)$ for $t \geq 0$, $\sigma \in R$, $\phi \in X^\alpha$. This defines a process on $R^+ \times R \times X^\alpha$. If there is an $\omega > 0$ such that

(2.6) $\qquad f(t+\omega, \phi) = f(t, \phi), \qquad (t, \phi) \in R \times X^\alpha,$

then the process is ω-periodic, $T(t, \sigma + \omega) = T(t, \sigma)$.

Since the resolvent of A is compact, $T(t, \sigma)$ is compact for each $t > 0$ if $T(t, \sigma)$ is a bounded map for each $t > 0$. We can now state the following consequence of Theorems 3.6.3 and 3.6.4.

THEOREM 4.2.6. (i) *For (2.4), (2.5), (2.6) suppose* $T(t, \sigma): X^\alpha \to X^\alpha$ *defined as above exists for $t \geq 0$, $\sigma \in R$, is a bounded map for each $t \geq 0$, $\sigma \in R$, and is point dissipative. Then there is a connected global attractor A_σ for the map $T(\omega, \sigma)$ and the set*

$$M = \{(t, \psi): \psi = T(t, 0)\phi, t \in R, \phi \in A_0\}$$

is invariant and a global attractor for (2.4). Finally, there is an ω-periodic solution of (2.4).

(ii) *If $f(t, u)$ is linear in u and there is a bounded solution of (2.4), then there is an ω-periodic solution of (2.4).*

4.3. A scalar parabolic equation.

4.3.1. *Existence and gradient.* Let us consider a scalar parabolic equation in one space dimension following Henry [1981, p. 118].

Suppose $a \in C([0, 1], R)$, $a(x) > 0$, $0 \leq x \leq 1$, $f \in C^2(R, R)$,

(3.1) $\qquad \varlimsup_{|u| \to \infty} f(u)/u \leq 0$

and consider the equation

(3.2) $\qquad u_t = (a(x)u_x)_x + f(u), \qquad 0 < x < 1,$

with the boundary condition

(3.3) $\qquad u = 0 \quad \text{at } x = 0, 1.$

Let $X = L^2(0, 1)$, $A\phi(x) = -(a\phi_x)_x$ for smooth ϕ which vanish at 0 and 1. Then A can be extended to a positive definite selfadjoint densely defined operator in X. Then $D(A) = H_0^1(0, 1) \cap H^2(0, 1)$, $D(A^{1/2}) = H_0^1(0, 1) = X^{1/2}$, A has compact resolvent and is a sectorial operator. For any $\phi \in X^{1/2}$, we have ϕ is continuous and $|\phi(x)| \leq |\phi|_{X^{1/2}}$. If $f^e: X^{1/2} \to X$ is defined by $f^e(\phi)(x) = f(\phi(x))$, $x \in [0, 1]$, and $\phi, \psi \in X^{1/2}$, $|\phi|_{X^{1/2}}, |\psi|_{X^{1/2}} \leq r$, then the fact that $|\phi(x)| \leq r$, $|\psi(x)| \leq r$ and f is C^1 implies that there is a constant $c = c_r$ such that

$$|f^e(\phi) - f^e(\psi)|_X^2 = \int_0^1 [f(\phi(x)) - f(\psi(x))]^2\, dx = \int_0^1 \left(\int_{\psi(x)}^{\phi(x)} f'(s)\, ds\right)^2 dx$$

$$\leq c \int_0^1 |\phi(x) - \psi(x)|^2\, dx \leq c|\phi - \psi|_{X^{1/2}}^2$$

and f^e is locally Lipschitz. Using the fact that f is C^2, one obtains that f^e is a C^1-function. Thus, (3.2), (3.3) define a local C^1-semigroup on $X^{1/2}$.

To show $T(t)$ is a C^1-semigroup on $H_0^1(0,1)$, we must show that solutions are defined for $t \geq 0$. If

(3.4) $$V(\phi) = \int_0^1 \left[\frac{a(x)}{2}\phi_x^2(x) - F(\phi(x))\right] dx, \qquad F(u) = \int_0^u f(s)\, ds,$$

then

(3.5) $$\frac{d}{dt} V(u(t, \cdot, \phi)) = -\int_0^1 u_t^2\, dx \leq 0.$$

Furthermore, for any $\varepsilon > 0$, condition (3.1) implies that

$$F(s) - \varepsilon s^2 = \int_0^s [f(\tau)/\tau - 2\varepsilon]\tau\, d\tau \leq C_\varepsilon \quad \text{for } s > 0.$$

If $s < 0$, one obtains a similar estimate so that

$$F(u) \leq \varepsilon u^2 + C_\varepsilon, \qquad u \in R.$$

Therefore, there is a constant $k > 0$ such that

$$V(\phi) \geq k|\phi|_{H_0^1}^2 - \int_0^1 [\varepsilon\phi^2(x) + C_\varepsilon]\, dx \geq k|\phi|_{H_0^1}^2 - \varepsilon|\phi|_{H_0^1}^2 - C_\varepsilon$$

from the fact that $|\phi(x)| \leq |\phi|_{H_0^1}$. For $\varepsilon = k/2$, we have

$$k|\phi|_{H_0^1}^2 \leq 2V(\phi) + 2C_{k/2}.$$

Since $V(u(t, \cdot, \phi))$ is nonincreasing in t, this implies $k|u(t, \cdot, \phi)|_{H_0^1}^2 \leq 2V(\phi) + 2C_{k/2}$, solutions are globally defined, and $T(t)$ defines a C^1-semigroup on H_0^1.

One can also show that $T(t)\colon H_0^1 \to H_0^1$ is one-to-one and $D_\phi T(t)\phi$ is an isomorphism (see, for example, Henry [1981]).

For any $r > 0$, there is a constant $k(r) > 0$, such that $|f^e(\phi)|_X \leq k(r)$ if $|\phi|_{X^{1/2}} \leq r$. Thus, there is another constant $c(r)$ such that $V(\phi) \leq c(r)$ if $|\phi|_{X^{1/2}} \leq r$. Using this fact and the fact that V is nonincreasing along solutions of (3.2), (3.3) implies that orbits of bounded sets under $T(t)$ are bounded. Since the resolvent of A is compact, $T(t)$ is therefore completely continuous for $t > 0$. In particular, the positive orbit $\gamma^+(\phi)$ of any point $\phi \in X^{1/2}$ is precompact and has a nonempty compact connected invariant ω-limit set $\omega(\phi)$ (Lemma 3.2.1). From the expression for dV/dt along solutions and the continuity of $T(t)$ this implies that $\omega(\phi)$ belongs to the set E of equilibrium points of $T(t)$, that is, solutions of the boundary value problem

(3.6)
$$(au_x)_x + f(u) = 0, \quad 0 < x < 1,$$
$$u = 0 \quad \text{at } x = 0, 1.$$

It actually can be shown with a nontrivial proof that the set $\omega(\phi)$ consists of exactly one point in E (see Zelenyak [1968], Matano [1978], and Hale and Massatt [1982]).

The fact that $\omega(\phi) \subset E$ for each $\phi \in X^{1/2}$ implies that $T(t)$ is point dissipative if E is bounded. Let us now prove that E is bounded. An equilibrium point $\phi \in H_0^1$ is an extreme value of the functional

$$V(\phi) = \int_0^1 \left[\frac{a(x)}{2}\phi_x^2(x) - F(\phi(x))\right] dx;$$

that is,

(3.7)
$$\int_0^1 [a(x)\phi_x\psi_x - f(\phi)\psi] dx = 0 \quad \text{for all } \psi \in H_0^1.$$

Since f satisfies (3.1), for any $\varepsilon > 0$, there is an $M > 0$ such that $f(u)/u \le \varepsilon$ if $|u| \ge M$. Choosing $\psi = \phi$ in (3.7), we have

$$\int_0^1 a(x)\phi_x^2 \, dx = \int_0^1 f(\phi)\phi \, dx = \int_{I_1} f(\phi)\phi \, dx + \int_{I_2} f(\phi)\phi \, dx,$$

where $I_1 = \{x \in [0,1]: |\phi(x)| \ge M\}$, $I_2 = [0,1]\setminus I_1$. Then there is a constant $K = K(\varepsilon)$ such that

$$\int_0^1 a(x)\phi_x^2 \, dx \le \varepsilon |\phi|_{L^2}^2 + K \le \varepsilon |\phi|_{H_0^1}^2 + K.$$

There is a constant $k > 0$ such that $a(x) \ge k$ on $[0,1]$. So $\int_0^1 a(x)\phi_x^2 \, dx \ge k|\phi|_{H_0^1}^2$. Thus, $|\phi|_{H_0^1}^2$ is bounded by $K(k-\varepsilon)^{-1}$ if $\varepsilon < k$ and the equilibrium set is bounded.

We can now prove the following result.

THEOREM 4.3.1. *If f satisfies (3.1), then (3.2), (3.3) is a gradient system and there exists a connected global attractor $A = W^u(E)$ in $H_0^1(0,1)$. If each element of E is hyperbolic, then*

$$A = \bigcup_{\phi \in E} W^u(\phi).$$

Finally, A is a bounded set in $C^2(0,1)$.

PROOF. We have shown above that V defined in (3.4) satisfies (ii)$_1$, (ii)$_2$ of definition (3.8.1). Also, (3.5) implies condition (ii)$_3$ is satisfied. To prove (ii)$_4$ is satisfied, suppose ϕ is such that $V(T(t)\phi) = \phi$ for $t \in R$. Then $dV(T(t)\phi)/dt = 0$ for all $t \in R$ and (3.5) implies $\phi \in E$. Thus (3.1) is a gradient system. We have shown above that E is a bounded set. Applying Theorem 3.8.5, one obtains the existence of the global attractor A and its representation as stated in the theorem.

Since the solutions of (3.1), (3.2) are defined on $(-\infty, \infty)$, it follows that they represent classical solutions of the equation (see, for example, Henry, §3.5). This completes the proof of the theorem.

If $\phi \in E$, then ϕ is *hyperbolic* if and only if there is no nontrivial solution of the equation
$$(au_x)_x + f'(\phi(x))u = 0, \qquad 0 < x < 1, \qquad u = 0 \text{ at } x = 0, 1.$$
This is a consequence of the selfadjointness of this problem.

In an appropriate function space X, one can generalize this example to

(3.8a) $$u_t = (a(x)u_x)_x + f(x, u), \qquad 0 < x < 1$$

with boundary condition

(3.8b)
$$\alpha_0 u(0, t) - (1 - \alpha_0)u_x(0, t) - \beta_0 = 0,$$
$$\alpha_1 u(1, t) + (1 - \alpha_1)u_x(1, t) - \beta_1 = 0,$$

where $0 \le \alpha_j \le 1$, $\beta_j, j = 0, 1$, are constants. If $F(x, u) = \int_0^u f(x, s)\, ds$,

$$V(\phi) = \int_0^1 \left[\frac{a(x)}{2}\phi_x^2(x) - F(x, \phi(x)) \right] dx + B(\phi),$$

$$B(\phi) = a(1)\left[\frac{\alpha_1}{2}\phi^2(1) + \frac{1-\alpha_1}{2}\phi_x^2(1) - \beta_1\phi(1) \right]$$
$$- a(0)\left[\frac{\alpha_0}{2}\phi^2(0) - \frac{1-\alpha_0}{2}\phi_x^2(0) - \beta_0\phi(0) \right],$$

then
$$\frac{d}{dt}V(u(t,\cdot,\phi)) = au_x u_t\big|_0^1 - \int_0^1 u_t^2\, dx + \frac{\partial}{\partial t}B(u(t,\cdot,\phi)).$$

On the boundary, with $\alpha = \alpha_1$ and $\beta = \beta_1$, we have $\alpha u + (1-\alpha)u_x = \beta$ and so
$$u_t u_x = \alpha u_t u_x + (1-\alpha)u_t u_x = \alpha u_t u_x + u_t(1-\alpha)u_x$$
$$= -(1-\alpha)u_{xt}u_x + u_t(-\alpha u + \beta) = -\frac{\partial}{\partial t}\left[\frac{\alpha}{2}u^2 + \frac{1-\alpha}{2}u_x^2 - \beta u \right].$$

A similar computation at the boundary point $x = 0$ shows that $dV(u(t,\cdot,\phi))/dt = -\int_0^1 u_t^2 \le 0$. Now one proceeds as before.

Similarly, for the case of the n space variables,

$$u_t = \sum_{i,j=1}^n (a_{ij}(x)u_{x_j})_{x_i} + f(x, u) \quad \text{in } \Omega,$$

$$\alpha(x)u + [1 - \alpha(x)]\frac{\partial u}{\partial \nu} = \beta(x) \quad \text{in } \partial\Omega,$$

where $\alpha(x)$ is continuous, $0 \le \alpha(x) \le 1$, Ω is a bounded domain in R^n with smooth boundary, and the matrix $(a_{ij}(x))$ is symmetric and positive definite, one can take

$$V(u(t,\cdot,\phi)) = \int_\Omega \left[\frac{1}{2}\sum a_{ij}(x)u_{x_i}u_{x_j} - F(x, u) \right] dx$$
$$+ \int_{\partial\Omega} a\left[\frac{\alpha}{2}u^2 + \frac{1-\alpha}{2}\left(\frac{\partial u}{\partial \nu}\right)^2 - \beta u \right] d\sigma,$$

where

$$a(y) = \left(\sum_i \left[\sum_j a_{ij}(y)N_j(y)\right]^2\right)^{1/2}$$

and $N(y)$ is the unit outer normal to $\partial\Omega$ at y and show that

$$\frac{d}{dt}V(u(t,\cdot,\phi)) = -\int_\Omega u_t^2 \le 0$$

(see Matano [1979]).

One also can take the boundary conditions for this last equation in a more canonical way as

$$\alpha(x)u + [1-\alpha(x)]\frac{\partial u}{\partial \mu} = \beta(x) \quad \text{in } \partial\Omega$$

taking $\partial u/\partial \mu$ to be the conormal direction; that is, $\mu_i = \sum_{j=1}^n a_{ij}N_j$ where $N = (N_1, \ldots, N_n)$ is the unit outward normal. With this choice of boundary conditions, the function $V(u(t,\cdot,\phi))$ is the same as before except the factor a in the integral over $\partial\Omega$ is omitted.

4.3.2. *Qualitative properties of the flow on the attractor.* Since (3.8a), (3.8b) is a gradient system, the compact attractor is given as the union of the unstable manifolds of the equilibria if they are all hyperbolic. Therefore, to determine the flow on the attractor, one must determine the limit sets of elements of the unstable manifolds. This involves first of all an understanding of the restrictions, if any, that are imposed on equilibria if there is an orbit connecting them. It turns out that the Sturm-Liouville theory and the maximum principle impose restrictions. The precise statement is the following result.

LEMMA 4.3.2. *If ϕ, ψ are hyperbolic equilibria of (3.8a), (3.8b), and there is an orbit γ through a point η_0 and $\alpha(\eta_0) = \phi$ and $\omega(\eta_0) = \psi$, then $\dim W^u(\phi) > \dim W^u(\psi)$.*

PROOF. We indicate the essential elements of the proof and refer the reader to Henry [1985], Hale [1985], or Angenent [1986] for more details. To simplify matters we also take $a = 1$, $f(x,u) = f(u)$ independent of x and assume Dirichlet conditions.

Let $\eta(t)$ be the solution of the equation through η_0 with $\eta(t) \to \phi$ as $t \to -\infty$ and $\eta(t) \to \psi$ as $t \to +\infty$. Then the linear variational equation about $\eta(t)$ is

$$v_t - v_{xx} - f'(\eta(t))v = 0 \quad \text{in } (0,1), \qquad v = 0 \quad \text{at } x = 0,1.$$

From Theorem 4.3.1, η is a classical solution with $\eta(t,\cdot)$ in a bounded set in $C^2(0,1)$ for $t \in R$. Therefore, η_t exists, is bounded on R, and satisfies the linear variational equation about η. Since ϕ is hyperbolic, the function $\eta \to \phi$ exponentially as $t \to -\infty$ along the tangent space $TW^u(\phi)$ of the unstable manifold $W^u(\phi)$ of ϕ. Likewise, $\eta \to \psi$ exponentially as $t \to \infty$ along the tangent space $TW^s(\psi)$ of the stable manifold $W^s(\psi)$ of ψ. Thus, $\eta_t \to 0$ as $t \to \pm\infty$.

Now, let us consider the behavior of η_t as $t \to -\infty$. The space $TW^u(\phi)$ is finite dimensional and is given by the span of the eigenvectors corresponding to the positive eigenvalues of the operator $\partial^2/\partial x^2 + f'(\phi)$. Since the function $f'(\eta(t)) \to f'(\phi)$ as $t \to -\infty$ exponentially, it is certainly reasonable to guess that η_t must approach zero along one of the eigenvectors of the above operator.

To make this statement precise, one can consider the linear variational equation for $t \leq -\tau$, where τ is a large positive number, as a perturbation of the corresponding equation with $\eta(t)$ replaced by ϕ. More precisely, rewrite this equation as

$$v_t = v_{xx} + f'(\phi)v + b(t,x)v, \qquad 0 < x < 1,$$
$$b(t,x) = f'(\eta(t,x)) - f'(\phi(x)).$$

Let μ be an eigenvalue of $\partial^2/\partial x^2 + f'(\phi)$ acting on functions satisfying homogeneous Dirichlet boundary conditions. Let e_μ be an eigenfunction for μ with $|e_\mu|_0 = 1$, where $|e_\mu|_0^2 = \langle e_\mu, e_\mu \rangle$ and $\langle \,,\, \rangle$ is the L^2-inner product. Let $M_\mu = \mathrm{sp}[e_\mu]$ and let

$$X^{1/2} = M_\mu \oplus M_\mu^\perp,$$

where $M_\mu^\perp = \{\varsigma \colon \langle \varsigma, e_\mu \rangle = 0\}$. If $v = e^{\mu t}[e_\mu y + w]$ then

$$\dot{y} = A(t,y,w), \qquad w_t = w_{xx} + (f'(\phi) - \mu)w + B(t,y,w).$$

The functions $A(t,y,w), B(t,y,w)$ are linear in y, w and there are positive constants k, α such that

$$|A(t,y,w)|_{L^2}, |B(t,y,w)|_{L^2} \leq k e^{\alpha t}(|y| + |w|_{L^2}), \qquad t \leq 0.$$

One can now show that there is a $\tau > 0$ such that for each $t \leq -\tau$ there is a bounded linear operator $H_\mu(t) \colon R \to M_\mu^\perp$ continuous in t in the operator topology such that $H(t) \to 0$ as $t \to -\infty$ and if

$$M_\mu^t = \{\varsigma \in X^{1/2} \colon \varsigma = y e_\mu + H_\mu(t)y,\ y \in R\}$$

and $T(t,s)$ is the solution operator of the linear system, then $T(t,s)M_\mu^s = M_\mu^t$ for $s \leq t \leq \tau$. This is equivalent to saying that the set S_μ in $R \times X^{1/2}$ given by

$$S_\mu = \{(t,\varsigma) \colon \varsigma \in M_\mu^t,\ t \leq -\tau\}$$

is an invariant set for the linear system.

The method for showing the existence of $H_\mu(t)$ is the classical approach of Lyapunov-Perron in the theory of integral manifolds for ordinary differential equations (see, for example, Hale [1969]). One makes a further decomposition of the space $M_\mu^\perp$ as

$$M_\mu^\perp = M_{\mu u}^\perp \oplus M_{\mu s}^\perp$$

where $M_{\mu u}^\perp = \mathrm{sp}\,\{\text{eigenfunctions corresponding to the eigenvalues of } \partial^2/\partial x^2 + f'(\phi) \text{ which are larger than } \mu\}$ and $M_{\mu s}^\perp$ is the closure of the span of the eigenvectors corresponding to the eigenvalues less than μ. Let $P_\mu \colon M_\mu^\perp \to M_{\mu u}^\perp$ be the projection along $M_{\mu s}^\perp$ and let $P_s = I - P_u$. For a given $H \colon (-\infty, -\tau] \times R \to M_\mu^\perp$, consider the mapping defined in the following way. Let $-A_\mu = \partial^2/\partial x^2 + (f'(\phi) - \mu)$ with Dirichlet conditions. Let $\Phi(t, t_0, y_0, H)$ be the solution of the equation

$$\dot{y} = A(t, y, H(t)y), \qquad y(t_0) = y_0,$$

and define

$$\mathcal{L}H(t_0)y_0 = \int_{-\infty}^{0} e^{As} P_s B(t_0 + s, \Phi(t_0 + s, t_0, y_0, H),$$

$$H(t_0 + s)\Phi(t_0 + s, t_0, y_0, H))\,ds$$

$$- \int_{0}^{\infty} e^{-As} P_u B(t_0 + s, \Phi(t_0 + s, t_0, y_0, H),$$

$$H(t_0 + s)\Phi(t_0 + s, t_0, y_0, H))\,ds.$$

In each of these integrals, the kernels e^{As} and e^{-As} are decaying exponentially. This fact, together with the exponential decay estimates on A, B, permits one to use the contraction principle to show that there is a fixed point of the operator $\mathcal{L}$ if τ is chosen sufficiently large. If $H_\mu(t)$ is this fixed point, then the function $(\Phi(t, t_0, y_0, H_\mu), H_\mu(t)\Phi(t, t_0, y_0, H)) = (y(t), w(t))$ is a solution of the original equation, as asserted. This allows one to define the sets M_μ^t and S_μ above.

Once $H_\mu(t)$ has been constructed, the manifold M_μ^t is one dimensional and one can select an e_μ^t with $|e_\mu^t|_0 = 1$ and $e_\mu^t \to e_\mu$ as $t \to -\infty$.

Since there are only a finite number of positive eigenvalues $\mu_1, \ldots, \mu_N$, we can carry out the above analysis for each such μ_j to obtain $e_{\mu_j}^t$ with $|e_{\mu_j}^t|_0 = 1$ and $e_{\mu_j}^t \to e_{\mu_j}$ as $t \to -\infty$. For each $t_0 \leq -\tau$ with τ sufficiently large, the vector $e_{\mu_j}^{t_0}$ forms a basis for $M_{\mu_j}^{t_0}$. Let

$$M^t = \mathrm{sp}[e_{\mu_1}^t, \ldots, e_{\mu_N}^t], \qquad S_\tau = \{(t, \varsigma) : \varsigma \in M^t,\ t \leq -\tau\}.$$

Since $\eta_t \to 0$ as $t \to -\infty$ along $TW^u(\phi)$, it is not difficult to show that $(t, \eta_t) \in S_\tau$ for $t \geq \tau$. Therefore, $\eta_t \to 0$ as $t \to -\infty$ along the direction of one of the eigenvectors of the operator $\partial^2/\partial x^2 + f'(\phi)$. The Sturm-Liouville theory therefore implies that η_t has at most $\dim W^u(\phi) - 1$ zeros in $(0, 1)$.

The same type of proof as used for $t \to -\infty$ can be applied to the case where $t \to \infty$. One considers the linear variational equation as $t \to \infty$ as a perturbation of the limiting equation at $t = \infty$. More precisely, we write it as

$$u_t = u_{xx} + f'(\psi)u + c(t, x)u, \qquad 0 < x < 1,$$

$$c(t, x) = f'(\eta(t, x)) - f'(\psi(x)).$$

For each negative eigenvalue μ with corresponding eigenvector e_μ, use the same type of decomposition as before and show that there is a $\tau > 0$ such that, for each $t \geq \tau$, there is a linear operator $K_\mu(t): R \to M_\mu^\perp$ continuous in the operator topology such that $K_\mu(t) \to 0$ as $t \to \infty$ and the set

$$N_\mu^t = \{\varsigma \in X^{1/2}, \varsigma = ye_\mu + K_\mu(t)y,\ y \in R\}$$

has the property that $T(t, s)N_\mu^s = N_\mu^t$ for $t \geq s \geq \tau$ for the solution operator $T(t, s)$ and the set $\tilde{S}_\mu = \{(t, \varsigma) : \varsigma \in N_\mu^t,\ t \geq \tau\}$ is invariant.

An additional difficulty occurs here because there are infinitely many negative eigenvalues. To reason as above for the case where $t \to -\infty$, we need to be able to choose τ independent of μ. This is possible in this case because the eigenvalues μ_n satisfy $\mu_{n+1} - \mu_n = O(n)$ as $n \to \infty$. This is more than enough to show that τ can be chosen independently of μ.

If μ_n is an eigenvalue define $e_{\mu_n}^t$, $K_{\mu_n}(t)$, $N_{\mu_n}^t$, $\tilde{S}_{\mu_n}$ as above. It is a consequence of the proof of existence that $K_{\mu_n}(t)$ approach zero as $t \to \infty$ uniformly in n. This allows one to select the $e_{\mu_n}^t$ so that $e_{\mu_n}^t \to e_{\mu_n}$ as $t \to \infty$ uniformly in n. If

$$N^t = \bigcup_n N_{\mu_n}^t, \quad \tilde{S}_\tau = \{(t,\varsigma), \varsigma \in N^t, t \geq \tau\}$$

then this latter fact implies that the $\{e_{\mu_n}^t\}$ for $t \geq \tau$, for τ sufficiently large, form a basis for N^t since the $\{e_{\mu_n}\}$ form a basis for $M_\mu^\perp$.

Since $\eta_t \to 0$ as $t \to \infty$ along $TW^s(\psi)$, it is not difficult to show that $(t, \eta_t) \in \tilde{S}_\tau$ for $t \geq \tau$. Since the $\{e_{\mu_n}^{t_0}\}$ form a basis for N^{t_0}, it follows that, if $\eta_t \neq 0$, there is an expansion of $\eta_t(t_0)$ in the form

$$\eta_t(t_0) = \sum_{j \geq k} \alpha_{\mu_j}(t_0) e_{\mu_j}^{t_0}, \quad \alpha_{\mu_k}(t_0) \neq 0.$$

Since $\eta_t(t) = T(t, t_0)\eta_t(t_0)$ and $T(t, t_0)N_{\mu_j}^{t_0} = N_{\mu_j}^t$, we have

$$\eta_t(t) = \sum_{j \geq k} \alpha_{\mu_j}(t_0) e^{\mu_j(t-t_0)}[1 + g_j(t, t_0)] e_{\mu_j}^t,$$

where $g_j(t, t_0) \to 0$ as $t \to \infty$ uniformly in j. Thus,

$$e^{-\mu_k t}\eta_t(t) \to \alpha_{\mu_k}(t_0) e^{-\mu_k t_0} e_{\mu_k} \neq 0,$$

and $\eta_t \to 0$ as $t \to \infty$ along the direction of an eigenvector corresponding to a negative eigenvalue of $\partial^2/\partial x^2 + f'(\psi)$.

The Sturm-Liouville theory together with the conclusion in the previous paragraph implies that η_t has at least $\dim W^u(\psi)$ zeros in $(0, 1)$.

The proof of Lemma 3.2 is completed by an application of the following result which was proved by Nickel [1962] as well as others mentioned in Walter [1970]. It also can be found in Matano [1982].

LEMMA 4.3.3. *If $v(t,x)$ is a solution of the linear variational equation and $N_v(t)$ is the number of zeros of $v(t,x)$ in $(0,1)$, then $N_v(t)$ is nonincreasing in t.*

The proof of this lemma uses the strong maximum principle and the Jordan curve theorem. In fact, if $\tau < t$, then there are $p \stackrel{\text{def}}{=} N_v(1) + 1$ points in $(0, 1)$, $Q_1, \ldots, Q_p$ such that $f(t, Q_j) \cdot v(t, Q_{j+1}) < 0$ for all j. For definiteness, suppose $v(t, Q_1) > 0$ and let S_1 be the connected component in $[\tau, t] \times (0, 1)$ such that $(t, Q_1) \in S_1$ and $v(s, x) > 0$ if $(s, x) \in S_1$. Then the strong maximum principle implies that $S_1 \cap (\{\tau\} \times (0, 1)) \neq \emptyset$. If we define S_j in an analogous way for Q_j, then $S_j \cap (\{\tau\} \times (0, 1)) \neq \emptyset$. If $(\tau, P_j) \in S_j$, then $v(\tau, P_j) \cdot v(\tau, P_{j+1}) < 0$ for all j from the Jordan curve theorem. This proves the lemma.

Lemma 4.3.2 allows one to give an interesting Morse decomposition (for the definition, see §4.1) of the attractor A in the case when all equilibrium points are hyperbolic. Define the Morse set S_N by the relations

$$S_N = \{\phi \in E \colon \dim W^u(\phi) = N\}.$$

Since the system is gradient, each α- and ω-limit set must belong to one of the S_N. Lemma 4.3.2 implies that the S_N, $N = 0, 1, 2, \ldots$, are Morse sets for A.

Using Lemma 4.3.2, one obtains the very surprising result of Henry [1985] (another proof has been given by Angenent [1986]).

THEOREM 4.3.4. *If ϕ, ψ are hyperbolic equilibria of (3.8), (3.9), then $W^u(\phi)$ is transversal to $W^s(\psi)$.*

The idea of the proof is very geometric. Let us introduce the adjoint of the linear variational equation,

$$w_t + w_{xx} + f'(\eta(-t))w = 0,$$

together with the boundary conditions $w = 0$ at $x = 0, 1$. If $w(t)$ is a solution of this equation and $v(t)$ is a solution of the linear variational equation, then $\langle w(t), v(t) \rangle$ is a constant on any interval I of common existence. This permits one to characterize $TW^s(\psi)$ in terms of all solutions which are orthogonal to the solutions of the adjoint equation which approach $TW^u(\phi)$ as $t \to \infty$.

Now suppose $\eta(t)$ as before is such that $\eta(t) \to \phi$ as $t \to -\infty$, $\eta(t) \to \psi$ as $t \to +\infty$. Then $\eta_t \in TW^u(\phi) \cap TW^s(\psi)$ for all $t \in R$. Let us take t near $-\infty$. Then $TW^u(\phi)$ is approximately the same linear subspace as the one corresponding to the above solutions of the adjoint equation for t large. If $W^u(\phi)$ is not transversal to $W^s(\psi)$ then there must be a nontrivial solution of the adjoint equation which approaches zero as $t \to -\infty$. Thus, the adjoint equation has a nontrivial solution which approaches zero as $t \to \pm\infty$. One now proves that Lemma 4.3.3 is valid for this equation. But this will imply $\dim W^u(\phi) < \dim W^u(\psi)$. This contradicts Lemma 4.3.2.

The details of the above argument as well as a proof of the fact that Theorem 4.3.4 is valid for equations where $f = f(x, u, u_x)$ can be found in Henry [1985]. One also can use exponential dichotomies to give a complete proof, but we do not pursue this further.

Theorem 4.3.4 has very interesting implications. In particular, for the gradient system (3.8a), (3.8b), it says that changes in the dynamical behavior on the global attractor due to changes in f can occur only when there is a local bifurcation of equilibria. No changes in the dynamics will occur if the equilibria remain hyperbolic. The only ordinary differential equation for which we know this to be true is a scalar equation!

To understand the remaining structure of the flow on A, one must determine the connecting orbits C_K^N between the Morse sets. Theorem 4.3.4 plays a major role in understanding these connecting orbits since it reduces the discussion to the analysis of the orbit structure near points of bifurcation near equilibria.

Further developments from these ideas are contained in Henry [1985], Brunovsky and Fiedler [1986], and Rocha [1987].

Now let us consider a one parameter family of equations

$$(3.9)_\lambda \qquad u_t = (a(x)u_x)_x + \lambda f(u), \qquad 0 < x < 1,$$

with the boundary conditions (3.3) and f satisfying (3.1). Here, $\lambda > 0$ is a real parameter. For each $\lambda > 0$, there is a global attractor A_λ. It is not too difficult to show that the equilibrium set E_λ is bounded uniformly in λ. Also, the corresponding semigroup $T_\lambda(t)$ satisfies the property that $T_\lambda(t)x$ is continuous in (t, x, λ) and the continuity in λ is uniform with respect to (t, x) in bounded sets. Therefore, Theorem 3.5.3 implies that A_λ is upper semicontinuous in λ. If the equilibrium points of $(3.9)_{\lambda_0}$ are hyperbolic at $\lambda = \lambda_0$, then Theorem 3.4 implies that the family of semigroups $T_\lambda(t)$ defined by $(3.9)_\lambda$ for λ in a neighborhood U of λ_0 is Morse-Smale; that is,

(i) $T_\lambda(t): H_0^1 \to H_0^1$ is one-to-one and $DT_\lambda(t): H_0^1 \to H_0^1$ is an isomorphism for each $t \geq 0$ and $\lambda \in U$;

(ii) the equilibrium points are hyperbolic for $\lambda \in U$;

(iii) the stable and unstable manifolds of equilibrium points are transversal;

(iv) A_λ is upper semicontinuous in λ for $\lambda \in U$.

We summarize this in

THEOREM 4.3.5. *Consider the system* $(3.9)_\lambda$, *(3.3) with f satisfying (3.1) and $\lambda > 0$ a real parameter. If the equilibrium points at $\lambda = \lambda_0$ are hyperbolic, then there is a neighborhood U of λ_0 such that the system is Morse-Smale for $\lambda \in U$.*

Since the flow on the global attractor for Morse-Smale systems is stable, it follows that the qualitative properties of the flow in $(3.9)_\lambda$, (3.2) are the same for all $\lambda \in U$, the set defined in Theorem 4.3.5. In particular, if ϕ_λ, ψ_λ are equilibrium points of $(3.9)_\lambda$, (3.3) which are continuous in λ and there is an orbit γ_{λ_0} connecting ϕ_{λ_0} to ψ_{λ_0}, then there is an orbit γ_λ connecting ϕ_λ to ψ_λ for all $\lambda \in U$. Therefore, to understand the change in the orbit structure of the flow on A_λ for any $\lambda > 0$, one can restrict attention to a small neighborhood of those values $\lambda = \lambda_1$ for which there is a nonhyperbolic equilibrium point of $(3.9)_{\lambda_1}$, (3.3). This reduces the global problem of orbit structure to one which is more local. A discussion of an example for which the complete orbit structure is analyzed in this way can be found in Henry [1985].

4.3.3. *Stability properties of equilibria.* In this section, we relate some of the structure of an equilibrium solution to the stability of the solution.

The following result is essentially due to Yanagida [1982]. Chipot and Hale [1983] gave the result under the condition $a'' \leq 0$. For the case where a in (3.2) is constant and Dirichlet boundary conditions, it is due to Chafee [1975]. For the other boundary conditions and a in (3.2) a constant, see Matano [1982].

THEOREM 4.3.6. *Consider equation (3.2) with*

$$(3.10) \qquad a(x) = b^2(x), \qquad b(x) > 0, \qquad b''(x) \leq 0.$$

For the boundary conditions

(3.11) $\quad \alpha u_x(0,t) - (1-\alpha)u(0,t) = 0, \quad \beta u_x(1,t) + (1-\beta)u(1,t) = 0$

with $0 \leq \alpha, \beta < 1$, every nonconstant equilibrium solution v such that $v_x = 0$ at two distinct points in $(0,1)$ is unstable. For the Neumann boundary condition $(\alpha = \beta = 1)$ every nonconstant equilibrium solution of (3.2) is unstable.

PROOF. The proof is based on Chipot and Hale [1983].

Neumann boundary conditions. If v is an equilibrium solution of (3.2) with Neumann boundary conditions, let $H: H^1(0,1) \to R$ be defined by

$$H(\phi) = \int_0^1 [a\phi_x^2 - f'(v)\phi^2]\,dx.$$

The first eigenvalue λ_1 of the operator

$$Lu = -(au_x)_x - f'(v)u$$

is given by

$$\lambda_1 = \min\{H(\phi): \phi \in H^1(0,1), |\phi|_{L^2(0,1)} = 1\}.$$

Furthermore, the equilibrium solution v is unstable if $\lambda_1 < 0$ (see, for example, Henry [1981]). Consequently, it is sufficient to show that, if $a = b^2, b > 0, b'' \leq 0$, then $\lambda_1 < 0$ if v is not a constant. This will be the case if we show that

$$H(bv_x) = \int_0^1 \{b^2[(bv_x)_x]^2 - f'(v)(bv_x)^2\}\,dx < 0$$

when v is not a constant. Let us first compute the last term in this expression.

$$\int_0^1 f'(v)(bv_x)^2 = \int_0^1 f'(v)v_x b^2 v_x = \int_0^1 [f(v)]_x b^2 v_x$$

$$= -\int_0^1 f(v)(b^2 v_x)_x = \int_0^1 [(b^2 v_x)_x]^2$$

$$= \int_0^1 [b(bv_x)_x + b_x(bv_x)]^2.$$

Therefore,

$$H(bv_x) = -\int_0^1 (b_x bv_x)^2 + bb_x \cdot 2bv_x(bv_x)_x$$

$$= -\int_0^1 (b_x bv_x)^2 + bb_x[(bv_x)^2]_x$$

$$= \int_0^1 [-(b_x bv_x)^2 + (bb_x)_x(bv_x)^2]$$

$$= \int_0^1 [-b_x^2 + (bb_x)_x](bv_x)^2$$

$$= \int_0^1 bb_{xx}(bv_x)^2 \leq 0.$$

If $H(bv_x) < 0$, then $\lambda_1 < 0$ and the solution v is unstable. If $\lambda_1 \geq 0$, then the definition of λ_1 and the fact that $H(bv_x) \leq 0$ imply $H(bv_x) = 0$. Thus, $\lambda_1 = 0$.

Since the first eigenvalue of L is simple, there is a ϕ such that $L\phi = 0$, $\phi_x = 0$ at $x = 0$, $x = 1$, $|\phi|_{L^2(0,1)} = 1$. Also, $H(bv_x) = 0$ implies there is a constant c such that $bv_x = c\phi$ on $[0, 1]$. If $c \neq 0$, then $v_x = 0$ at $x = 0$, $x = 1$ and $b > 0$ imply that $\phi = 0$ at $x = 0$, $x = 1$. But this would imply that $\phi = 0$ on $[0, 1]$ which is a contradiction. Thus, $c = 0$ and $bv_x = 0$, $v_x = 0$, and v is a constant. This completes the proof for the Neumann case.

General boundary conditions. Suppose v is an equilibrium point of (3.2), (3.11) which is nonconstant and such that $v_x = 0$ at $x = x_0$ and $x_1 > x_0$ in $(0, 1)$. Then v is an equilibrium solution of the Neumann problem on $[x_0, x_1]$. As in the proof of the first part of the theorem, one shows that the first eigenvalue λ_1 of the linear variational operator is negative if v is not a constant on $[x_0, x_1]$. Let μ_1 be the first eigenvalue of the linear variational operator about v for the boundary condition (3.11). By the characterization of μ_1 as the minimum of the functional $H(u)$ it follows that μ_1 is negative. This proves the theorem.

For Neumann boundary conditions, Yanagida [1982] has shown that Theorem 4.3.6 is the best possible in the following sense: if there is an x_0 in $[0, 1]$ such that $b''(x_0) > 0$, then there is a nonlinear function f such that (3.2) has a stable nonconstant equilibrium. The proof of this result is rather difficult. However, it is intuitively possible to see why some condition on the diffusion a must be imposed. For example, if $f(u) = u(1 - u^2)$ and a is a diffusion coefficient with a deep well in the interval $[1/4, 3/4]$ and almost one outside this interval, then one would expect that there should be a stable solution which is almost one in $(0, 1/2)$ and almost -1 in $(1/2, 1)$ with a smooth transition in between. Exploiting this fact, examples of equations (3.2) with stable nonconstant equilibria have been given by Matano [1979], Fife and Peletier [1981], and Fusco and Hale [1985].

For Ω a bounded domain in R^n, consider the equation

$$(3.12) \qquad u_t = \Delta u + f(u) \quad \text{in } \Omega, \qquad \partial u/\partial \nu = 0 \quad \text{in } \partial\Omega.$$

If Ω is convex, then Casten and Holland [1978], Matano [1979] have shown that every nonconstant equilibrium is unstable. If Ω is not convex, it is not known if there is a nonlinear function f such that the equation has a stable nonconstant equilibrium solution. However, there are nonconvex domains and nonlinear functions f such that there are stable nonconstant equilibria (Matano [1979]). For the nonlinear function f having two distinct stable zeros, the domain Ω can be chosen such that there are two convex islands with a small causeway between. If the causeway is sufficiently narrow, then one can get a solution which is approximately u_0 on one of the islands and u_1 on the other. Hale and Vegas [1984] and Vegas [1983] have discussed all of the solutions of such an equation as a bifurcation problem for a special type of domain using the size of the causeway as the bifurcation parameter, with the limiting domain being the islands connected by a line. These authors imposed conditions on the magnitude of the derivative of the nonlinear function f. Jimbo [1988] has shown that a limitation on the magnitude is necessary in order to be able to determine the nature of the flow by only considering the solutions of the limiting problem on the islands. If the

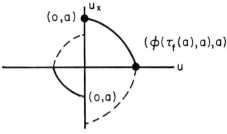

FIGURE 3.1

derivative is too large then a boundary value problem on the line connecting the islands plays an important role. Jimbo [1986] and Vegas [1986] have other results for more general domains. For Ω convex, Kishimoto and Weinberger [1985] have generalized Theorem 4.3.2 to the case of variable coefficients and the analogue of the Newmann boundary conditions. No general results seem to be known for boundary conditions other than Neumann in (3.12).

4.3.4. *A bifurcation problem—Dirichlet conditions.* In this section, we consider two bifurcation problems for the equation

(3.13) $\qquad u_t = u_{xx} + \lambda^2 f(u), \qquad 0 < x < 1,$

(3.14) $\qquad u = 0 \quad \text{at } x = 0, \ x = 1,$

(3.15) $\qquad f(u) = u(u - b)(1 - u),$

where $\lambda > 0$ is the bifurcation parameter. This system is a special case of (3.1), (3.2), and (3.9). Thus, there is a connected global attractor. We discuss the flow on the attractor for every λ.

The equilibrium points for this equation are solutions of

(3.16) $\qquad u_{xx} + \lambda^2 f(u) = 0, \qquad 0 < x < 1,$

with the boundary conditions (3.14). It is convenient to rescale the x variable as $x \to \lambda x$ so that the equilibrium solutions coincide with the solutions of the boundary value problem

(3.17) $\qquad u_{xx} + f(u) = 0,$

(3.18) $\qquad u = 0 \quad \text{at } x = 0, \ x = \lambda.$

In the (u, u_x)-phase plane, this means that we need to determine a solution $\phi(x, a)$ of (3.17) with $\phi(0, a) = 0$, $\phi_x(0, a) = a$ such that $\phi(\lambda, a) = 0$ (see Figure 3.1). Because of the symmetry of solutions with respect to the u axis, this means that we need to find the solution $\phi(x, a)$ with $\phi(0, a) = 0$, $\phi_x(0, a) = a$, and $\phi_x(\lambda/2, a) = 0$.

To discuss this problem, it is therefore natural to discuss the "time" map from the u_x axis to the u axis. More specifically, for any $a \in R$, let $\phi(x, a)$ be the solution of (3.17) with $\phi(0, a) = 0$, $\phi_x(0, a) = a$. For some $a \in R$, there may be a $\tau_f(a) > 0$ such that $\phi_x(x, a) \neq 0$, $0 < x < \tau_f(a)$, $\phi_x(\tau_f(a), a) = 0$. Let $D(\tau_f) = \{a \in R: \tau_f(a) \text{ is defined}\}$ be the domain of τ_f. The function $\tau_f(a)$ is

APPLICATIONS

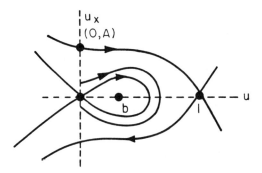

FIGURE 3.2

the "time" map for (3.17) and this function determines the number of solutions of the boundary value problem that can be obtained. In fact, the solutions with no zeros in $(0, \lambda)$ coincide with the solutions of the equation $2\tau_f(a) = \lambda$. The solutions with one zero in $(0, \lambda)$ coincide with the ones which are periodic of period λ; that is, $2\tau_f(a) + 2\tau_f(-a) = \lambda$. The solutions with two zeros satisfy $4\tau_f(a) + 2\tau_f(-a) = \lambda$, etc.

Let us now compute the formula for $\tau_f(a)$. The solution $\phi(x, a)$ satisfies

(3.19) $$\phi_x^2(x, a) + 2F(\phi(x, a)) = a^2,$$

where $F(u) = \int_0^u f(s)\, ds$. On the interval $(0, \tau_f(a))$ the function $\phi(x, a)$ is implicitly defined as a function of x, a by the integral equation

$$\int_0^{\phi(x,a)} \frac{da}{[a^2 - 2F(s)]^{1/2}} = x.$$

If $a \in D(\tau_f)$, let $\alpha(a) = \phi(\tau_f(a), a)$. Since $a^2 = 2F(\alpha(a))$, we have

(3.20) $$2^{1/2}\tau_f(a) = \int_0^{\alpha(a)} \frac{ds}{[F(\alpha(a)) - F(s)]^{1/2}}.$$

Case 1. $0 < b < 1/2$. In this case, the phase portrait for (3.17) is shown in Figure 3.2. The constant solution $u = 0$ always satisfies the boundary value problem and $D(\tau_f) = (0, A)$ where A is the value of u_x where the stable manifold of the equilibrium point $u = 1$ intersects the u_x-axis. The basic lemma is

LEMMA 4.3.7. *If $0 < b < 1/2$ in (3.15), then $D(\tau_f) = (0, A)$, $\tau_f(a) \to \infty$ as $a \to 0$, $a \to A$ and $\tau_f(a)$ has a unique critical point at a_0 which corresponds to a minimum, that is, $\tau_f''(a_0) > 0$.*

We do not give the proof, but refer the reader to Smoller [1983, pp. 188–190].

Lemma 4.3.7 implies that there are at most three solutions of (3.17), (3.18). In fact, there is one solution if $0 < \lambda < 2\tau_f(a_0)$, two solutions at $\lambda = 2\tau_f(a_0)$, and three solutions if $\lambda > 2\tau_f(a_0)$. Any nonzero solution is positive on $(0, 1)$. The bifurcation diagram is shown in Figure 3.3.

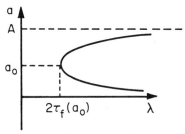

FIGURE 3.3

LEMMA 4.3.8. *An equilibrium solution v of (3.13), (3.14) with $v(0) = 0$, $v_x(0) = a_0$ is hyperbolic if and only if $d\tau_f(a_0)/da \neq 0$.*

PROOF. See Smoller [1983].

Let us now discuss the attractor A_λ of (3.13), (3.14). If $0 < \lambda < 2\tau_f(a_0)$, then zero is the only equilibrium point of (3.13), (3.14) and $A_\lambda = \{0\}$. Furthermore, the linear variational equation for the zero solution is

$$u_t = u_{xx} - \lambda^2 bu, \qquad 0 < x < 1,$$
$$u = 0 \quad \text{at } x = 0, \ x = 1.$$

The largest eigenvalue μ of the operator $\partial^2/\partial x^2 - \lambda^2 b$ with Dirichlet boundary data is $\mu = -\pi^2 - \lambda^2 b < 0$. Thus, zero is always exponentially asymptotically stable.

The point $\lambda_0 = 2\tau_f(a_0)$ is a bifurcation point for equilibrium solutions which corresponds to the solution $\phi(\lambda, x, a_0)$ of (3.13) for $\lambda = \lambda_0$ where $\phi(\lambda, x, a_0)$ is the solution of (3.17) with $\phi(\lambda, 0, a_0) = 0$, $\phi_x(\lambda, 0, a_0) = a_0$. To complete the global picture of the flow on the attractor, we need

LEMMA 4.3.9. *Let f be any C^2-function in (3.13) and suppose v is an equilibrium solution of (3.13), (3.14) which is positive on $(0,1)$. Then $\dim W^u(v) \leq 1$. Also, if $\dim W^u(v) = 1$, the tangent space $T_v W^u(v)$ is equal to span $[v_0]$ where $v_0(x) > 0$ on $(0,1)$.*

PROOF. We show that the eigenvalue problem

(3.21)
$$u'' + \lambda^2 f'(v(x))u = \mu u, \qquad 0 < x < 1,$$
$$u = 0 \quad \text{at } x = 0, x = 1$$

has at most one positive eigenvalue. Suppose $\mu > 0$ is an eigenvalue and v_0 is a corresponding eigenfunction. The function $v'(x)$ is a solution of $u'' + \lambda^2 f'(v(x))u = 0$ and phase plane analysis implies $v'(x)$ has exactly one zero in $(0,1)$. The Sturmian comparison theorem implies that v_0 has no zeros in $(0,1)$. Thus, μ corresponds to the maximal eigenvalue and so there is at most one positive eigenvalue of (3.21). Thus, $\dim W^u(v) \leq 1$. If $\dim W^u(v) = 1$, then there is a positive eigenvalue of (3.21) with corresponding eigenfunction $v_0(x)$ positive in $(0,1)$. Thus, $T_v W^u(v) = [v_0]$ and the lemma is proved.

Now suppose that $\lambda > 2\tau_f(a_0)$ and let $u_0 = 0$, u_1, u_2 be the three equilibrium solutions with $0 < u_1(x) < u_2(x)$ on $(0,1)$. From Lemma 4.3.8, u_0, u_1, and

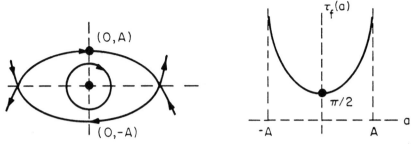

FIGURE 3.4

u_2 are hyperbolic. The attractor $A_\lambda = \bigcup_j W^u(u_j)$ and $\dim W^u(u_j) \leq 1$ by Lemma 4.3.9. Also, if $\dim W^u(u_j) = 1$ for some j, then Lemma 4.3.9 implies that $T_{u_j} W^u(u_j) = [v_0]$ where $v_0(x) > 0$, $0 < x < 1$. The maximum principle implies that, if $\phi \in W^u(u_j)$, $\phi \neq u_j$, then either $\phi(x) < u_j(x)$ for $x \in (0,1)$ or $\phi(x) > u_j(x)$ for $x \in (0,1)$. Since the system is gradient this implies there are equilibrium points v_1, v_2 with $v_1(x) < u_j(x) < v_2(x)$ such that if $\phi \in W^u(u_j)$ then $\omega(\phi) = v_1$ if $\phi(x) < u_j(x)$ in $(0,1)$ and $\omega(\phi) = v_2$ if $\phi(x) > u_j(x)$ in $(0,1)$. This obviously implies that $j = 1$, $v_1 = u_0$, and $v_2 = u_2$. Thus, the attractor is a one dimensional set ·—←—·—→—·. If $\lambda = \tau_f(a)$, the center manifold theorem and the maximum principle imply that A_λ is given by ·—←—·.

The argument that has been used in this example also proves a more general result. We say an equilibrium point ϕ is *connected to* an equilibrium point ψ if there is an orbit γ such that $\alpha(\gamma) = \phi$, $\omega(\gamma) = \psi$.

THEOREM 4.3.10. *Suppose* (3.1), (3.2) *has a global attractor A which contains only the hyperbolic equilibrium points* $u_1, u_2, \ldots, u_p$ *with* $\dim W^u(u_j) \leq 1$ *for all j and* $u_1(x) < u_2(x) < \cdots < u_p(x)$ *for x in* $(0,1)$. *Then* $p = 2k+1$ *for some integer k, the equilibrium points* u_{2j+1}, $j = 0, 1, 2, \ldots, k$, *are u.a.s. and* u_{2j}, *are unstable,* $\dim W^u(u_{2j}) = 1$, $j = 1, 2, \ldots, k$, *and there is an orbit connecting* u_{2j} *to* u_{2j-1} *and an orbit connecting* u_{2j} *to* u_{2j+1} *for* $j = 1, 2, \ldots, k$. *The attractor A is given by*

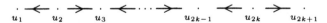

The flow on the attractor in Theorem 4.3.10 preserves the natural order of the equilibrium points. This is a consequence of the maximum principle for scalar parabolic equations. Contrast this with the gradient scalar FDE of Theorem 4.1.7.

Case 2. $b = -1$. In this case, $f(u) = u - u^3$ and the phase portrait for the solutions of (3.17) are shown in Figure 3.4. In this case $D(\tau_f) = (-A, A) \setminus \{0\}$ and τ_f can be extended to $(-A, A)$ by defining $\tau_f(0) = \pi/2$. Also, $\tau_f(a) = \tau_f(-a)$. One can then prove the following lemma (see Chafee and Infante [1974]).

THEOREM 4.3.11. *If* $b = -1$ *in* (3.15), *then* $D(\tau_f) = (-A, A)$, $\tau_f(a) \to \infty$ *as* $a \to \pm A$, *and* $\tau_f(a)$ *has a unique critical point which is a minimum at zero,* $\tau_f(0) = \pi/2$, $\tau_f''(0) > 0$.

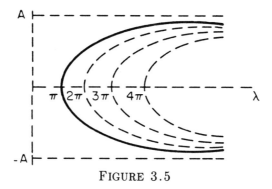

FIGURE 3.5

Lemma 4.3.11 and the complete symmetry in the phase plane diagram allow one to discuss the bifurcation diagram for the equilibrium solutions of (3.13), (3.14) with $f(u) = u - u^3$. In fact, all solutions bifurcate from $u = 0$ at the values $\lambda_n = n\pi$, $n = 1, 2, 3, \ldots$. At λ_n, the solutions ϕ_n^+, ϕ_n^-, $d\phi_n^+(0)/dx > 0$, $d\phi_n^-(0)/dx < 0$, that bifurcate from zero and exist for $\lambda > \lambda_n$ are hyperbolic, have $n - 1$ zeros in $(0, 1)$, $\dim W^u(\phi_n^\pm) = n - 1$, and $T_{\phi_n^\pm} W^u(\phi_n^\pm)$ is the span of the n eigenfunctions $\psi_1^\lambda, \ldots, \psi_{n-1}^\lambda$ corresponding to the $n-1$ positive eigenvalues $\mu_1^\lambda, \ldots, \mu_{n-1}^\lambda$ of the boundary value problem

$$v_{xx} + \lambda^2[1 - 3(\phi_n^\pm)^2]v = \mu v, \qquad v(0) = v(1) = 0.$$

The constants $\mu_j^\lambda \to (n-j)^2\pi^2$, $\psi_j^\lambda(x) \to \xi_{n-j}(x) \stackrel{\text{def}}{=} \sin(n-j)\pi x$ as $\lambda \to n\pi$. The equilibrium point zero has $\dim W^u(0) = n$ and $T_0 W^u(0)$ is spanned by the functions $\xi_j(x)$, $j = 1, 2, \ldots, n$. The bifurcation diagram is in Figure 3.5 with the unstable branches labeled with dashed lines.

Let us summarize these results in

THEOREM 4.3.12. *If $\lambda \in (n\pi, (n+1)\pi]$ where n is an integer, then there are $2n + 1$ equilibrium points $\phi_0 = 0$, ϕ_j^+, ϕ_j^-, $j = 1, 2, \ldots, n$, of (3.13), (3.14) with $f(u) = u - u^3$, the points $\phi_j^\pm$ are hyperbolic with $\dim W^u(\phi_j^\pm) = j - 1$, $j = 1, 2, \ldots, n$. If $\lambda \in (n\pi, (n+1)\pi)$, then $\phi_0 = 0$ is hyperbolic, $\dim W^u(0) = n$, and the attractor A_λ is given*

$$A_\lambda = W^u(0) \cup \left(\bigcup_{j=1}^n W^u(\phi_j^\pm) \right).$$

Chafee and Infante [1974] proved the existence of the equilibria and their stability properties. The other statements are taken from Henry [1981].

From Lemma 4.3.11, if $\lambda \in (0, \pi)$, then $A_\lambda = \{\phi_0\}$. If $\lambda \in (\pi, 2\pi)$, then the same type of argument as used in the proof of Theorem 4.3.10 shows that A_λ is given as $\text{Cl } W^u(0)$ and looks like ·——←·—→——·. Since the next bifurcation occurs from ϕ_0, one would guess that A_λ for $\lambda \in (2\pi, 3\pi)$ looks like the deformed disk in Figure 3.6. Furthermore, the next bifurcation should simply increase the dimension. This can be proved, but it is far from trivial. It uses in a significant

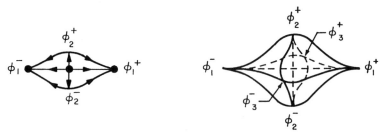

FIGURE 3.6

way Theorem 4.3.4 stating that the stable and unstable manifolds for hyperbolic equilibria are always transversal and the fact the semigroup $T_\lambda(t)$ generated by the equation is Morse-Smale for $\lambda \in (n\pi, (n+1)\pi)$ (see the remarks after Theorem 4.3.4). This reduces the discussion to λ in a small neighborhood of $n\pi$. To complete the discussion, one uses the gradient structure of the equation and a transversality result for the stable and unstable sets at $\lambda = n\pi$, a bifurcation point (see Henry [1985]).

It is an interesting exercise to discuss (3.13), (3.15) with the boundary condition $u = \tau$ at $x = 0$, $x = 1$, where τ is a positive constant.

In (3.13), if we replace t by $\lambda^{-2} t$ then the equation becomes

$$u_t = \lambda^{-2} u_{xx} + f(u), \qquad 0 < x < 1,$$

and the boundary conditions are not changed. Therefore, the bifurcation diagrams that have been obtained can be considered as bifurcation diagrams with the diffusion coefficient as the parameter.

4.3.5. A bifurcation problem—other boundary conditions. Let us consider (3.13), (3.15) with Neumann boundary conditions

(3.22) $$u_x = 0 \quad \text{at } x = 0, 1.$$

In the phase plane, one must find a solution of (3.17) which goes from the u-axis at $x = 0$ to the u-axis at $x = \lambda$. The "time" map $\tau_f(a)$ will be solutions $\phi(x, a)$ (if they exist) which satisfy $\phi(0, a) = a$, $\phi_x(0, a) = 0$, $\phi_x(x, a) \neq 0$, $0 < x < \tau_f(a)$, $\phi_x(\tau_f(a), a) = 0$.

To discuss the solutions of this boundary value problem, it again is convenient to rescale the variables $x \to \lambda x$ obtaining (3.17) with the boundary conditions

(3.23) $$u_x(0) = 0, \qquad u_x(\lambda) = 0.$$

We now define the "time" map for the Neumann problem (3.17), (3.23). For any real number a, let $\phi(x, a)$ be the solution of (3.17) with $\phi(0, a) = a$, $\phi_x(0, a) = 0$. Let $D(\tau_f)$ be the set of a for which there is a $\tau_f(a) > 0$ such that $\phi_x(x, a) \neq 0$ for $0 < x < \tau_f(a)$ and $\phi_x(\tau_f(a), a) = 0$. The "time" map $\tau_f \colon D(\tau_f) \to (0, \infty)$ is the mapping taking a into $\tau_f(a)$.

This map τ_f determines the number of solutions of the boundary value problem (3.17), (3.23) that can be obtained. In fact, the solutions with no zero of u_x in $(0, \lambda)$ coincide with the solutions of the equation $\tau_f(a) = \lambda$. The solutions with

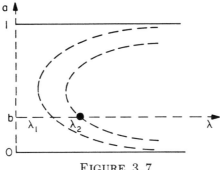

FIGURE 3.7

one zero of u_x in $(0, \lambda)$ coincide with the solutions of the equation $2\tau_f(a) = \lambda$, etc.

We discuss the bifurcation problem only for the case $0 < b \leq 1/2$. It is clear that $D(\tau_f) = (0, b) \cup (b, 1)$. If we define $\tau_f(b) = \pi[b(1-b)]^{-1/2}$, then we can take $D(\tau_f) = (0, 1)$. Since the orbits of (3.17) are symmetric with respect to the u-axis, it follows that if $a \in D(\tau_f)$, then $\phi(\tau_f(a), a) \in D(\tau_f)$ and $\tau_f(\phi(\tau_f(a), a))$ is defined. The symmetry of the orbits of (3.17) about the u-axis implies that $\tau_f(\phi(\tau_f(a), a)) = \tau_f(a)$.

As for Dirichlet conditions, one shows that $\tau_f(a)$ has a unique critical point which is a minimum at some point a_0, $d^2\tau_f(a_0)/da^2 > 0$. This allows one to obtain the bifurcation diagram shown in Figure 3.7. The nth branch of nonconstant solutions intersects the branch corresponding to the constant function b at λ_n. The solutions ϕ_n^+, ϕ_n^- on this branch have n zeros for u_x in $(0, 1)$, $\dim W^u(\phi_n^\pm) = n$. The constant solutions $u = 0$, $u = 1$ are u.a.s. and $u_0 = b$ is unstable with $\dim W^u(0) = n + 1$ if $\lambda_n < \lambda < \lambda_{n+1}$. The attractor A_λ is $\cdot\!\!\leftarrow\!\!\cdot\!\!\rightarrow\!\!\cdot$ for $0 < \lambda < \lambda_1$ and can be shown to resemble the Dirichlet problem for $\lambda_n < \lambda < \lambda_{n+1}$.

Now consider the boundary conditions

(3.23) $(1-\theta)\lambda^{-1}u_x - \theta u = 0$ at $x = 0$, $\quad (1-\theta)\lambda^{-1}u_x + \theta u = 0$ at $x = 1$,

where $0 \leq \theta \leq 1$. The factor λ^{-1} is inserted here for convenience. At $\theta = 0$, we have Neumann conditions and at $\theta = 1$, we have Dirichlet conditions. One now seeks solutions of (3.17) which go from the line $L_0 = \{(u, u_x): u_x = \lambda\theta(1-\theta)^{-1}u\}$ at $x = 0$ to the line $L_1 = \{(u, u_x): u_x = -\lambda\theta(1-\theta)^{-1}u\}$ at $x = \lambda$ (see Figure 3.8). The slope of the stable manifold $W^s(0)$ of zero at zero is $-\lambda b^{1/2}$; the slope of the unstable manifold $W^u(0)$ of zero at zero is $\lambda b^{1/2}$. Thus, if $\theta > \theta_b = b^{1/2}(1 + b^{1/2})^{-1}$, there can be no solution of the boundary value problem inside the homoclinic orbit.

In the rescaled variables $x \to \lambda x$, the problem becomes

(3.24) $\qquad\qquad u_{xx} + f(u) = 0, \qquad 0 < x < \lambda,$

(3.25) $\quad (1-\theta)u_x - \theta u = 0$ at $x = 0$, $\quad (1-\theta)u_x + \theta u = 0$ at $x = \lambda$.

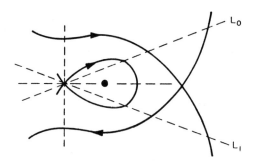

FIGURE 3.8

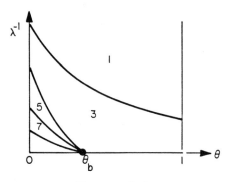

FIGURE 3.9

The analysis of the "time" map from L_0 to L_1 in this case is very complicated and not complete. It is conjectured by Hale and Rocha [1987b] that the bifurcation diagram is the one shown in Figure 3.9 where the number of solutions in each region is designated. Inside each of the regions bounded by two curves in the diagram, all equilibrium points are hyperbolic. As in the discussion for the previous examples, the attractor $A_{d,\theta}$ is a point in region 1, ·—←·—→·— in region 3. The situation in the others is not known. Gardner [1985] needed the fact that region 3 extended from $\theta = 0$ to $\theta = 1$ in his study of multidimensional traveling waves.

4.3.6. *Equations whose flow is equivalent to an ODE.* The purpose of this section is to summarize the results of Fusco [1987] showing that the flow on the attractor for equations (3.2) with boundary conditions (3.11) for a large class of diffusion coefficients is equivalent to the flow on the attractor for a carefully constructed system of ODE. Once this equivalence is established, one can discuss the influence of diffusion and boundary conditions on the structure of the attractor and, in particular, the appearance of stable equilibria.

To motivate the construction of the ODE, suppose in (3.2) that $a = \nu b$ where $b > 0$, ν are real parameters, and $f(u)$ is such that $\dot{u} = f(u)$ has at least two stable equilibria. The attractor A_ν undergoes drastic transformations as ν goes from zero to ∞. In fact, from our discussion of bifurcation problems in the previous sections, one expects that the number of equilibria approaches

∞ as $\nu \to 0$ and, thus, the number of bifurcations experienced by A_ν grows unboundedly. On the other hand, if ν is sufficiently large, the flow on A_ν should be equivalent to the flow defined by $\dot u = f(u)$ in the case of Neumann boundary conditions. When $\nu \to 0$, there is no obvious way to obtain an ODE whose dynamics would reflect the dynamics of the PDE.

On the basis of the above discussion, it is natural to ask if it is possible to choose more general deformations of the diffusion function a which would allow one to show that the flow on the attractor does not change its topological structure and becomes equivalent to an ODE for which the dynamics is more easily ascertained. For example, the deformations a_ν, $\nu > 0$, of the diffusion could have the property that, as $\nu \to 0$, a_ν approaches zero at a finite number of points in $[0, 1]$ and diverges to ∞ otherwise. It turns out that this can be done if the idea is exploited in the proper way.

To avoid excessive notation, we only describe the detailed construction for a special case. Suppose $\nu > 0$, $x_1 \in (0, 1)$, $b_0 > 0$, $b'_0 > 0$, $l_0 > 0$, $l'_0 > 0$, $e_0 > 0$ are given constants and let $x_0 = 0$, $x_2 = 1$. Consider the class of C^2-functions a on $[0, 1]$ satisfying

$$(3.26) \quad \begin{cases} a(x) \geq e_0 \nu^{-1} & \text{for } x \in [0, x_1 - \nu l'_0] \cup [x_1 + \nu l'_0, 1], \\ a(x) \geq \nu b_0 & \text{for } x \in [x_1 - \nu l'_0, x_1 + \nu l'_0], \\ a(x) \leq \nu b'_0 & \text{for } x \in [x_1 - \nu l_0, x_1 + \nu l_0]. \end{cases}$$

The parameter in the problem will be ν and the constants and functions a considered are required to satisfy the following technical hypotheses:

$$(3.27) \quad \begin{aligned} & l'_0 - l_0 = O(\nu^q) \quad \text{for some } 0 < q < 1, \\ & (b'_0 - b_0)/(l'_0 - l_0) \to 0 \quad \text{as } \nu \to 0, \\ & \int_0^1 \frac{a_x^2}{a^4} \, dx = O(\nu^{-(3+q+\varepsilon)}) \quad \text{for some } \varepsilon \geq 0. \end{aligned}$$

For any small $\nu > 0$ and a as above, let us consider the equation

$$(3.28) \quad u_t = (a u_x)_x + f(u), \quad 0 < x < 1,$$

with the Neumann boundary conditions

$$(3.29) \quad u_x = 0 \quad \text{at } x = 0, 1.$$

If we formally identify dependent variables z_1, z_2 in R with the "averages" of the solution u of (3.28), (3.29) by the relations

$$z_1 = x_1^{-1} \int_0^{x_1} u \, dx, \quad z_2 = (1 - x_1)^{-1} \int_{x_1}^1 u \, dx,$$

then, as $\nu \to 0$, it is natural to expect that their averages should satisfy the ODE

$$(3.30) \quad \dot z_1 = \mu_1(z_2 - z_1) + f(z_1), \quad \dot z_2 = -\mu_2(z_2 - z_1) + f(z_2),$$

where

$$(3.31) \quad \mu_1 = b_0/2l_0 x_1, \quad \mu_2 = b_0/2l_0(1 - x_1).$$

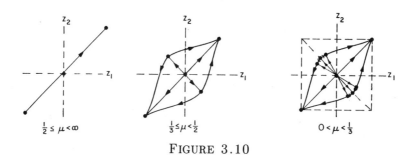

FIGURE 3.10

If we assume that

(3.32) $\quad \overline{\lim\limits_{|u|\to\infty}} f(u)/u \leq -\beta < 0, \quad \int_0^u f(s)\,ds \to -\infty \quad \text{as } u \to \pm\infty,$

then the equation (3.32) has a global attractor $A_{\mu,f}$.

With a_ν defined as in (3.26), (3.2) with Neumann boundary conditions also has a global attractor $A_{\nu,\omega,f}$ where $\omega = (b_0, l_0, e_0, x_1)$. From the manner in which the function a_ν is constructed, one would expect that, for ν small, the flow on the attractor $A_{\nu,\omega,f}$ in function space would be closely related to the flow on the attractor $A_{\mu,f}$ for the ODE (3.30) provided that the latter flow is stable under perturbations of the vector field in (3.30). To begin to show that this is the case, one must embed the solutions of (3.30) into the function space in which solutions of (3.2) are to be considered. Using center manifold techniques, one then would show that there is a global attractor for the PDE close to the one for the embedded one from the ODE. The next step is to compare the flows on the attractors. Fusco [1987] has shown that this program can be successfully completed, but the proof requires several delicate steps and is omitted.

Many authors (for example, Matano [1979], Yanagida [1982], Fusco and Hale [1985], and Hale and Rocha [1985]) have shown that, for a deep well in the conductivity function a and a nonlinear function f with two stable equilibria, the equation (3.2) with Neumann boundary conditions can have a stable nonconstant solution. Fusco and Hale [1985] have shown how these solutions can occur through secondary bifurcations. Using the ODE (3.30), one can recover these qualitative results. In fact, consider the example where $f(u) = u(1 - u^2)$, $x_1 = 1/2$ and Neumann conditions. Equations (3.30) become

(3.33) $\quad \dot{z}_1 = \mu(z_2 - z_1) + z_1(1 - z_1^2), \qquad \dot{z}_2 = -\mu(z_2 - z_1) + z_2(1 - z_2^2)$

where $\mu = b_0/l_0$. We take the parameter μ as the bifurcation parameter and note that a deep well in a_ν corresponds to μ small. It is easy to show that the attractor as a function of μ is the one depicted in Figure 3.10.

As $\mu \to 0$, the outer boundary of the attractor approaches the boundary of the unit square and the three nonconstant equilibria have a transition layer at $x = 0$.

It should be noted that the example considered above is only a very special case of Fusco [1987]. Firstly, several deep wells can be created in a_ν. If there are

$n - 1$ wells, then the number of ordinary differential equations has dimension n and has the form

$$\dot{z}_i = \mu_i(z_{i+1} - z_i) - \nu_i(z_i - z_{i-1}) + f_i(z_i), \qquad 1 \leq i \leq n,$$

where each μ_i, ν_i is a positive constant. Secondly, the theory applies to arbitrary boundary conditions of the form (3.8b). See Fusco [1987] for the details. As a final remark, the ordinary differential equations obtained through the above process have the property that stable and unstable manifolds of hyperbolic equilibria always intersect transversally. This reflects the fact that transversality holds for parabolic equations in one space dimension (see Theorem 4.3.4). It happens that the basic reason that transversality holds for the ODE is that the linear part is a positive Jacobi matrix (tridiagonal with positive elements off the diagonal) and the nonlinearity in the ith equation depends only on z_i (see Fusco and Oliva [1986]).

4.3.7. *A method for determining stability.* In this section, we present the method in Fusco and Hale [1985], Hale and Rocha [1985], and Rocha [1985] for determining the stability of equilibrium solutions of the equation

(3.34) $$u_t = (a^2 u_x)_x + f(x, u), \qquad 0 < x < 1,$$

with the Neumann boundary conditions

(3.35) $$u_x = 0 \quad \text{at } x = 0, 1.$$

Similar methods have been used by C. Jones [1984] in the study of stability of traveling waves.

We assume that $f(x, u)$ is continuous in x, u with continuous first and second derivatives in u, and

(3.36) $$f(x, 0) = 0.$$

Along with (3.34) and (3.35), we consider the initial value problem

(3.37) $$\begin{aligned} a^2 u_x &= v, & v_x &= -f(x, u), & 0 < x < 1, \\ u(0) &= u_0, & v(0) &= 0, \end{aligned}$$

where $u_0 \in R$. Let $(u(x, u_0), v(x, u_0))$ be the solution of (3.37) and suppose it is defined for $0 \leq x < \alpha_{u_0}$. Let $\Sigma \subset R^3$ be defined by $\Sigma = \{(u(x, u_0), v(x, u_0), x): 0 \leq x < \alpha_{u_0}, u_0 \in R\}$. The set E of equilibrium points of (3.34), (3.35) coincides with the functions $u(x, u_0)$ which are defined for $0 \leq x \leq 1$ and have $u_x(1, u_0) = v(1, u_0) = 0$; that is, $u(\cdot, u_0) \in E$ if and only if

$$(u(1, u_0), v(1, u_0), 1) \in \Sigma_1 \cap \{(u, 0, 1) \in R^3\},$$

where

$$\Sigma_x = \{(u, v, x) \in \Sigma\}$$

is the curve representing the cross-section of Σ at x. The set Σ_x is the image of the solutions of (3.37) at x that began on the u-axis.

For the study of stability of equilibria, an important role is played by the angle t (measured clockwise) that the tangent vector to the curve Σ_x makes with the

u-axis. The tangent vector to Σ_x is given by $(\partial u(x, u_0)/\partial u_0, \partial v(x, u_0)/\partial u_0)$ and satisfies the linear variational equation

(3.38)
$$a^2 \bar{u}_x = \bar{v}, \qquad \bar{v}_x = -f_u(x, u(x, u_0))\bar{u},$$
$$\bar{u}(0) = 1, \qquad \bar{v}(0) = 0.$$

If we let $\bar{u} = p\cos t$, $\bar{v} = -p\sin t$, then

(3.39) $\qquad t_x = (\sin^2 t)/a^2 + f_u(x, u(x, u_0))\cos^2 t, \qquad t(0, u_0) = 0.$

A principal result on stability is

THEOREM 4.3.13. *Let $\theta(u_0) = t(1, u_0)$ where $t(x, u_0)$ is the solution of (3.39). Then*

(i) $\theta: R \to (-\pi/2, \infty)$.

(ii) *An equilibrium point $u = u(\cdot, u_0)$ is hyperbolic if and only if $\theta(u_0) \neq k\pi$ for any integer $k \geq 0$; that is, Σ_1 is transversal to $\{(u, 0, 1): u \in [0, 1]\}$ at u_0.*

(iii) *If $u(\cdot, u_0)$ is a hyperbolic equilibrium and $W^u(u)$ is the unstable manifold, then $\dim W^u(u) = 1 + [\theta(u_0)/\pi]$ where $[\cdot]$ denotes the integer part.*

PROOF. (i) Note that t_x in (3.38) is positive for $t = (k - 1/2)\pi$ where $k \geq 0$ is an integer. Thus, the range of θ is in $(-\pi/2, \infty)$.

(ii) If $u(\cdot, u_0)$ is an equilibrium point, then $u(\cdot, u_0)$ is hyperbolic if and only if $\lambda = 0$ is not an eigenvalue of the problem

$$(a^2(x)w_x)_x + f_u(x, u(x, u_0))w = \lambda w, \qquad 0 \leq x \leq 1,$$
$$w_x(0) = w_x(1) = 0.$$

The number λ is an eigenvalue of this problem if and only if

$$w = \eta \cos \xi, \qquad a^2 w_x = -\eta \sin \xi,$$

where ξ satisfies the equation

(3.40)
$$\xi_x = (\sin^2 \xi)/a^2 + [f_u(x, u(x, u_0)) - \lambda]\cos^2 \xi, \qquad 0 \leq x \leq 1,$$
$$\xi(0) = 0, \qquad \xi(1) = k\pi \quad \text{for some integer } k \geq 0.$$

For $\lambda = 0$, this equation is the equation for the angle to the tangent vector of Σ_1. The result in (ii) now is immediate.

To prove (iii), let $\xi(x, \lambda)$ be the solution of (3.40) with $\xi(0) = 0$. Note that $\xi(1, \lambda)$ is strictly decreasing in λ, $\xi(1, \lambda) \to -\pi/2$ as $\lambda \to +\infty$ and $\xi(1, \lambda) \to \infty$ as $\lambda \to -\infty$. Also, note that $\xi(1, 0) = \theta(u_0)$ and λ is an eigenvalue if and only if $\xi(1, \lambda) = k\pi$. This gives the proof of the assertion in (iii).

It also is possible to obtain a characterization of the equilibrium solutions and their stability by integrating the equations backward from $x = 1$. In fact, consider the initial value problem

(3.41)
$$a^2 u_x = v, \qquad v_x = -f(x, u), \qquad 0 < x < 1,$$
$$u(1) = u_0, \qquad u_x(1) = 0.$$

If $(u(x, u_0, 1), v(x, u_0, 1))$ is the solution of (3.41) defined on a maximal interval $\beta_{u_0} < x \leq 1$, let

$$\Sigma^1 = \{(u(x, u_0, 1), v(x, u_0, 1), x): \beta_{u_0} < x \leq 1, \ u_0 \in R\}$$

and define the cross-sections Σ_x^1 as before. The equilibrium solutions are then in one-to-one correspondence with those u_0 which belong to $\Sigma_0^1 \cap \{(u, 0, 0) \in R^3\}$. More generally, u_0 is an equilibrium point if and only if $u_0 \in \Sigma_x \cap \Sigma_x^1$ for $0 \le x \le 1$. One can introduce the angle t' for the linear variational equation of (3.41) about an equilibrium and then prove the following

THEOREM 4.3.14.
(i) *The set of equilibria of* (3.34), (3.35) *is in one-to-one correspondence with the set* $\Sigma_x \cap \Sigma_x^1$ *for* $0 \le x \le 1$.
(ii) *An equilibrium point corresponding to* $u_0 \in [0, 1]$ *is hyperbolic if and only if* Σ_x *is transversal to* Σ_x^1 *at* u_0 *for every* $x \in [0, 1]$.
(iii) *If* $u_0 \in [0, 1]$ *corresponds to an hyperbolic equilibrium point and* $W^u(u_0)$ *is the unstable manifold, then*

$$\dim W^u(u_0) = 1 + [\phi(u_0, x)/\pi], \qquad 0 \le x \le 1,$$

where $\phi(u_0, x) = t(x, u_0) - t'(x, u_0)$ *is the angle between* Σ_x *and* Σ_x^1 *measured clockwise.*

We have proved part (i) above. The proof of part (ii) is an easy consequence of Theorem 4.3.13 since the statement is true for $x = 1$ and the sets Σ, Σ^1 correspond to solutions of (3.37), (3.41), respectively. Property (iii) holds for $x = 1$. The formula in (iii) must therefore hold for all x by (ii).

4.3.8. Stable solutions in a singularly perturbed equation. In this section, for $\varepsilon > 0$ a constant, we consider the equation

$$(3.42) \qquad u_t = \varepsilon^2 u_{xx} + f(x, u), \qquad 0 < x < 1,$$

with the Neumann boundary conditions

$$(3.43) \qquad u_x = 0 \quad \text{at } x = 0, 1$$

and the function $f(x, u)$ being the cubic function

$$(3.44) \qquad f(x, u) = u(1 - u)[u - b(x)].$$

If $b(x)$ is independent of x, we have seen that the only stable equilibrium solutions of (3.42), (3.43) are constant functions. For $b(x)$ depending upon x, this may not be the case. The purpose of this section is to characterize those solutions which can be stable. This characterization is given in terms of the transition layer behavior of the solutions as the parameter $\varepsilon \to 0$. For the case in which b is a C^1-function, the results are due to Angenent, Mallet-Paret, and Peletier [1987] and, when a is a step function, they are due to Rocha [1987]. The results may be loosely summarized in the following way: Suppose $b(x)$ crosses $1/2$ exactly N times in the interval $(0, 1)$. Then there are an integer k_N and an $\varepsilon_0 > 0$ such that, for $0 < \varepsilon < \varepsilon_0$, there are exactly k_N stable equilibria. Furthermore, the equilibrium point is stable if and only if the solution crosses $1/2$ in a direction opposite to the direction that the function b crosses $1/2$.

Let us now state the result of Rocha [1987].

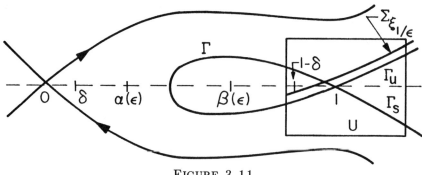

FIGURE 3.11

THEOREM 4.3.15. *Let $b: [0, 1] \to (0, 1)$ be a step function with $b(x) \neq 1/2$ for all x. Let b jump across $1/2$ at the N points $0 < \xi_1 < \xi_2 < \cdots < \xi_N < 1$ and let $\xi_0 = 0$, $\xi_{N+1} = 1$. Then there is an $\varepsilon_0 > 0$ such that, for $0 < \varepsilon \leq \varepsilon_0$, the number of stable solutions of (3.42), (3.43) is exactly the Nth Fibonacci number. Furthermore, a stable solution $u(x, \varepsilon) \to 0$ or 1 on each interval $(\bar{\xi}_j, \bar{\xi}_{j+1})$, $j = 0, 1, 2, \ldots, N$, and is monotone in a neighborhood of each ξ_j with*

$$[b(\xi_j^+) - b(\xi_j^-)]u_x(\xi_j, \varepsilon) < 0, \qquad j = 1, 2, \ldots, N.$$

The Fibonacci numbers are defined recursively by the relations $k_0 = 2$, $k_1 = 3$, $k_N = k_{N-1} + k_{N-2}$.

Let us outline the essential elements of the proof of Theorem 4.3.15. The equilibrium points are the solutions of the equation

(3.45) $$\varepsilon^2 u_{xx} + f(x, u) = 0$$

with the boundary conditions (3.43).

We use the method discussed in §4.3.7 for determining the stability of equilibria. More specifically, we use (3.37), (3.38), (3.39), and (3.41) with $a^2 = \varepsilon^2$ and the notation $u(x, u_0)$, $t(x, u_0)$, $\theta(u_0)$, $t'(x, u_0)$ of that section.

LEMMA 4.3.16. *For any $0 < \delta < 1/2$, there is an $\varepsilon_0 > 0$ such that, for $0 < \varepsilon < \varepsilon_0$, all the equilibria $u = u(\cdot, u_0)$ of (3.42), (3.43) with $u_0 \in [\delta, 1 - \delta]$ are unstable.*

PROOF. Let $b(x) = b_1$ for $x \in [0, \xi)$ with $b_1 > 1/2$. The proof will be similar for $b_1 < 1/2$. If we let $x = \varepsilon y$ in (3.37) with $a^2 = \varepsilon^2$, then

(3.46) $$\begin{aligned} u_{yy} + u(1 - u)(u - b_1) &= 0, \qquad 0 < y < \xi_1/\varepsilon, \\ u(0) = u_0, \qquad u_y(0) &= 0. \end{aligned}$$

The solutions (u, v), $v = u_x$, of this equation lie on level curves of the function $H(u, v) = v^2/2 + \int_0^u s(1 - s)(s - b_1)\, ds$. If $b_1 > 1/2$, the phase portrait is shown in Figure 3.11. Let Γ be the level curve of H passing through $(1, 0)$. It contains an orbit homoclinic to $(1, 0)$. Let U be a small square neighborhood of $(1, 0)$ and let Γ_s, Γ_u be the subsets of $\Gamma \cap U$ corresponding to connected components of the stable and unstable manifolds of $(1, 0)$.

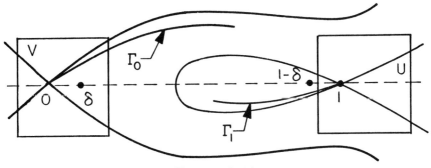

FIGURE 3.12

Let $\Sigma_y = \{(u(y,u_0), v(y,u_0)),\ 0 \le u_0 \le 1\}$ be the curve in the phase space defined by the solutions $(u(y,u_0), v(y,u_0))$ of (3.46), $0 \le u_0 \le 1$. For ε small enough, the curve $\Sigma_{\xi_1/\varepsilon}$ has a nonempty intersection with U and $\Sigma_{\xi_1/\varepsilon} \cap U$ and Γ_u are C^k-close manifolds for any $k \ge 1$ (this is the λ-lemma; see Palis [1969]). If we let $(\alpha(\varepsilon), \beta(\varepsilon))$ be the interval of initial values corresponding to the set $\Sigma_{\xi_1/\varepsilon} \cap U$, then the tangent vector to any point of $\Sigma_{\xi_1/\varepsilon} \cap U$ makes an angle $t(\xi_1/\varepsilon, u_0) > \pi/2$ with the u axis for all $u_0 \in (\alpha(\varepsilon), \beta(\varepsilon))$. Since $t_x > 0$ when $t = \pi/2$ this clearly implies that $t(1, u_0) > 0$. From Theorem 4.3.13, part (iii), any equilibrium solution corresponding to $u_0 \in (\alpha(\varepsilon), \beta(\varepsilon))$ is unstable. Since every equilibrium solution must correspond to a $u_0 \in [0, 1]$, we conclude from the phase portrait that there are no equilibrium solutions in the interval $[\delta, \alpha(\varepsilon)]$ for any $\delta > 0$. If $u_0 \in [\beta(\varepsilon), 1 - \delta]$, then one concludes also that $t(\xi_1/\varepsilon, u_0) > \pi/2$ for ε small enough and so these equilibria will be unstable. This proves Lemma 4.3.16.

To complete the proof of Theorem 4.3.15, it is only necessary to discuss the equilibrium points corresponding to $u_0 \in [0, \delta] \cup [1 - \delta, 1]$ for δ small. Let us continue to use the notation in the proof of Lemma 4.3.16 and also define a corresponding square V around the saddle point $(0,0)$ for (3.46) (see Figure 3.12). Let $\Gamma_0(y)$ be that part of the section curve Σ_y corresponding to $u_0 \in [0, \delta]$ and $\Gamma_1(y)$ that corresponding to $u_0 \in [1 - \delta, 1]$. For ε small, $\Gamma_0(\xi_1/\varepsilon)$ and $\Gamma_1(\xi_1/\varepsilon)$ become C^k-close respectively to the unstable manifolds of $(0,0)$ and $(1,0)$. If $\xi_1 = 1$, then we conclude as in the proof of Lemma 4.3.16 that there are no stable equilibria except 0 and 1; that is, there are exactly two stable equilibria.

Now suppose $\xi_1 < 1$ and the function b jumps to a value $b_2 > 1/2$. One can use the same type of argument as above to show that no new stable equilibria are introduced. Thus, we suppose $\xi_1 < 1$ and $b_2 < 1/2$. The phase portrait of the solutions of the equation for x in (ξ_1, ξ_2) is given from the solutions of the equation

(3.47) $\quad u_{yy} + u(1-u)(u - b_2) = 0, \quad \xi_1/\varepsilon < y < \xi_2/\varepsilon$

and is shown in Figure 3.13. The origin now has a homoclinic orbit and the new stable and unstable manifolds of $(1, 0)$ have a transversal intersection respectively

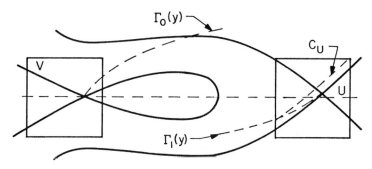

FIGURE 3.13

with the old unstable and stable manifolds of $(0,0)$. This implies there is an $\varepsilon_0 > 0$ such that the curve $\Gamma_0(y)$ defined before has a transversal intersection with the stable manifold of $(1,0)$ if $\varepsilon y \in [\xi_1, \xi_2]$, $0 < \varepsilon \leq \varepsilon_0$. Furthermore, the λ-lemma implies that we can choose ε_0 so that the set $\Gamma_0(y) \cap U \neq \emptyset$ and there is a curve C_U as shown in Figure 3.13 for $\varepsilon y \in [\xi_1, \xi_2]$ which is C^k-close to the unstable manifold of $(1,0)$ in U. Thus, if $\xi_2 = 1$ there is a unique equilibrium point with initial data $u_0 \in (0, \delta)$ if ε is small enough. It is also clear that the angle of the tangent vector C_U at this point is negative. Theorem 4.3.13 implies it is a stable equilibrium. For the case $\xi_2 = 1$, one can use Theorem 4.3.13 to show that no stable equilibrium point can be obtained with initial data in $(1 - \delta, 1)$. Thus, the number of stable equilibria is three.

If $\xi_2 < 1$, we repeat the above process using the λ-lemma and studying carefully the behavior of the flow in the sets U, V. At each ξ_k, a new transversal intersection is introduced between the unstable manifold of one equilibrium, either 0 or 1, in the previous phase portrait and the stable manifold of the other in the new phase portrait, producing a number of segments of Σ_y for $y = \xi_j/\varepsilon$ transversal to the stable manifold. This then produces a number of segments of Σ_y for $y = \xi_{j+1}/\varepsilon$ approaching the unstable manifold of the last equilibrium and whose tangent curves intersect the u-axis at an angle < 0. Also, all of the previous segments in U and V are preserved.

If k_j^U, k_j^V denote the number of such intersections existing at the jump ξ_j in the neighborhoods U, V respectively, and assuming the jump across $1/2$ is positive, we obtain $k_{j+1}^V = k_j^U + k_j^V$ and $k_{j+1}^U = k_j^U$. Moreover, since the jump at ξ_{j-1} has the opposite sign, we have $k_j^V = k_{j-1}^V$ and $k_j^U = k_{j-1}^U + k_{j-1}^V$. Since $k_j = k_j^U + k_j^V$, we have $k_{j+1} = k_j + k_{j-1}$, the Fibonacci sequence.

The shape of the monotone transition layers are obtained from the phase portraits and the level curves of the function $H(u,v)$.

We now state the result of Angenent, Mallet-Paret, and Peletier [1987].

THEOREM 4.3.17. *Let $b \in C^1[0,1]$, $\Sigma_0 = \{\xi \in [0,1]: b(\xi) = 1/2\}$, and suppose that $0, 1 \notin \Sigma_0$ and $b'(\xi) \neq 0$ for $\xi \in \Sigma_0$. If $\Sigma \subset \Sigma_0$ is the sequence $0 < \xi_1 < \xi_2 < \cdots < \xi_M < 1$, then there is an $\varepsilon_0 > 0$ such that, for $0 < \varepsilon \leq \varepsilon_0$, there is a stable solution $u(x, \varepsilon)$ of (3.42), (3.43), and $u(x, \varepsilon)$ is monotone in a* neighborhood of each ξ_j, $b'(\xi_j)u'(\xi_j,\varepsilon) < 0$ for each j and $u(x,\varepsilon) \to 0$ or 1 as $\varepsilon \to 0$ uniformly on any closed interval not containing Σ. Furthermore, all stable solutions are obtained in this way.

The proof of this result is more difficult than the proof of Theorem 4.3.15 and we merely outline the ideas in the proof, referring the reader to the original paper for the interesting details.

To avoid excessive notation, let us assume that Σ contains only one element ξ and $b'(\xi) < 0$. An equilibrium solution must satisfy the equation

$$\varepsilon^2 u_{xx} + f(x,u) = 0 \tag{3.48}$$

together with the boundary conditions (3.43). To understand the role of $b(\xi) = 1/2$, let us introduce the stretched variables $x - \xi = \varepsilon y$ to obtain

$$u_{yy} + f(\xi + \varepsilon y, u) = 0. \tag{3.49}$$

For $\varepsilon = 0$, this equation has a heteroclinic orbit $u_0(y)$ which satisfies $u_0(y) \to 0$ exponentially as $y \to -\infty$ and $u_0(y) \to 1$ exponentially as $y \to \infty$.

The first step in the proof is to show that there exists a solution with a monotone transition layer at ξ as $\varepsilon \to 0$. A proof of this fact could be supplied using the ideas of Fife [1976] (see also Ito [1984]). This method of proof also allows one to obtain a solution with a monotone transition layer which is not stable. Since more information is contained in the case of stable transition layers in Angenent, Mallet-Paret, and Peletier [1987], we outline their approach.

Using the heteroclinic orbit $u_0(y)$ of (3.49) for $\varepsilon = 0$ and a small $\delta > 0$, they construct a lower solution $\underline{u}$ of (3.51) with $\underline{u}(x) = 0$ on $[0, \xi - \delta]$, $\underline{u}(x) = \beta$ on $[\xi + \delta, 1]$ and $\underline{u}(x)$ is essentially a scaling of the above heteroclinic orbit $u_0((x - \xi)/\varepsilon)$ on $(\xi - \delta, \xi + \delta)$. In the same way, they construct an upper solution $\overline{u}$ of (3.48) with $\overline{u}(x) = \alpha$ on $(0, \xi - \delta]$, $\overline{u}(x) = 1$ on $[\xi + \delta, 1]$ and a scaled version of the heteroclinic orbit on $[\xi - \delta, \xi + \delta]$. Let

$$[\underline{u}, \overline{u}] = \{\phi \in C^2 : \phi(x) \in [\underline{u}(x), \overline{u}(x)], 0 \le x \le 1\}.$$

If one now considers any initial function $\phi \in [\underline{u}, \overline{u}]$, then the solution $u(t,x)$ of (3.42) satisfies $u(t, \cdot) \in [\underline{u}, \overline{u}]$ for all $t \ge 0$. A result of Matano [1979] implies there is a stable equilibrium $u_0 \in [\underline{u}, \overline{u}]$. The next observation which is far from trivial is that, for ε small, every equilibrium solution in $[\underline{u}, \overline{u}]$ is monotone increasing as long as it has its values in $[\alpha, \beta]$. The next step in the proof is to obtain an estimate on the principal eigenvalue u_ε of the linear variational equation about any equilibrium solution $u_0 \in [\underline{u}, \overline{u}]$ as

$$\mu_\varepsilon = -c\varepsilon \int_0^1 |f_x(\xi, s)|\, ds + O(\varepsilon) \quad \text{as } \varepsilon \to 0$$

where c is a positive constant. This shows that u_0 is stable and is the only equilibrium solution with a transition layer at ξ as $\varepsilon \to 0$.

To show that these are the only possible stable solutions, one shows that every stable solution must have a transition layer at ξ as $\varepsilon \to 0$ and then uses

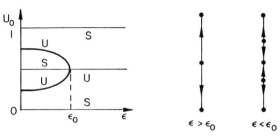

FIGURE 3.14

the previous computations on the principal eigenvalue to conclude that it must be monotone increasing.

Let us mention some examples of Rocha [1988] in which the complete flow on the global attractor is discussed by using the angles introduced in §4.3.7 and the transversality theory of §4.3.2. Some notation is needed. If $0 < c < 1/2$, the equation

$$u_{yy} + u(1-u)(u - 1/2 + c) = 0$$

has a homoclinic orbit at the origin $(0,0)$ in the phase space. Let $(\gamma(c), 0)$ be the intersection of this orbit with the u-axis.

PROPOSITION 4.3.18. *If*

$$b(x) = \begin{cases} 1/2 + c_1, & x < c_3, \\ 1/2 - c_2, & x \geq c_3, \end{cases}$$

where $c_1, c_2 \in (0, 1/2)$ and $\gamma(c_1) + \gamma(c_2) \leq 1$, then the bifurcation diagram and the flow on the global attractor for (3.42), (3.43), (3.44) are the ones shown in Figure 3.14.

For $\varepsilon > \varepsilon_0$, there are two stable equilibria and one unstable equilibrium. For $\varepsilon < \varepsilon_0$, there are three stable equilibria and two unstable equilibria. In both cases, the unstable equilibria have one dimensional unstable manifolds.

PROPOSITION 4.3.19. *If*

$$b(x) = \begin{cases} 1/2 + c_1, & 0 \leq x < c_3, \\ 1/2 - c_2, & c_3 \leq x < c_5, \\ 1/2 + c_4, & c_5 \leq x \leq 1, \end{cases}$$

where $0 < c_3 < c_5 < 1$, $c_1, c_2, c_4 \in (0, 1/2)$, and $\gamma(c_1) + \gamma(c_2) \leq 1$, $\gamma(c_2) + \gamma(c_4) \leq 1$, without both being equal, then there is an open set G in $(0,1) \times (0,1)$ containing the point $(1/4, 3/4)$ such that if $(c_3, c_5) \in G$, the bifurcation diagram and the flow on the attractor for (3.42), (3.43), (3.44) are shown in Figure 3.15. Note that two different types of bifurcation diagrams are possible and there is a maximum of five stable equilibria.

As a final remark, we should mention some differences that occur between the case when b is a step function and the case when b is a C^1-function. The previous examples of Rocha [1987] show that it is possible in some cases to have an upper

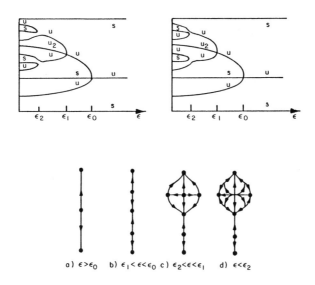

FIGURE 3.15

bound on the total number of equilibria independently of $\varepsilon > 0$. For the case in which b is a C^1-function, one does not expect that such an upper bound exists. This is a consequence of results of Kurland [1983] where he shows that for any integer n_0, there is an ε_0 such that, for $0 < \varepsilon \leq \varepsilon_0$, the equation has at least n_0 equilibrium solutions that oscillate around the function $b(x)$. All of these solutions except the one discussed in Theorem 4.3.17 are unstable. However, they play an important role in understanding the flow on the attractor.

From the remarks in the previous paragraph, it is clear that an attempt should be made to understand how the highly oscillatory solutions obtained by Kurland behave as the function $b(x)$ approaches a step function.

4.3.9. *Quenching as a dynamic problem.* H. Kawarada [1975] posed the following problem. For a fixed $a > 0$, let $u(t, x, a)$ be the solution of the boundary value problem

$$u_t = u_{xx} + (1-u)^{-1}, \quad -a < x < a, \ t > 0,$$
$$u(t,-a) = u(t,a) = 0, \quad t \geq 0,$$
$$u(0,x) = 0, \quad -a < x < a.$$

Let $[0, t_a)$ be the maximal interval of existence of the solution. If $t_a = \infty$, then we have global existence and if $t_a < \infty$, then $\max\{u(t,x,a): |x| \leq a\} \to 1$ as $t \to t_a^-$. Kawarada refers to this latter phenomenon by saying that the solution "quenches." The solution remains bounded, but the derivatives blow up in finite time. The reason for such behavior is that the nonlinear term has a singularity at $u = 1$. If $a^* = \sup\{a > 0 : t_a = \infty\}$, then there is global existence for $a < a^*$ and quenching for $a > a^*$. Kawarada proved that $a^* \leq 2^{1/2}$.

Acker and Walter [1976] proved the following interesting result: let a_0 be the supremum of those $a > 0$ such that there is a solution of the boundary value

problem
$$\phi_{xx} + (1-\phi)^{-1} = 0, \quad -a < x < a,$$
$$\phi(-a) = \phi(a) = 0$$
with $0 \leq \phi(x) < 1$. Then $a^* = a_0$ and $a_0 \sim 0.765$. Acker and Walter draw the same conclusion that $a^* = a_0$ when $(1-u)^{-1}$ is replaced by a C^1-function $f\colon [0,1) \to R$ with $f(0) > 0$ and $f(u) \to \infty$ as $u \to 1^-$.

Let us consider the same problem from the point of view of the dynamics of the flow defined by the equation

(3.50) $$u_t = u_{xx} + a^2 f(u), \quad -1 < x < 1,$$

(3.51) $$u = 0 \quad \text{at } x = -1, \; x = 1,$$

where $f\colon [0,1) \to R$ is a C^1-function such that $f(0) > 0$, $f(u) \to \infty$ as $u \to 1^-$. We have rescaled the variables $x \to ax$, $t \to a^2 t$. We choose the initial data $\phi \in \tilde{H}_0^1 = \{\phi \in H_0^1 \colon 0 \leq \phi(x) < 1, -1 \leq x \leq 1\}$.

With $F(u) = \int_0^u f(s)\,ds$ and
$$V(\phi) = \int_{-1}^{1} \left[\frac{1}{2}\phi_x^2 - F(\phi)\right] dx$$
we know that
$$dV(u(t,\cdot,\phi,a))/dt = -\int_{-1}^{1} u_t^2(t,x,\phi,a)\,dx,$$
where $u(t,\cdot,\phi,a)$ is the solution of (3.50) and (3.51) with $u(0,\cdot,\phi,a) = \phi$. The maximum principle implies that $u(t,\cdot,\phi,a) \in \tilde{H}_0^1$ if $\phi \in \tilde{H}_0^1$ for all $t \in [0, t_{\phi,a})$, the maximal interval of existence of $u(t,\cdot,\phi,a)$.

If $\gamma_{\phi,a}(t) = \max\{u(t,x,\phi,a), -1 \leq x \leq 1\}$ and $\sup\{\gamma_{\phi,a}(t) \colon t \in [0, t_{\phi,a})\} < 1$, then $t_{\phi,a} = \infty$ and the ω-limit set of ϕ belongs to the set E_a of equilibrium points of (3.50), (3.51), that is, the solutions of

(3.52) $$\phi_{xx} + a^2 f(\phi) = 0, \quad -1 < x < 1,$$
$$\phi(-1) = \phi(1) = 0$$

in $\tilde{H}_0^1$.

Suppose there is a $\phi \in \tilde{H}_0^1$ such that $t_{\phi,a} = \infty$ and $\gamma_{\phi,a}(t) \to 1$ as $t \to \infty$. For any $\varepsilon > 0$ consider the solution $u(t,\cdot,\phi,a+\varepsilon)$ and the corresponding $t_{\phi,a+\varepsilon}$. We claim that $t_{\phi,a+\varepsilon} < \infty$. In fact, if $w(t,x) = u(t,x,\phi,a+\varepsilon) - u(t,x,\phi,a)$, then
$$w_t = w_{xx} + (a+\varepsilon)^2[f(u+w) - f(u)] + (2\varepsilon a + \varepsilon^2)f(u)$$
and $w(t,-1) = w(t,1) = 0$, $w(0,\cdot) = 0$. If $t_{\phi,a+\varepsilon} = \infty$, then the strong maximum principle implies that there are $t_0 > 0$, $\delta > 0$, such that $\max\{w(t,x), |x| \leq 1\} \geq \delta$ for $t \geq t_0$. Since $\gamma_{\phi,a}(t) \to 1$ as $t \to \infty$ this implies there is a $t_1 > t_0$ such that $\gamma_{\phi,a+\varepsilon}(t) > 1$ for $t \geq t_1$. This is a contradiction and shows that $t_{\phi,a+\varepsilon} < \infty$.

Let $a_0 = \sup\{a\colon E_a \neq \varnothing\}$ and, for $a < a_0$, let $\phi_{m,a}, \phi_{M,a}$ be the minimal and maximal elements of E_a. The above analysis shows in particular that if $\phi \leq \phi_{m,a}$, then $t_{\phi,a} = \infty$ and $\omega(\phi) = \phi_{m,a}$. If $a > a_0$, then $t_{\phi,a} < \infty$ for every

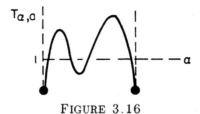

FIGURE 3.16

$\phi \in \tilde{H}_0^1$ and $\gamma_{\phi,a}(t) \to 1$ as $t \to t_{\phi,a}^-$. This is a slight improvement on the results of Acker and Walter [1976].

With a little more analysis, one can obtain a better picture of the flow defined by (3.50), (3.51). Let $\phi(x, a, \alpha)$, $\alpha > 0$, be the solution of $\phi_{xx} + a^2 f(\phi) = 0$ with $\phi(0, a, \alpha) = 0$, $\phi_x(0, a, \alpha) = \alpha$ and let $T_{\alpha,a}$, if it exists, be such that $\phi_x(x, a, \alpha) > 0$, $0 < x < T_{\alpha,a}$, $\phi_x(T_{\alpha,a}, a, \alpha) = 0$. A typical graph for $T_{\alpha,a}$ is shown in Figure 3.16. The set $E_a = \{\phi(1 + \cdot, a, \alpha): T_{\alpha,a} = 1\}$. A point $\psi \in E_a$ is hyperbolic if $T'_{\alpha,a} \neq 0$, and is stable if $T'_{\alpha,a} > 0$, unstable if $T'_{\alpha,a} < 0$ with $\dim W^u(\psi) = 1$. For the picture shown and $a \leq a_0$, let $\psi^* = \psi^*(a)$ be the maximal equilibrium point and consider the set $\tilde{H}_0^{1^*} = \{\phi \in \tilde{H}_0^1: \phi(x) \leq \psi^*(x), -1 \leq x \leq 1\}$. The differential equation (3.50), (3.51) defines a flow on $\tilde{H}_0^{1^*}$ and there is an attractor A_a. As a increases to a_0, the flow on the attractor A_a changes from $\cdot\!\leftarrow\!\cdot\!\rightarrow\!\cdot\!\leftarrow\!\cdot\!\rightarrow$ to $\cdot$. For $a > a_0$, the equilibrium set is empty.

We remark that Acker and Walter [1978] have characterized a^* in the same way as above for the more general equation with $f(u)$ replaced by $f(u, u_x)$ with $f(0,0) > 0$ and $f(u,p) \to \infty$ uniformly over all p in a bounded interval as $u \to 1^-$. One possibly could use the ideas of Zelenyak [1968] together with the previous remarks to give another proof and extension of the results of Acker and Walter [1978].

The example discussed in this section shows that it is possible to obtain information about the flow by using ideas from dynamical systems even when no global attractor exists. A recent paper of Rothe [1987] shows that the same fact is true for the general scalar parabolic equation (3.12).

4.4. The Navier-Stokes equation.

Suppose Ω is a bounded open set in R^n, $n = 2$ or 3, with smooth boundary $\partial \Omega$. The form of the Navier-Stokes equation for a homogeneous incompressible fluid is

(4.1) $$\partial u / \partial t - \nu \Delta u + (u \cdot \text{grad})u + \text{grad } p = f \quad \text{in } \Omega,$$

(4.2) $$\text{div } u = 0,$$

where $u = u(x,t) \in R^n$ is the velocity of fluid which is at point x at time t, p is the pressure, and f is the density of force per unit volume. The Reynolds number $\text{Re} = \nu^{-1} > 0$ and $f \in L^2(\Omega, R^n)$.

The first equation (4.1) represents Newton's law balancing the forces, and (4.2) expresses conservation of mass and is often referred to as the equation of continuity. For boundary conditions, we take

(4.3) $$u = 0 \quad \text{on } \partial\Omega.$$

At first glance, system (4.1), (4.2) does not appear to be the usual type of evolutionary system since (4.2) has no time derivative and the pressure term appears with no time derivative. As we shall see, if we choose the space of functions u so that $\operatorname{div} u = 0$, the pressure term disappears.

To motivate the space, let us observe that $\operatorname{div}(\phi u) = u \cdot \operatorname{grad} \phi + \phi \operatorname{div} u$ and use Gauss's theorem to obtain

$$\int_\Omega u \cdot \operatorname{grad} \phi \, dx = -\int_\Omega \phi \cdot \operatorname{div} u \, dx + \int_{\partial\Omega} \phi u \cdot n \, ds$$

where n is the outward normal to $\partial\Omega$. Thus, if $\operatorname{div} u = 0$ in Ω and $u \cdot n = 0$ on $\partial\Omega$, then u is orthogonal to all gradients. This suggests the following decomposition of $L^2(\Omega, R^n)$. Let

$$H_\pi = \text{closure in } L^2(\Omega, R^n) \text{ of } \{\operatorname{grad} \phi \colon \phi \in C^1(\Omega)\},$$
$$H_\sigma = \text{closure in } L^2(\Omega, R^n) \text{ of } \{u \in C^1(\Omega, R^n) \colon \operatorname{div} u = 0 \text{ in } \Omega,$$
$$u \cdot n = 0 \text{ on } \partial\Omega\}.$$

One can then prove (Ladyzenskaya [1963, p. 28] or Henry [1981, p. 80]).

LEMMA 4.4.1. *H_π and H_σ are closed orthogonal subspaces in $L^2(\Omega, R^n)$ and $L^2(\Omega, R^n) = H_\pi \oplus H_\sigma$.*

PROOF. The fact that H_π and H_σ are orthogonal follows from the previous computation. To show their sum is $L^2(\Omega, R^n)$, it is sufficient to show that every smooth $u \colon \Omega \to R^n$ with compact support (since these are dense in $L^2(\Omega, R^n)$) has the form $u = v + \operatorname{grad} \phi$ with $v \in H_\sigma$ and $\operatorname{grad} \phi \in H_\pi$. Let ϕ be the solution of

$$\Delta \phi = \operatorname{div} u \quad \text{in } \Omega, \qquad \partial \phi / \partial n = u \cdot n = 0 \quad \text{on } \partial\Omega$$

and let $v = u - \operatorname{grad} \phi$. The $\operatorname{div} v = 0$ in Ω and $v \cdot n = 0$ on $\partial\Omega$ since u has compact support. This completes the proof.

One can show that H_σ is the completion in $L^2(\Omega, R^n)$ of divergence-free vectors of compact support in Ω (see Temam [1979]). Let V be the completion in $H^1(\Omega, R^n)$ of the space of divergence-free vectors of compact support in Ω. It is known that $V = \{u \in H_0^1(\Omega, R^n) \colon \operatorname{div} u = 0\}$ (see Temam [1979]). If we let P be the orthogonal projection of $L^2(\Omega, R^n)$ onto H_σ, then (4.1), (4.2) is formally equivalent to the equation

(4.4) $$dv/dt + Av = B(v) + g$$

on V, where $A = -P\Delta$, $B(v) = -P(v \cdot \operatorname{grad})v$, and $g = Pf$.

Some properties of A are given in the following

LEMMA 4.4.2. *The operator A may be considered as a selfadjoint positive definite densely defined operator in H_σ, the inverse of A is compact, and $D(A) = H^2(\Omega, R^n) \cap V$.*

PROOF. If $u, v \in C_0^3(\Omega, R^n) \cap H_\sigma$ (subscript zero means compact support) then $Au \in H_\sigma$. If (u, v) is the inner product in $L^2(\Omega, R^n)$, then integration by parts yields

$$(Au, v) = (u, Av), \qquad (Au, u) = (-\Delta u, u) = \int_\Omega |\text{grad } u|^2 \, dx \geq 0.$$

Thus A is formally selfadjoint and nonnegative. By Friedrich's extension theorem, the operator A may be considered as selfadjoint and nonnegative.

To obtain the other properties of the operator A, consider the equation $Av = g$ for $g \in H_\sigma$, $v \in V$, that is, the equation

(4.5)
$$\begin{aligned} -\nu \Delta v + \text{grad } p &= f \quad \text{in } \Omega, \\ \text{div } v &= 0 \quad \text{in } \Omega, \\ v &= 0 \quad \text{on } \partial \Omega, \end{aligned}$$

where $f \in L^2(\Omega, R^n)$, $g = Pf$.

By a *generalized solution* of (4.5), we mean a function $v \in V$ satisfying

(4.6)
$$\nu \int_\Omega \sum_{i=1}^n \frac{\partial v}{\partial x_i} \cdot \frac{\partial \phi}{\partial x_i} \, dx = \int_\Omega f \cdot \phi \, dx = \int_\Omega g \cdot \phi \, dx$$

for all $\phi \in V$. Here, we have used the fact that $g = Pf$. Multiplying the first equation in (4.5) by $\phi \in V$ and integrating by parts, one observes that any classical solution of (4.5) is a solution of (4.6) since $f = \text{grad } \psi + g$ for some function ψ and the gradient terms drop out by Lemma 4.4.1. Conversely, if v satisfies (4.6) and belongs to $H^2(\Omega, R^n) \cap V$, then $f \in L^2(\Omega, R^n)$ implies

$$\int_\Omega (\nu \Delta v + f) \cdot \phi \, dx = 0$$

for all $\phi \in V$ with $\phi = 0$ on $\partial \Omega$. Since this set is dense in H_σ it follows from Lemma 4.4.1 that $\nu \Delta v + f = \text{grad } p$ for some function p. Thus, the solution of (4.6) satisfies (4.5).

The equation (4.6) has at most one solution. In fact, $\int_\Omega f \cdot \phi \, dx$ is a continuous linear functional on V equipped with the inner product

$$[u, v] = \int_\Omega \text{grad } u \cdot \text{grad } v \, dx.$$

From the Riesz representation theorem, there is a unique $u \in V$ such that $\int_\Omega f \cdot \phi \, dx = [u, \phi]$ and $u = u(f)$ is continuous in f. The fact that $f \mapsto u(f)$ is continuous from $L^2(\Omega, R^n)$ to $H_0^1(\Omega, R^n)$ implies A^{-1} is compact.

We do not prove the last statement of the lemma concerning regularity and refer the reader to Ladyzenskaya [1963, p. 38] or Temam [1979].

From Lemma 4.4.2, A is a sectorial operator on H_σ and one can define the fractional power spaces H_σ^α for any $\alpha \in R$. For $0 \leq \alpha < 1$, A generates an analytic semigroup e^{-At} on H_σ^α which is compact for each $t > 0$. Furthermore, we know from §4.2 that any solution of (4.4) must be a solution of the integral equation

$$v(t) = e^{-At}v_0 + \int_0^t e^{-A(t-s)}[B(v(s)) + g(s)]\,ds. \tag{4.7}$$

To show existence and uniqueness by using (4.7), one must estimate the function B as a map from H_σ^α to H_σ. From the form of B and the Sobolev embedding theorems, the function B will be Lipschitzian if $\alpha > 1/2$ for $\Omega \subset R^2$ or $\alpha > 3/4$ for $\Omega \subset R^3$ since $H_\sigma^\alpha \subset W^{1,q}(\Omega, R^3) \cap L^\infty(\Omega, R^3)$ for any $q \geq 2$ if $\Omega \subset R^2$ and any $q^{-1} > (5-4\alpha)/6$ if $\Omega \subset R^3$. Thus, Theorem 4.2.1 implies the local existence and uniqueness of the solutions of (4.7). Under a further smoothness assumption on f, one obtains a classical solution of (4.1).

After local existence has been proved, one must discuss the global existence of solutions. This requires obtaining a priori bounds on the solutions in H_σ^α for $\alpha > 1/2$ if $\Omega \subset R^2$ and $\alpha > 3/4$ if $\Omega \subset R^3$. Since these a priori bounds are not known, this limits the success of the "soft" approach using the spaces H_σ^α, $\alpha > 1/2$. The best that one can do is to obtain a priori bounds in $H_\sigma^{1/2}$. These estimates are very delicate and depend upon the fact that the L^1-norm of B can be estimated in terms of $|u|$, $|\operatorname{grad} u|$, and $|Au|$. We will not give all of these estimates but only summarize the results that are known and show how they imply the existence of a global attractor. The presentation follows Temam [1983] and Ladyzenskaya [1972].

Let V' be the dual space of V and suppose $f \in L^2(0,T;V')$, $u_0 \in H_\sigma$. A *weak* (or *generalized*) *solution* of (4.4) is a function $u \in L^2(0,T;V)$ such that, for all $v \in V$,

$$d(u,v)/dt + \nu[u,v] + b(u,u,v) = (f,v), \tag{4.8}$$

$$u(0) = u_0, \tag{4.9}$$

where

$$(u,v) = \int_\Omega u \cdot v\,dx,$$

$$[u,v] = \int_\Omega \operatorname{grad} u \cdot \operatorname{grad} v\,dx,$$

$$b(u,v,w) = \sum_{i,j=1}^n \int_\Omega u_i \frac{\partial v_j}{\partial x_i} w_j\,dx.$$

It is shown in Temam [1983] that any solution of (4.8) is continuous almost everywhere on $[0,T]$ so that (4.9) is meaningful. According to Ball [1976], weak solutions also satisfy the integral equation.

For $f \in L^2(0,T;H_\sigma)$, $u_0 \in V$, a *strong solution* of (4.4) is a function $u \in L^2(0,T;D(A)) \cap L^\infty(0,T;V)$ which satisfies (4.8), (4.9).

A typical existence and uniqueness result for $\Omega \subset R^2$ obtained by using Galerkin methods is the following (for proofs, see Temam [1983] and Ladyzenskaya [1963]).

THEOREM 4.4.3. ($\Omega \subset R^2$). For any $T > 0$, if $f \in L^2(0,T;V')$, $u_0 \in H_\sigma$, then there is a unique weak solution u of (4.4) and, in addition, $u \in C([0,T], H_\sigma)$, $du/dt \in L^2(0,T;V')$. In addition, if $f \in L^2(0,T;H_\sigma)$, $u_0 \in V$, then there is a unique strong solution u of (4.4) and $u \in L^2(0,T;D(A)) \cap C([0,T], V)$, $du/dt \in L^2(0,T;H)$.

In particular, solutions exist for all $t > 0$. Therefore, one can define the mapping $T(t): V \to V$ by the weak solution $T(t)u_0 = u(t)$ of (4.8), (4.9). One also can prove that $T(t)$ is continuous in u_0 and so $T(t)$ is a C_0-semigroup on V.

For the case where $\Omega \subset R^3$, the results are not as good as Theorem 4.4.3 and are contained in the following result. The notation $H_{\sigma,w}$ denotes H_σ with the weak topology.

THEOREM 4.4.4. ($\Omega \subset R^3$). If $f \in L^2(0,T;V')$, $u_0 \in H_\sigma$, then there is a weak solution u of (4.4) and, in addition, $u \in L^\infty(0,T;H_\sigma) \cap C([0,T]; H_{\sigma,w})$, $du/dt \in L^{4/3}(0,T;V')$. If, in addition, $f \in L^\infty(0,T;H_\sigma)$, $u_0 \in V$, then there is a $T^* = T^*(u_0)$ such that there is a unique strong solution u of (4.4) on $[0,T^*]$, $u \in L^2(0,T^*;D(A)) \cap C([0,T^*];V)$, $du/dt \in L^2(0,T;H_\sigma)$.

Theorem 4.4.4 for $\Omega \subset R^3$ is weaker than the case $\Omega \subset R^2$ in two significant ways:

(1) It is not known if weak solutions are unique.
(2) It is not known if strong solutions exist for all $t > 0$.

As long as a strong solution exists, it is known that it is unqiue in the class of weak solutions. Without an answer to these questions, one cannot associate semigroups with (4.4) for $\Omega \subset R^3$.

Using Theorem 3.4.8, the following theorem is a consequence of the results of Ladyzenskaya [1972]. Some parts of the proof had been obtained previously by Foiaş and Prodi [1967].

THEOREM 4.4.5. Let $\Omega \subset R^2$ and let $T(t): V \to V$ be the semigroup defined by (4.4), and $f \in V$. Then $T(t)$ is compact for $t > 0$, point dissipative, and there is a connected global attractor A. Also, there is at least one steady state solution of (4.4). Finally, $T(t)|A$ is a group.

If we denote by A_ν the global attractor for $T_\nu(t): V \to V$ where ν is the viscosity, then, for $\nu_0 > 0$, A_ν is upper semicontinuous at $\nu = \nu_0$. The proof is exactly the same as the approximation result of §6 of Hale, Lin, and Raugel [1985].

We emphasize again that our definition of an attractor A implies that A attracts bounded sets of V; that is, for any $\varepsilon > 0$ and any bounded set B in V, there is a $t_0 = t_0(\varepsilon, B, A)$ such that $T(t)B \subset N_\varepsilon(A)$ for $t \geq t_0$. The universal attractor in Temam [1984] is the global attractor in Theorem 4.4.5.

We indicate the ideas of the proof, but make none of the estimates, referring the reader to Ladyzenskaya [1972]. Let $|u(t)|^2 = (u(t), u(t))$, $\|u(t)\|^2 = [u(t), u(t)]$. It is known that u is a sufficiently regular solution so that (4.8) becomes

(4.10) $\quad (u'(t), v) + \nu[u(t), v] + b(u(t), u(t), v) = (f, v)$

for all $v \in V$. Since $b(u, v, w) = -b(u, w, v)$ for all u, v, w, it follows that $b(u, u, u) = 0$. If $v = u$ in (4.10), then

$$\frac{1}{2}\frac{d}{dt}|u(t)|^2 + \nu\|u(t)\|^2 = (f, u(t)) \leq |f| \cdot \|u(t)\| \leq \frac{\nu}{2}\|u(t)\|^2 + \frac{1}{2\nu}|f|^2$$

and

(4.11) $\quad \dfrac{d}{dt}|u(t)|^2 + \nu\|u(t)\|^2 \leq \dfrac{1}{\nu}|f|^2.$

If λ_1 is the first eigenvalue of $-\Delta$ on Ω, then, for every $u \in V$, the Poincaré inequality implies that $|u|^2 \leq \lambda_1^{-1}\|u\|^2$. Therefore, from (4.11),

$$\frac{d}{dt}|u(t)|^2 + \nu\lambda_1|u(t)|^2 \leq \frac{1}{\nu}|f|^2$$

and

(4.12) $\quad |u(t)|^2 \leq e^{-\nu\lambda_1 t}|u_0|^2 + \dfrac{1}{\nu^2\lambda_1}[1 - e^{-\nu\lambda_1 t}]|f|^2$

for all $t \geq 0$. This implies that, for every solution $u(t)$, there is a $t_0(u_0)$ such that

(4.13) $\quad |u(t)|^2 \leq \dfrac{2}{\nu^2\lambda_1}|f|^2 \stackrel{\text{def}}{=} R^2 \quad \text{for } t \geq t_0(u_0);$

that is, the L^2-norm of u is ultimately bounded by a constant $R = R(f)$ independently of the initial data.

It is also clear from (4.12) that, if $|u_0| \leq R$, then $|u(t)| \leq R$ for all $t \geq 0$; that is, if $K_R = \{u \in V : |u| \leq R\}$, then $T(t)K_R \subset K_R$ for $t \geq 0$. The next step is to discuss the behavior of the orbit $\gamma^+(K_R) \subset V$. Ladyzenskaya [1972] proves that, for any $\sigma > 0$, the set $\gamma^+(T(\sigma)K_R)$ is bounded in V and $T(t)$ is completely continuous on V for each $t > 0$. Therefore, for any $\varepsilon > 0$, $\sigma > 0$, $T(\varepsilon)\gamma^+(T(\sigma)K_R) = \gamma^+(T(\sigma + \varepsilon)K_R)$ belongs to a compact set and has a nonempty compact invariant ω-limit set $\omega(K_R)$ and $\omega(K_R)$ attracts $T(\varepsilon + \sigma)K_R$. From (4.13), this implies that $\omega(K_R)$ attracts each point of V and $T(t)$ is point dissipative. Since $T(t)$ is completely continuous, Theorem 3.4.8 implies the conclusions of the theorem since $T(t)$ is one-to-one on V.

The proofs in Ladyzenskaya [1972] can be extended to the case where $f = f(t)$ in (4.4) belongs to $C(R, V)$ and is ω-periodic in t. One then obtains an ω-periodic process. One can use the same arguments as in the proof of Theorem 4.4.5 and apply Theorem 3.6.3 to obtain the following result.

THEOREM 4.4.6. *Let $\Omega \subset R^2$ and $f \in C(R, V)$ be ω-periodic. Then the ω-periodic process $U(t, \sigma)$ generated by (4.8) has a connected global attractor A in $R \times V$ for which the cross-section $A_\sigma = \{u \in V : (\sigma, u) \in A\}$ is a connected global attractor for $U(\omega, \sigma)$. Also, there is an ω-periodic solution of (4.8).*

We mention that the existence of an ω-periodic solution of (4.8) is an easy consequence of the Schauder fixed point theorem. In fact, if $|f(t)|^2 \leq \sigma^2$ for all

t and $R^2 = 2\sigma^2/\nu^2\lambda_1$, then the ball B_R in L^2 of radius R is mapped into itself under $U(\omega, 0)$. For $\omega > 0$, $U(\omega, 0)B_R$ is precompact in L^2 and so $U(\omega, 0)$ has a fixed point.

From the work of Constantin, Foiaş, and Temam [1985] and the situation when $\Omega \subset R^3$, it should be possible to modify the proof of Theorem 4.4.5 to show that, if each solution of (4.8) is defined for all $t \geq 0$, then $T(t)$ is point dissipative. Temam [1984] has similar remarks.

Even though one does not know if (4.8) and (4.9) have unique global solutions for arbitrary (u_0, f), it is generically true as stated in the following result of Furskiov [1980].

THEOREM 4.4.7. *For $\Omega \subset R^3$ and given ν and $u_0 \in V$, there exists a set $F \subset L^2(0, T; H_\sigma)$ dense in $L^q(0, T; V')$ for all q, $1 \leq q < 4/3$ such that, for every $f \in F$, equations (4.8), (4.9) possess a unique solution (strong solution).*

One can use the above procedure to obtain the existence of global attractors for other types of fluids, for example, the so-called Oldroit fluids whose dynamics are described by the Navier-Stokes equation coupled with an ordinary differential equation (see Kotsiclis and Oskolkov [1986]).

4.5. Neutral functional differential equations.

In this section, we discuss a special class of neutral functional differential equations. These equations arise naturally in the theory of transmission lines where the hyperbolic partial differential equations are linear and the boundary conditions are nonlinear. Other problems of nonlinear vibrations can be formulated in terms of these equations.

4.5.1. *Properties of the semigroup.* To define the class of equations, we use the notation of §4.1 and introduce some additional terminology. Suppose $g: C \to R^n$ is a bounded linear operator. We say g is *nonatomic at zero* if there is an $n \times n$ matrix $\mu: [-\delta, 0] \to R^{n^2}$ of bounded variation such that

(5.1) $$g(\phi) = \int_{-\delta}^{0} [d\mu(\theta)]\phi(\theta), \qquad \mu(0) - \mu(0^-) = 0.$$

Let $D: C \to R^n$ be defined by

(5.2) $$D\phi = \phi(0) - g(\phi),$$

where g is nonatomic at zero; for example, D could be $D\phi = \phi(0) - A\phi(-1)$ where A is an $n \times n$ matrix.

With D as in (5.1), (5.2) and $f: C \to R^n$, a neutral functional differential equation, sometimes referred to as NFDE or NFDE(D, f), is the equation

(5.3) $$(d/dt)Dx_t = f(x_t).$$

If $\phi \in C$ is a given function, a solution of (5.3) with initial value ϕ at $t = 0$ is a continuous function $x: [-\delta, \alpha) \to R^n$ for some $\alpha > 0$ such that $x_0 = \phi$, and $D(x_t)$ is continuously differentiable on $[0, \alpha)$ and satisfies (5.3) on $[0, \alpha)$. Note that $x(t)$ need not be differentiable in t.

If $f \in C^r(C, R^n)$, $r \geq 1$, then, for any $\phi \in C$, there is a unique solution $x(t, \phi)$ through ϕ at $t = 0$, defined on a maximal interval $[-\delta, \alpha_\phi)$. If $\alpha_\phi < \infty$, then $\lim_{t \to \alpha_\phi} |x_t| = +\infty$. The solution $x(t, \phi)$ is continuous in t, ϕ together with derivatives up through order r in ϕ. If we suppose that all solutions are defined for all $t \geq -\delta$ and let $T(t)\phi = x_t(\cdot, \phi)$, then $T(t): C \to C$, $t \geq 0$, is a C^r-semigroup (see Hale [1977]).

One of our main objectives is to obtain a representation of the semigroup $T(t)$ as a sum $S(t) + U(t)$ where $U(t)$ is completely continuous and $S(t)$ is closely related to the linear functional equation

(5.4) $$Dy_t = 0.$$

To do this, we need detailed properties of (5.4). If $C_D = \{\phi \in C: D\phi = 0\}$ and $\psi \in C_D$, then there is a unique solution $y(t, \psi)$ of (5.4) defined for all $t \geq 0$. If $T_D(t)\psi = y_t(\cdot, \psi)$, then $T_D(t): C_D \to C_D$, $t \geq 0$, is a bounded linear semigroup of operators and there are constants k_D, ω_D such that

(5.5) $$|T_D(t)| \leq k_D e^{\omega_D t}, \qquad t \geq 0.$$

One can actually relate the modulus ω_D of the semigroup $T_D(t)$ to the solutions of the characteristic equation

(5.6) $$\det \Delta_D(\lambda) = 0, \qquad \Delta_D(\lambda) = I - \int_{-\delta}^{0} e^{\lambda \theta} d\mu(\theta)$$

if the function μ has no singular part; that is,

(5.7) $$D\phi = D_0\phi - \int_{-\delta}^{0} B(\theta)\phi(\theta)\, d\theta,$$
$$D_0\phi = \phi(0) - \sum_{j=1}^{\infty} B_j \phi(-\delta_j),$$
$$\sum_{j=1}^{\infty} |B_j| < \infty, \qquad \int_{-\delta}^{0} |B(\theta)|\, d\theta < \infty.$$

One can show that ω_D can be chosen as any real number satisfying

(5.8) $$\omega_D > \sup\{\operatorname{Re} \lambda: \det \Delta_D(\lambda) = 0\}$$

(see, for example, Hale [1977]). The general case has been considered by Staffans [1982].

Since $D(\phi) = \phi(0) - g(\phi)$ and g is nonatomic at zero, there are n functions $\phi_1^D, \ldots, \phi_n^D$ in C such that

(5.9) $$D(\Phi_D) = I, \text{ the identity}, \qquad \Phi_D = (\phi_1^D, \ldots, \phi_n^D).$$

In fact, for any $s \in [0, \delta]$, let $\psi \in C[-r, 0]$ be defined by $\psi(s) = 0$, $-r \leq \theta \leq -s$, $\psi(s) = 1 + \theta/s$, $-s < \theta \leq 0$. For s sufficiently small, $\det D\psi I \neq 0$ and $D\psi I$ forms a basis for R^n. If $\Phi_D = \psi I(D\psi I)^{-1}$, then $D(\Phi_D) = I$.

LEMMA 4.5.1. *Let $T_D(t): C_D \to C_D$ be the semigroup defined by (5.4) with D satisfying (5.7). Let $\Phi_0 = (\phi_1, \ldots, \phi_n)$ be n functions in C such that $D_0(\Phi_0) = I$ and let $\Psi_0 = I - \Phi_0 D_0: C \to C_{D_0}$. Then*

(5.10) $$T_D(t) = T_{D_0}(t)\Psi_0 + U_D(t), \qquad t \geq 0,$$

where $U_D(t)$ is completely continuous.

PROOF. We give the main ideas with the details being in Hale [1977]. If
$$U_D(t)\phi = T_D(t)\phi - T_{D_0}(t)\Psi_0\phi,$$
$$v(t,\tau,\phi) = U_D(t+\tau)\phi - U_D(t)\phi$$
$$h(t,\tau,\phi) = \int_{t+\tau-\delta}^{t+\tau} B(s-t-\tau)(T_D(s)\phi)(0)\,ds - \int_{t-\delta}^{t} B(s-t)(T_D(s)\phi)(0)\,ds,$$

then
$$D_0 v(t,\tau,\phi) = h(t,\tau,\phi);$$

that is, $v(t,\tau,\phi)$ satisfies a nonhomogeneous functional equation. Also, $v(0,\tau,\phi) = U_D(\tau)\phi - \Phi_0 D_0 \phi$. Since $U_D(0) = \Phi_0 D_0 \phi$ varies over a finite dimensional space as ϕ varies over C, and $U_D(\tau)$, $\tau \geq 0$, is a continuous family of bounded linear operators, the initial data for v with ϕ in a bounded set can be made arbitrarily small by making τ sufficiently small. Since the nonhomogeneous term uniformly approaches zero as $\tau \to 0$ for t, ϕ in a bounded set, estimates on the nonhomogeneous linear equation complete the proof of the lemma. For more details, see Hale [1977].

Lemma 4.5.1 says that the noncompact part of $T_D(t)$ comes from the semigroup $T_{D_0}(t)$, $t \geq 0$, for the difference equation

(5.11) $$D_0 y_t = y(t) - \sum_{k=1}^{m} B_k y(t-\delta_k) = 0.$$

Also, from the remarks made above concerning (5.5) and (5.8), if

(5.12) $$a_{D_0} = \sup\left\{\operatorname{Re}\lambda: \det\left[I - \sum_{k=1}^{\infty} B_k e^{-\lambda \delta_k}\right] = 0\right\},$$

then, for any $\varepsilon > 0$, there is a k_ε such that

(5.13) $$|T_{D_0}(t)| \leq k_\varepsilon e^{(a_{D_0}+\varepsilon)t}, \qquad t \geq 0.$$

We say the linear operator $D_0: C \to R^n$ is *asymptotically stable* if $a_{D_0} < 0$. With $S_D(t) = T_{D_0}(t)\Psi_0$ in (5.10), we have

LEMMA 4.5.2. *If D is given in (5.7) and D_0 is asymptotically stable, then*
$$T_D(t) = S_D(t) + U_D(t),$$
where $U_D(t)$ is completely continuous and $|S(t)| \to 0$ exponentially as $t \to \infty$. In particular, there is an equivalent norm in C_D such that $T_D(t)$, for each $t > 0$, is an α-contraction on C_D.

Let us now consider the general equation (5.3). For any given $\phi \in C$, the solution of (5.3) through ϕ at $t = 0$ is given by

$$Dx_t = D\phi + \int_0^t f(x_s)\, ds.$$

Let Φ_0 be defined as before with $D_0 \Phi_0 = I$; let $\Psi_0 = I - \Phi_0 D_0 \colon C \to C_{D_0}$ and $x_t = T(t)\phi = T_{D_0}(t)\Psi_0 \phi + U(t)\phi$. Then

$$D_0(U(t)\phi) = \int_{t-\delta}^t B(s-t)(T(s)\phi)(0)\, ds + \int_0^t f(T(s)\phi)\, ds.$$

Using the same type of reasoning as in the proof of Lemma 4.5.2, one obtains

THEOREM 4.5.3. *Suppose D is given in (5.7) and $T(t) \colon C \to C$ is the C^r-semigroup defined by (5.3). If, for each $t > 0$, the set $\{T(s)B,\ 0 \leq s \leq t\}$ is bounded for $B \subset C$ bounded, then*

(5.14) $$T(t) = S(t) + U(t)$$

where $U(t)$ is completely continuous for each $t \geq 0$, and $S(t)$, $t \geq 0$, is a family of bounded linear operators (actually a semigroup on C_{D_0}) such that, for any $\varepsilon > 0$, there is a k_ε such that

$$|S(t)| \leq k_\varepsilon e^{(a_{D_0}+\varepsilon)t}, \qquad t \geq 0,$$

$$a_{D_0} = \sup\left\{\operatorname{Re}\lambda \colon \det\left[I - \sum_{k=1}^\infty B_k e^{-\lambda \delta_k}\right] = 0\right\}.$$

In particular, if D_0 is asymptotically stable ($a_{D_0} < 0$) there is an equivalent norm in C such that $T(t)$ is an α-contraction for each $t > 0$.

4.5.2. Global attractor in the space of continuous functions.

THEOREM 4.5.4. *Suppose D is given in (5.7) with D_0 asymptotically stable, $T(t)$ is the semigroup for (5.3) and, for each $t \geq 0$, satisfies $\{T(s)B,\ 0 \leq s \leq t\}$ bounded if B is bounded.*

(i) If (5.3) is compact dissipative, then there is a maximal compact invariant set $A = A_{D,f}$ that attracts neighborhoods of compact sets of C.

(ii) If (5.3) is point dissipative and orbits of bounded sets are bounded, then (5.3) has a global attractor $A = A_{D,f}$.

In either case, there is an equilibrium point of (5.3). Finally

(iii) if $\phi \in A$ and $f \in C^k$ (or analytic), then $\phi(\theta)$ is a C^k (or analytic) function of θ.

The proof of all parts of the theorem except (iii) follows from Theorem 4.5.3 and Theorems 3.4.2 and 3.4.7. The proof of part (iii) will not be given and is contained in Hale and Scheurle [1985].

The functions D, f in (6.3) may also depend on t. In this case, the solution operator $T(t,\sigma) \colon C \to C$, $t \geq \sigma$, will have the same properties as stated in Theorem 4.5.4 if there is a continuous $n \times n$ matrix function $A(t,\theta)$ such that

(5.15) $$D(t)\phi = D_0(t)\phi + \int_{-\delta}^0 A(t,\theta)\phi(\theta)\, d\theta$$

and the equation

(5.16) $$D_0(t)y_t = 0$$

has the solution zero uniformly asymptotically stable (u.a.s). One can state

THEOREM 4.5.5. *Suppose $D(t)$, $f(t,\cdot)$ are ω-periodic in t with D satisfying (5.15) and the zero solution of (5.16) being u.a.s. If the solution operator $T(t,\sigma)$, $t \geq \sigma$, for*

(5.17) $$(d/dt)D(t)x_t = f(t,x_t)$$

is point dissipative and $\{T(t,\sigma)B,\ t \geq \sigma\}$ is bounded for each bounded set B, then there is a global attractor A_σ for $T(\omega+\sigma,\sigma)$, $T(\sigma,0)A_0 = A_\sigma$ for all $\sigma \in R$ and there is an ω-periodic solution of (5.17). Finally, if $\phi \in A_\sigma$ and $D, f \in C^k$ (or analytic), then $\phi(\theta)$ is a C^k (or analytic) function in θ.

4.5.3. **Global attractor in $W^{1,\infty}$.** Theorem 4.5.5 is the analogue of Theorem 4.1.11 for retarded FDE's. However, there is the additional hypothesis that orbits of bounded sets are bounded. For retarded FDE's, it was shown that point dissipative implies this latter property. This leaves open the possibility that perhaps there are additional properties of neutral FDE's that are being overlooked. In particular, it is interesting to try to determine how much information about the asymptotic behavior of the solutions can be obtained by assuming only that the associated semigroup is point dissipative. For example, is there some type of an attractor and does there exist an ω-periodic solution? Massatt [1981] has shown that this is the case by exploiting, in a nontrivial way, the fact that solutions of neutral functional differential equations can be defined on $W^{1,\infty}([-\delta,0], R^n)$, the space of absolutely continuous functions with derivatives essentially bounded. For $\phi \in W^{1,\infty}$, let

$$|\phi|_{W^{1,\infty}} = \sup_{[-\delta,0]} |\phi(\theta)| + \operatorname{ess\,sup}_{[-\delta,0]} |\phi'(\theta)|.$$

The space $W^{1,\infty}$ is compactly embedded in C and is dense in C.

We are going to use the results of §2.9 to prove the following result.

THEOREM 4.5.6. *Suppose $D(t)$, $f(t,\cdot)$ are ω-periodic in t with D satisfying (5.15) and the zero solution of (5.16) being uniformly asymptotically stable. If $D(t)$ is C^1 in t, then point dissipative and compact dissipative are equivalent. Thus, if (5.17) is point dissipative,*

  (i) *there is an ω-periodic solution of (5.17).*
  (ii) *There is a maximal compact invariant set $A_\sigma \subset C$ for $T(\omega+\sigma,\sigma)$, $T(\sigma,0)A_0 = A_\sigma$, for all $\sigma \in R$, and A_σ is u.a.s. and attracts a neighborhood of any compact set of C.*
  (iii) $A_\sigma \subset W^{1,\infty}$ *and A_σ is a global attractor for $T(\omega+\sigma,\sigma)$ in $W^{1,\infty}$.*
  (iv) *If $\phi \in A_\sigma$, then $\phi(\theta)$ is as smooth in θ as $D(t), f(t,\psi)$ are in t, ψ.*

For simplicity in notation in the proof of Theorem 4.5.6, we assume that D is independent of t, D is stable, and the solution operator $T_{D,f}(t,\sigma): C \to C$ takes bounded sets into bounded sets. The other situations are proved in an analogous way. We first prove

LEMMA 4.5.7. *Suppose the solutions of* (5.17) *are uniquely defined for* $t \geq \sigma$ *by the initial data in* C *at* σ *and let* $T_{D,f}(t,\sigma) \colon C \to C$ *be the solution operator. Then* $T_{D,f}(t,\sigma) \colon W^{1,\infty} \to W^{1,\infty}$ *and* $T_{D,f}(t,\sigma)\phi$ *is continuous. It is also continuous with respect to* f *in the topology of uniform convergence on bounded sets of* $[0,\infty) \times C$.

PROOF. Recall that we are assuming for simplicity in notation that D is independent of t and is stable. It is also sufficient to give the proof for $T(t,0)$ for $0 \leq t \leq \tau$ with $\tau > 0$ small, since the other situation can be obtained by translation of the initial time.

We first use the Schauder fixed point theorem to prove that there is a $\tau > 0$ such that $T_{D,f}(t,0) \colon W^{1,\infty} \to W^{1,\infty}$ for $0 \leq t \leq \tau$. For any $\tau > 0$, and any continuous function $x \colon [-\delta, \tau] \to R^n$, let

$$z(t) = \begin{cases} x(t) - \phi(0), & t \geq 0, \\ 0, & t \leq 0, \end{cases}$$

$$\tilde{\phi}(t) = \begin{cases} \phi(0), & t \geq 0, \\ \phi(t), & -\delta \leq t \leq 0. \end{cases}$$

If $D\phi = \phi(0) - L\phi$, where L is nonatomic at zero, then $x \colon [-\delta, \tau] \to C$ is a solution of (5.17) through ϕ at zero if and only if $z = Mz$ where

$$(Mz)(t) = Lz_t + D(\phi - \tilde{\phi}_t) + \int_0^t f(s, z_s + \tilde{\phi}_s)\, ds, \qquad 0 \leq t \leq \tau.$$

If $\phi \in W^{1,\infty}$, then M can be considered as a map of the set $Z = \{z \in W^{1,\infty}([-\delta, \tau], R^n) \colon z(t) = 0 \text{ for } t \leq 0\}$ into itself. Since f is completely continuous, $D(\phi - \tilde{\phi}_t) = 0$ for $t = 0$, and L is a contraction on Z for τ small, there is a $\tau > 0$ and a closed, bounded convex set $B \subset Z$ such that $MB \subset B$. Since B is compact in C and M is continuous in C, there must be a fixed point of T in M. This shows that $T(t,0) \colon W^{1,\infty} \to W^{1,\infty}$ for $0 \leq t \leq \tau$.

The continuous dependence in $W^{1,\infty}$ is proved in a similar manner. Let $\phi_0 \in W^{1,\infty}$ be given and let $x^0(t)$ be the solution through ϕ_0 at zero. If $x(t)$ is the solution of (5.17) through $\phi_0 + \phi$ at zero and $x_t = x_t^0 + z_t + \tilde{\phi}_t$, $t \geq 0$, then $z = Nz$ where, for $0 \leq t \leq \tau$,

$$(Nz)(t) = Lz_t + D(\phi - \tilde{\phi}_t) + \int_0^t [f(s, x_s^0 + z_s + \tilde{\phi}_s) - f(s, x_s^0)]\, ds.$$

For τ small enough, L is a contraction on Z. Furthermore, there are constants $\tau > 0$, $c = c(|\phi|_{W^{1,\infty}})$, $c(r) \to 0$ as $r \to 0$ such that, if $B_c = \{z \in Z \colon |z|_{W^{1,\infty}} \leq c\}$, then $NB_c \subset B_c$. Since B_c is compact in C, we have a fixed point of N in B_c. Since $c(r) \to 0$ as $r \to 0$, we have $z \to 0$ as $|\phi|_{W^{1,\infty}} \to 0$, that is, continuous dependence of $T_{D,f}(t)\phi$ on ϕ. The proof for continuous dependence on f is analogous and so the proof is omitted.

From the theory in §4.5.1, the solution operator $T_{D,f}(t,\sigma)$ can be written as

(5.18) $$T_{D,f}(t,\sigma) = S_D(t-\sigma) + U_{D,f}(t,\sigma),$$

where $U_{D,f}(t,\sigma)$ is conditionally completely continuous for $t \geq \sigma$ and $S_D(t)$ is linear and there are constants $k > 0$, $\alpha > 0$, such that

(5.19) $$|S_D(t)| \leq ke^{-\alpha t}, \qquad t \geq 0.$$

The operator $S_D(t)$ is the solution operator of the equation

$$Dy_t = 0, \qquad y_0 = \phi, \qquad D\phi = 0.$$

If $V = U_{D,f}(\omega+\sigma,\sigma)$, then $V: W^{1,\infty} \to W^{1,\infty}$ is continuous and $V: C \to C$ is continuous. We shall prove

(i) $V: C \to W^{1,\infty}$ is a continuous map and, if B and $V(B)$ are bounded in C, then $V(B)$ is bounded in $W^{1,\infty}$.

(ii) V is conditionally completely continuous in $W^{1,\infty}$.

As remarked before, we assume for simplicity that D is independent of t, asymptotically stable, and $T_{D,f}(t,\sigma)$ takes bounded sets in C to bounded sets in C. For any $B \subset C$ bounded, the set $\{T_{D,f}(s,\sigma), \sigma \leq s \leq t\}$ is bounded in C. Thus, the set $\{f(s, T_{D,f}(s,\sigma)\phi), \sigma \leq s \leq t, \phi \in B\}$ is bounded in R^n. Also, the integral operator $\int_0^t f(s, T_{D,f}(s,\sigma)\phi)\,ds$, $\sigma \leq t \leq \sigma+\omega$, is a continuous bounded map from C to $W^{1,\infty}$. For any $\phi \in B$, the function $V\phi = y_{\sigma+\omega}$, where y is the solution of the equation

$$dDy_t/dt = f(t,x_t), \qquad x_t = T_{D,f}(t,\sigma)\phi,$$

with $y_\sigma = \Phi_D D\phi$ where Φ_D is the matrix constructed in Lemma 4.5.1. We may choose the columns of Φ_D in $W^{1,\infty}$. Then $y_t \in W^{1,\infty}$ and the function $\dot{y}(t) = z(t)$ satisfies a.e. the equation $Dz_t = f(t,x_t)$ with initial data $\Phi'_D D\phi$. Since $\Phi'_D D\phi$ is a bounded set and D is stable, it follows that the set $VB = \{V\phi, \phi \in B)\}$ is bounded in $W^{1,\infty}$. This proof also shows that $V: C \to W^{1,\infty}$ is a continuous map.

To show V is completely continuous, suppose B is a bounded set in $W^{1,\infty}$. Since Cl B is compact in C, we have Cl$\{x_t: 0 \leq t \leq \omega, x_0 = \phi \in B\}$ is compact in C. Thus, the set of functions in $C([0,\omega], R^n)$ defined by $t \to f(t,x_t)$, $0 \leq t \leq \omega$, $x_0 = \phi \in B$, is in a compact set. By the continuous dependence result in Lemma 4.5.7 and the fact that $V\phi$, $\phi \in B$, is the solution at time $\omega + \sigma$ of the equation $dDy_t/dt = f(t,x_t)$ with $y_\sigma = \Phi_D D\phi$ varying over a bounded set in R^n, we obtain VB is precompact. Thus, V is completely continuous.

Since $S_D(t)$ satisfies (5.19), we can define an equivalent norm so that $S_D(\omega)$ is a contraction in C. The operator $S_D(\omega)$ also will be a contraction on $W^{1,\infty}$. Thus, $T(\sigma+\omega,\sigma)$ is an α-contraction on $W^{1,\infty}$ and C.

Using the above results, Theorem 2.9.3 implies that point dissipative in C implies compact dissipative in C. The conclusion in part (i) of the theorem is implied by Theorem 2.6.1 applied to the map $T(\omega,0)$. Part (ii) is implied by Theorem 2.4.2 applied to the map $T(\sigma+\omega,\sigma)$ and using the periodicity in the equation. If (5.17) is point dissipative in C, then Theorem 2.9.1 implies that (5.17) is bounded dissipative in $W^{1,\infty}$ under the α-contracting map $T(\sigma+\omega,\sigma)$. Corollary 2.9.2 implies there is a global attractor $\tilde{A}_\sigma$ for $T(\sigma+\omega,\sigma)$ in $W^{1,\infty}$. We can apply Theorem 2.9.4 to show that the invariant set A_σ in part (ii) of Theorem 4.5.6 belongs to $W^{1,\infty}$. Thus, $A_\sigma \subset \tilde{A}_\sigma$. Since $\tilde{A}_\sigma$ is invariant in C and precompact in C, we have $\tilde{A}_\sigma \subset A_\sigma$. Thus, $A_\sigma = \tilde{A}_\sigma$ and the proof of part (iii) of the theorem is complete. Part (iv) will not be proved and the reader is referred to Hale and Scheurle [1985] for this result.

4.6. Some abstract evolutionary equations.

In this section, we give some examples of abstract evolutionary equations for which we can show that the solution operator is an α-contraction. These results will be applied to more specific problems in subsequent sections. The proofs of the results follow standard arguments that have appeared in the literature (see Webb [1979a]).

THEOREM 4.6.1. *Let $T(t)$, $t \geq 0$, be a C^0-semigroup on a Banach space X satisfying*

(6.1) $$T(t)x = S(t)x + \int_0^t S(t-s)BT(s)x\,ds$$

for $t \geq 0$ and all $x \in X$, where $S(t)$ is a C^0-semigroup of linear mappings and B is a mapping from X to X satisfying

(6.2) $\qquad B$ *is completely continuous.*

(6.3) $S(t) = S_1(t) + S_2(t)$ *for all $t \geq 0$ where $S_1(t), S_2(t)$ are linear mappings from X to X with $|S_1(t)| \leq c(t)$, $c\colon \mathbf{R}^+ \to \mathbf{R}^+$, $\lim_{t\to\infty} c(t) = 0$, and $S_2(t)$ is completely continuous for t sufficiently large.*

Then
$$T(t) = S(t) + U(t)$$
where $U(t)$ is conditionally completely continuous for $t > 0$.

REMARK 4.6.2. Suppose $T(t)$, $t \geq 0$, is a linear C^0-semigroup on a Banach space X and let $\alpha(T(t)) = \inf\{k\colon \alpha(T(t)B) \leq k\alpha(B)$, for all bounded sets $B \subset X\}$ where α is the Kuratowski measure of noncompactness. The *linear semigroup* $T(t)$, $t \geq 0$, is an α-*contraction* if $\alpha(T(t)) \to 0$ as $t \to \infty$.

If $T(t)$, $t \geq 0$, is an α-contraction, then there is an integer $n > 0$ and a constant $\beta > 0$, such that $e^{-\beta n} \stackrel{\text{def}}{=} \alpha(T(n)) < 1$. For any $t > 0$, there is an integer p such that $0 \leq t - pn < n$. Let $k = \sup\{|T(t)|, 0 \leq t \leq n\}$. Then
$$\alpha(T(t)) = \alpha(T(t-pn)T^p(n)) \leq k\alpha(T^p(n))$$
$$\leq ke^{-\beta pn} = ke^{\beta(t-pn)}e^{-\beta t} \leq ke^{\beta n}e^{-\beta t} \stackrel{\text{def}}{=} c(t);$$
that is, $\alpha(T(t))$ is bounded above by a decaying exponential.

If $T(t)$ satisfies (6.3), then $T(t)$ is an α-contraction.

Suppose $T(t)$, $t \geq 0$, is a linear C^0-semigroup which is an α-contraction with $\alpha(T(t)) \leq k\exp(-\beta t)$. Using the method of proof in Deimling [1985] for Lemma 2.3.3, one can show that, for any $\varepsilon > 0$, there exist a decomposition of $T(t)$ as $T(t) = T_1(t) + T_2(t)$ where $T_2(t)$ is compact for $t \geq 0$ and there is an equivalent norm in X such that $|T_1(t)| \leq e^{-(\beta+\varepsilon)t}$ for $t \geq 0$.

Thus, the hypothesis (6.3) in Theorem 4.6.1 can be replaced by

(6.3)′ $\qquad\qquad S(t)$, $t \geq 0$, is an α-contraction on X,

since (6.3) implies $S(t)$ is an α-contraction.

REMARK 4.6.3. If $S_1(t)$ is a contraction for $t > 0$ and $S_2(t)$ is completely continuous for $t > 0$, then (6.3) is satisfied.

PROOF OF THEOREM 4.6.1. If we let

$$S_3(t)x = \int_0^t S(s)BT(t-s)x\,ds$$

we need only show that $S_3(t)$ is conditionally completely continuous for $t > 0$. Let M be a bounded subset of X for which $M_1 = \{T(t-s)x : x \in M,\ 0 \leq s \leq t\}$ is bounded. By (6.2), $M_2 = B(M_1)$ is precompact. We will show that $M_3 = \{S(s)x : x \in M_2,\ 0 \leq s \leq t\}$ is precompact by showing that it is totally bounded. Let k be a constant such that $|S(s)| \leq k$ for $0 \leq s \leq t$. Let $\varepsilon > 0$. There exist $x_1, \ldots, x_n \in M_2$ such that, if $x \in M_2$, there is an $i = i(x)$ such that $|x - x_i| < \varepsilon/2k$. Fix i, $1 \leq i \leq n$. Since $S(s)$, $s \geq 0$, is strongly continuous, there exist $s_1^i, \ldots, s_{k_i}^i \in [0,t]$, such that if $s \in [0,t]$, then $|S(s_j^i)x_i - S(s)x_i| < \varepsilon/2$ for some j. Then, for any $x \in M_2$, $s \in [0,t]$, there exist i and j such that

$$|S(s)x - S(s_j^i)x_i| \leq |S(s)x - S(s)x_i| + |S(s)x_i - S(s_j^i)x_i| \leq \varepsilon/2k + \varepsilon/2.$$

Thus, M_3 is totally bounded and hence precompact. By Mazur's theorem, the closed convex hull $\overline{\mathrm{co}}\,M_3$ is compact. Then, $\int_0^t S(s)BT(t-s)x\,ds \in t(\overline{\mathrm{co}}\,M_3)$ for all $x \in M$ and so $S_3(t)$ is compact. This proves the theorem.

THEOREM 4.6.4. *Let $T(t)$, $t \geq 0$, be a C^0-semigroup in a Banach space X satisfying (6.1) and*

(6.4) $\qquad\qquad S(t)$ *is compact for $t > 0$,*

(6.5) $\qquad\qquad B$ *is bounded.*

Then $T(t)$ is conditionally completely continuous for $t > 0$.

PROOF. Suppose M is a bounded subset of X for which $M_1 = \{T(t-s)x, x \in M,\ 0 \leq s \leq t\}$ is bounded and define $M_2 = \{\int_0^t S(s)BT(t-s)x\,ds : x \in M\}$ where $t > 0$ is fixed. We show that M_2 is compact. Let K be a constant such that $|S(s)| \leq K$ for $0 \leq s \leq t$ and let L be a constant such that $|BT(t-s)x| \leq L$ for $0 \leq s \leq t$, $x \in M$. If $0 < \varepsilon < t$, then for all $x \in M$,

$$\int_0^t S(s)BT(t-s)x\,ds = \int_0^\varepsilon S(s)BT(t-s)x\,ds + S(\varepsilon)\int_\varepsilon^t S(s-\varepsilon)BT(t-s)x\,ds.$$

Since $|\int_0^\varepsilon S(s)BT(t-s)x\,ds| \leq \varepsilon KL$, $|\int_\varepsilon^t S(s-\varepsilon)BT(t-s)x\,ds| \leq tKL$, and $S(\varepsilon)$ is compact, we have $\alpha(M_1) \leq \varepsilon KL + 0$. Since ε is arbitrary, this implies $\alpha(M_1) = 0$ and so $T(t)$ is compact for $t > 0$. This proves the theorem.

In the applications of Theorem 4.6.1 to hyperbolic equations, the operator B and the semigroup $S(t)$ have special properties which are convenient to have in an abstract form. Let X_1, X_2 be Banach spaces with norms $|\cdot|_1, |\cdot|_2$, respectively and suppose that X_1 is compactly embeddable into X_2, and X_1 is dense in X_2.

Let $X = X_1 \times X_2$. If the elements of X are denoted by (u_1, u_2), then $|(u_1, u_2)|_X = |u_1|_1 + |u_2|_2$. Let A be a closed linear operator on X_2 with $D(A)$ dense in X_1. Let C be an operator on X with $D(C) = D(A) \times X_2$ having the matrix representation

$$(6.6) \qquad C = \begin{bmatrix} 0 & I \\ -A & -2\alpha I \end{bmatrix};$$

that is, $C(u_1, u_2) = (u_2, -Au_1 - 2\alpha u_2)$, and suppose that

(6.7) $\quad$ C is the infinitesimal generator of a C_0-semigroup e^{Ct}, $t \geq 0$, in X and there exist real constants $M, \gamma > 0$ such that $|e^{Ct}| \leq Me^{-\gamma t}$, $t \geq 0$.

Let $F: X_1 \to X_2$ be an operator (perhaps nonlinear) satisfying the following conditions:

(6.8) $\qquad F: X_1 \to X_2$ is C^r for some $r \geq 1$,

(6.9) $\qquad F: X_1 \to X_2$ is completely continuous.

Let $\tilde{F}: X \to X$ be defined by $\tilde{F}(u_1, u_2) = (0, F(u_1))$. Then $\tilde{F}$ is continuous and locally Lipschitz. Consider the abstract evolutionary equation

$$(6.10) \qquad u_t = Cu + \tilde{F}(u), \qquad t > 0$$

and the corresponding integral equation

$$(6.11) \qquad u(t) = e^{Ct}u_0 + \int_0^t e^{C(t-s)}\tilde{F}(u(s))\,ds.$$

From Pazy [1983, p. 185], for any $u_0 \in X$, there is a unique solution of (6.11) defined on a maximal interval $[0, t_{u_0})$, and $u(t, u_0)$ is continuous in t, u_0 with continuous derivatives in u_0 of order r. If $t_{u_0} < \infty$, then $\lim_{t \to t_{u_0}} |u(t)| = \infty$. If $u_0 \in D(C)$, then the solution of (6.11) also satisfies (6.10) (Pazy [1983, p. 187]).

If we suppose that $t_{u_0} = \infty$ for all $u_0 \in R$ and also that (6.9) is satisfied, then Theorem 4.6.1 implies

THEOREM 4.6.5. *If* (6.6)–(6.9) *are satisfied and all solutions of* (6.11) *are defined for* $t \geq 0$, *then the semigroup* $T(t): X \to X$ *defined by* (6.11) *satisfies*

$$T(t) = e^{Ct} + U(t)$$

where $U(t)$ *is completely continuous for* $t \geq 0$ *provided that the set* $\{T(s)K, 0 \leq s \leq t\}$ *is bounded for any* $t > 0$ *and any bounded set* K *in* X. *Furthermore, there is an equivalent norm in* X *and* $\beta > 0$, *such that* $|e^{Ct}| \leq e^{-\beta t}$, $t \geq 0$.

We now give some additional conditions to ensure that C generates a C^0-group on X. Suppose the additional hypotheses:

(6.12) there is a linear operator B on X_2 with $D(B)$ dense in X_1 such that $B^2 = A$ and B^{-1} is a continuous operator from X_2 to X_1.

(6.13) $\pm iB$ are closed linear operators which are generators of C_0-semigroups on X_2.

We can prove

PROPOSITION 4.6.6. *If (6.12), (6.13) are satisfied, then e^{Ct} is a C_0-group on X. If $|e^{\pm iBt}| \leq K\exp\omega t$, $t \geq 0$, then $e^{Ct} = S_1(t) + S_2(t)$ where $S_2(t)$ is compact for $t \geq 0$ and, for any $\varepsilon > 0$, there is a $K(\varepsilon) > 0$ such that $|S_1(t)| \leq K(\varepsilon)e^{(\omega-\alpha+\varepsilon)t}$, $t \geq 0$.*

PROOF. Let $(u_1, u_2) = e^{Ct}(u_1^0, u_2^0)$. If $u_1 = e^{-\alpha t}v_1$, $u_2 = e^{-\alpha t}[v_2 - \alpha v_1]$, then

(6.14) $$v_t = D_\alpha v, \quad D_\alpha = \begin{bmatrix} 0 & I \\ -B^2 + \alpha^2 I & 0 \end{bmatrix}.$$

For the time being, let us suppose $\alpha = 0$. Condition (6.12) permits us to consider the transformation of variables from $X = X_1 \times X_2$ to $X_1 \times X_1$ given by

$$w_1 = v_1 + iB^{-1}v_2, \quad w_2 = v_1 - iB^{-1}v_2.$$

For $-w_1 + w_2$ in $D(B)$, the inverse transformation

$$v_1 = (w_1 + w_2)/2, \quad v_2 = iB(-w_1 + w_2)/2$$

is well defined. Since $D(B)$ is dense in X_1, this inverse transformation is defined on a dense set. By (6.13), the operator $\text{diag}(e^{-iBt}, e^{iBt})\colon X_2 \times X_2 \to X_2 \times X_2$ is a C_0-semigroup. Define the operator

(6.15)
$$e^{D_0 t}\colon X_1 \times X_2 \to X_1 \times X_2,$$
$$e^{D_0 t}(v_1, v_2) = ((e^{-iBt}w_1 + e^{iBt}w_2)/2, iB(-e^{-iBt}w_1 + e^{iBt}w_2)/2).$$

This is a C_0-semigroup on $X_1 \times X_2$ with generator D_0.

Since $D(B)$ is dense in X_1, it is not difficult to show that $(e^{D_0 t})^{-1}$ exists for each $t \geq 0$ and is a C^0-semigroup of bounded linear operators with infinitesimal generator $-D_0$. Therefore, by Pazy [1983, Lemma 6.4, p. 62], the operator

$$V(t) = \begin{cases} e^{D_0 t} & \text{for } t \geq 0, \\ (e^{-D_0 t})^{-1} & \text{for } t \leq 0 \end{cases}$$

is a C_0-group on X. This notation is cumbersome and we write $V(t) = e^{D_0 t}$, $t \in R$. This proves $e^{D_0 t}$ is a group on X. The fact that the same result holds for $e^{D_\alpha t}$ follows from S. G. Krein [1971, p. 242ff] or from the fact that $D_\alpha - D_0\colon X \to X$ is compact.

From (6.15), if $|e^{\pm iBt}| \leq Ke^{\omega t}$, $t \geq 0$, then there is a constant $K_1 > 0$ such that $|e^{D_0 t}| \leq K_1 e^{\omega t}$, $t \geq 0$. Since $D_\alpha - D_0\colon X \to X$ is compact, the variation of constants formula

$$e^{D_\alpha t}u_0 = e^{D_0 t}u_0 + \int_0^t e^{D_0(t-s)}(D_\alpha - D_0)e^{D_\alpha s}u_0 \, ds$$

and the fact that $e^{Ct} = e^{-\alpha t}e^{D_\alpha t}W$, where W is a 2×2 constant matrix, imply that $e^{Ct} = S_1(t) + S_2(t)$ where $S_1(t)$, $S_2(t)$ satisfy the conditions stated in the proposition. The proof is complete.

As an example, let us consider the linear equation

$$\begin{aligned} u_{tt} + 2\alpha u_t - \Delta u &= 0 \quad \text{in } \Omega \subset \mathbf{R}^n, \\ u &= 0 \quad \text{on } \partial\Omega, \end{aligned} \tag{6.16}$$

where $\alpha > 0$ is constant, and Ω is a bounded open set in $\mathbf{R}^n$ with smooth boundary. Equation (6.16) is equivalent to

$$\begin{aligned} u_{1t} &= u_2, \\ u_{2t} &= \Delta u_1 - 2\alpha u_2 \quad \text{in } \Omega, \\ u_1 &= 0 \quad \text{on } \partial\Omega. \end{aligned} \tag{6.17}$$

In terms of an abstract evolutionary equation $u_t = Cu$ with C in (6.6), we have $A = -\Delta$, $B = A^{1/2}$, $X_1 = H_0^1(\Omega) = D(B)$, $X_2 = L^2(\Omega)$, $X = H_0^1(\Omega) \times L^2(\Omega)$ and we can take $B > 0$. Furthermore, if λ_n^2 are the eigenvalues of A and ϕ_n are the normalized eigenfunctions, then each λ_n is real and can be taken > 0,

$$\phi = \sum_{n=1}^\infty (\phi, \phi_n)\phi_n, \qquad \phi \in L^2(\Omega),$$

$$A\phi = \sum_{n=1}^\infty \lambda_n^2 (\phi, \phi_n)\phi_n, \qquad \phi \in D(A) = H^2(\Omega) \cap H_0^1(\Omega),$$

$$B\phi = \sum_{n=1}^\infty \lambda_n (\phi, \phi_n)\phi_n, \qquad \phi \in D(B) = H_0^1(\Omega),$$

$$B^{-1}\psi = \sum_{n=1}^\infty \lambda_n^{-1} (\psi, \phi_n)\phi_n \qquad \text{for any } \psi \in L^2(\Omega).$$

From this formula, $B^{-1}: L^2(\Omega) \to H_0^1(\Omega)$ is continuous. Also, for any $\lambda \in \rho(B)$, the resolvent set of B,

$$(\pm iB - \lambda I)^{-m}\psi = \sum_{n=1}^\infty (\pm i\lambda_n - \lambda)^{-m}(\psi, \phi_n)\phi_n \quad \text{for any } \psi \in L^2(\Omega).$$

For any $\omega > 0$ and $\lambda < -\omega$, we have

$$|(\pm iB - \lambda I)^{-m}\psi|_{L^2(\Omega)} \le \max_n |\pm i\lambda_n - \lambda|^{-m}|\psi|_{L^2(\Omega)} \le \omega^{-m}|\psi|_{L^2(\Omega)}.$$

Thus, $\pm iB$ generate C^0-semigroups on $L^2(\Omega)$ and for any $\omega > 0$, $|e^{\pm iBt}| \le e^{\omega t}$, $t \ge 0$. Since (6.6), (6.13) are satisfied, e^{Ct} is a group on X. From Proposition 4.6.6, we have $e^{Ct} = S(t) + U(t)$ where $U(t)$ is compact for $t \ge 0$ and $|S(t)| \le K(\varepsilon)\exp(\omega - \alpha + \varepsilon)$ for any $\varepsilon > 0$. We can choose $\omega + \varepsilon < \alpha$. Thus, there are constants K_2, $\delta > 0$, such that $|e^{Ct}| \le K_2 \exp(-\delta t)$ for $t \ge 0$. This is summarized in

THEOREM 4.6.7. *Equation* (6.17) *generates a* C^0-*group* e^{Ct} *on* $X = H_0^1(\Omega) \times L^2(\Omega)$ *and there are constants* $K_2, \delta > 0$, *such that*
$$|e^{Ct}| \le K_2 e^{-\delta t}, \qquad t \ge 0.$$

In a later section, we need to consider (6.17) in a different space. More specifically, let $X_2 = H_0^1(\Omega)$, $X_1 = H^2(\Omega) \cap H_0^1(\Omega)$, and define the operators A and B as before, but take $D(B) = H_0^2(\Omega)$, the set of functions in $H^2(\Omega)$ which vanish together with their first derivatives on $\partial\Omega$. Then $D(B)$ is dense in X_1. Furthermore, one can use the formulas and the procedure in the proof of Theorem 4.6.7 to obtain the following result.

THEOREM 4.6.8. *Equation* (6.17) *generates a* C^0-*group* e^{Ct} *on* $X = (H^2(\Omega) \cap H_0^1(\Omega)) \times H_0^1(\Omega)$ *and the same estimate as in Theorem* 4.6.7 *holds.*

4.7. A one dimensional damped wave equation.

4.7.1. *Linear damping.* In this section, we discuss an example due to Webb [1979b]. Although more general results will be given in the next section, it is instructive to do the one dimensional case first because all Sobolev estimates are very simple.

Let $L^2 = L^2(0, \pi, R)$, $H^k = H^k([0, \pi], R)$, $H_0^k = H_0^k([0, \pi], R)$. For a given $\alpha > 0$ consider the equation for a function $u(x, t)$:

(7.1) $$u_{tt} + 2\alpha u_t - u_{xx} = f(u) - g$$

with the boundary conditions

(7.2) $$u(0, t) = u(\pi, t) = 0$$

or, equivalently, the system

(7.3) $$u_t = v, \qquad v_t = u_{xx} - 2\alpha v + f(u)$$

with the boundary conditions (7.2). We take the initial data (ϕ, ψ) for u, v at $t = 0$ to be in $H_0^1 \times L^2$. Observe that we can take $|\phi|_{H_0^1}^2 = |\phi'|_{L^2}$.

The function f will be required to satisfy

(7.4) $$f \in C^2(R, R), \qquad \varlimsup_{|u| \to \infty} f(u)/u \le 0, \qquad f(0) = 0,$$

and g is assumed to be in $L^2(0, \pi)$.

The main result of this section is

THEOREM 4.7.1. *Suppose f satisfies* (7.4). *Then* (7.2), (7.3) *defines a $C^{1,1}$ gradient semigroup $T(t)$ on $H_0^1 \times L^2$, the equilibrium set E is bounded, and there is a connected global attractor $A_f \subset H_0^1 \times L^2$. Furthermore, if each element $(\phi, 0) \in E$ is hyperbolic, then*
$$A_f = \bigcup_{(\phi,0) \in E} W^u(\phi, 0).$$

REMARK 4.7.2. Later in this section, we show that $A_f \subset (H^2 \cap H_0^1) \times H_0^1$ and that A_f is a global attractor for $T(t)$ in this smaller space.

To begin the proof, let us write (7.3) as an abstract evolutionary equation
$$w_t = Cw + F(w),$$

(7.5) $$w = \begin{bmatrix} u \\ v \end{bmatrix}, \quad C = \begin{bmatrix} 0 & I \\ -A & -2\alpha I \end{bmatrix}, \quad F = \begin{bmatrix} 0 \\ f^e - g \end{bmatrix}.$$

(7.6) $$A: H^2 \cap H_0^1 \to L^2, \quad A\phi = -\phi'',$$

(7.7) $$f^e(\phi)(x) = f(\phi(x)), \quad 0 \le x \le \pi, \quad \phi \in H_0^1.$$

LEMMA 4.7.3. *The operator C generates a C^0-group e^{Ct} on $H_0^1 \times L^2$ and there are positive constants K, δ such that*

(7.8) $$|e^{Ct}| \le Ke^{-\delta t}, \quad t \ge 0.$$

PROOF. The operator $A: H^2 \cap H_0^1 \to L^2$ is selfadjoint and positive definite. Therefore, $A^{1/2} = B$ is well defined and $D(B) = H_0^1$. In the previous section, we have shown that the operator $B: D(B) \to L^2$ has a compact inverse B^{-1}, $\pm iB$ generates a C^0-semigroup, e^{Ct} is a C^0-group, and the estimate 4.7.8 holds (see Theorem 4.6.7).

The constant $\delta = \delta(\alpha)$ is positive because α is positive but the dependence on α is rather complicated. Since x is in R, one can compute the group e^{Ct} explicitly to see how δ depends on α. The eigenvalues of A are n^2 and corresponding eigenfunctions can be chosen to be $\phi_n(x) = (2/\pi)^{1/2} \sin nx$. The following explicit representation of the group e^{Ct} is taken from Weinberger [1965, pp. 112–115].

$$e^{Ct}(\phi, \psi) = (u(t, \phi, \psi), v(t, \phi, \psi)),$$

$$u(t, \phi, \psi) = \sum_{n=1}^{\infty}[(\phi, \phi_n)U_n(t) + (\psi, \phi_n)V_n(t)]\phi_n,$$

$$v(t, \phi, \psi) = \sum_{n=1}^{\infty}[(\phi, \phi_n)U_n'(t) + (\psi, \phi_n)V_n'(t)]\phi_n,$$

$$U_n(t) = \begin{cases} e^{-\alpha t}[\cosh(\alpha^2 - n^2)^{1/2}t + \alpha(\alpha^2 - n^2)^{-1/2}\sinh(\alpha^2 - n^2)^{1/2}t], \\ \quad n^2 < \alpha^2, \\ e^{-\alpha t}(1 + \alpha t), \quad n^2 = \alpha^2, \\ e^{-\alpha t}[\cos(n^2 - \alpha^2)^{1/2}t + \alpha(n^2 - \alpha^2)^{-1/2}\sin(n^2 - \alpha^2)^{1/2}t, \\ \quad n^2 > \alpha^2, \end{cases}$$

$$V_n(t) = \begin{cases} e^{-\alpha t}(\alpha^2 - n^2)^{-1/2}\sinh(\alpha^2 - n^2)^{1/2}t, & n^2 < \alpha^2, \\ te^{-\alpha t}, & n^2 = \alpha^2, \\ e^{-\alpha t}(n^2 - \alpha^2)^{-1/2}\sin(n^2 - \alpha^2)^{1/2}t, & n^2 > \alpha^2. \end{cases}$$

For $0 < \alpha^2 \le n^2$, there are constants $k > 0$, $0 < \gamma \le \alpha$, such that
$$|U_n(t)|, |V_n(t)| \le ke^{-\gamma t}, \quad t \ge 0.$$
Furthermore, for $n^2 < \alpha^2$, there is a constant k_1 such that
$$|U_n(t)| \le k_1 e^{-\alpha t}, \quad |V_n(t)| \le e^{-\alpha t}(n^2 - \alpha^2)^{-1/2}.$$
Thus, for $(\phi, \psi) \in H_0^1 \times L^2$, one obtains the estimate (7.4) for an appropriate constant K and $\delta = \min(\gamma, \alpha)$.

LEMMA 4.7.4. *If $f \in C^2(R, R)$, then the function f^e in (7.7) satisfies*
(i) $f^e \colon H_0^1 \to L^2$ *is a local $C^{1,1}$-function,*
(ii) $f^e \colon H_0^1 \to L^2$ *is compact, in fact, $f^e \colon H_0^1 \to H_0^1$.*

PROOF. The proof of (i) is essentially contained in §4.3.
We next show that $f^e \colon H_0^1 \to L^2$ is compact. If $\phi \in H_0^1$, then ϕ is continuous, $|\phi(x)| \leq \sqrt{\pi}|\phi|_{H_0^1}$, the function $f^e\phi$ is absolutely continuous, and the derivative $(f^e\phi)'$ satisfies

$$\int_0^\pi [(f^e\phi)'(x)]^2 \, dx = \int_0^\pi [f'(\phi(x))\phi'(x)]^2 \, dx \leq \sup_x |f'(\phi(x))| |\phi|_{H_0^1}^2$$
$$\leq c(|\phi|_{H_0^1})|\phi|_{H_0^1}^2.$$

This shows that $f^e \colon H_0^1 \to H^1$ is a bounded map. Since $f(0) = 0$, $f^e \colon H_0^1 \to H_0^1$. Since bounded sets of H^1 are precompact in L^2, we have $f^e \colon H_0^1 \to L^2$ is compact. This completes the proof of the lemma.

Since $f^e \colon H_0^1 \to L^2$ is $C^{1,1}$, we obtain local existence and uniqueness of solutions of (7.2), (7.3) with initial data in $H_0^1 \times L^2$ and, thus, a local $C^{1,1}$-semigroup $T(t)$. If each solution is defined for $t \geq 0$, then $T(t)$ becomes a semigroup on $H_0^1 \times L^2$. In addition, if, for any $t > 0$ and any bounded set K in $H_0^1 \times L^2$, the set $\{T(s)K, 0 \leq s \leq t\}$ is bounded, then $T(t) = e^{Ct} + U(t)$ where $U(t)$ is compact from Theorem 4.6.5. Our next objective is to show that these latter properties are satisfied if (7.4) is satisfied.

For any $(\phi, \psi) \in H_0^1 \times L^2$, let

(7.9)
$$V(\phi, \psi) = \frac{1}{2}|\phi|_{H_0^1}^2 + \frac{1}{2}|\psi|_{L^2}^2 - \int_0^\pi F(\phi(x)) \, dx + \int_0^\pi g(x)\phi(x) \, dx,$$
$$F(u) = \int_0^u f(s) \, ds :$$

From §4.3, we know that, for any $\varepsilon > 0$, there is a constant c_ε such that

$$F(u) \leq \varepsilon u^2 + c_\varepsilon \quad \text{for all } u \in \mathbf{R}.$$

Therefore, for $(\phi, \psi) \in H_0^1 \times L^2$ and $\varepsilon = 1/4$, $C = C_{1/4}$, we have

(7.10)
$$V(\phi, \psi) \geq \frac{1}{2}|\phi|_{H_0^1}^2 + \frac{1}{2}|\psi|_{L^2}^2 - \int_0^\pi \left[\frac{1}{4}(\phi(x))^2 + c\right] dx$$
$$\geq \frac{1}{4}|\phi|_{H_0^1}^2 + \frac{1}{2}|\psi|_{L^2}^2 - \pi c.$$

If $(\phi, \psi) \in D(C)$ and $T(t)(\phi, \psi) = (u(\cdot, t), v(\cdot, t))$, then

(7.11)
$$\frac{d}{dt}V(u(\cdot, t), v(\cdot, t)) = \int_0^\pi [u_{xt}u_x + v_t v - f(u)u_t + gu_t] \, dx$$
$$= -2\alpha \int_0^\pi u_t^2 \, dx = -2\alpha \int_0^\pi v^2 \, dx \leq 0.$$

Thus,

(7.12)
$$V(T(t)(\phi, \psi)) \leq V(\phi, \psi).$$

Since $D(C)$ is dense in X, relation (7.12) is valid for all $(\phi, \psi) \in X$.

Now suppose that $|\phi|_{H_0^1} \leq r$, $|\psi|_{L^2} \leq r$. Then $|\phi(x)| \leq \sqrt{\pi} r$ and $|f(\phi(x))| \leq c(r)$ and $V(\phi, \psi) \leq r^2 + c(r) r \pi^{3/2}$. This fact, and relations (7.10), (7.12), imply that there is a constant $c_1(r)$ such that

(7.13) $$|T(t)(\phi, \psi)| \leq c_1(r),$$

and orbits of bounded sets are bounded. In particular, each orbit is bounded and $T(t)$ is a $C^{1,1}$-semigroup.

If $(u(t), v(t)) = T(t)(\phi, \psi)$ is such that $V(u(t), v(t)) = V(u(0), v(0))$ for all $t \in R$, then $dV(u(t), v(t))/dt = 0$ for $t \in R$ and (7.11) implies that $v(t) = 0$ for $t \in R$. From (7.3), this implies that $u(x, t) = u(x, 0) = \phi(x)$ for $t \in R$; that is, ϕ satisfies

(7.14) $$\phi'' + f(\phi) - g = 0, \qquad \phi(0) = \phi(\pi) = 0,$$

and $(\phi, 0)$ is an equilibrium point. Thus, (7.2), (7.3) is a gradient system.

We have shown in the proof of Theorem 4.3.1 that the set E is bounded. Since orbits of bounded sets are bounded under $T(t)$, Lemmas 4.7.3, 4.7.4 and Theorem 4.6.5 imply that $T(t)$ is an α-contraction. Since e^{Ct} is a group, it is not difficult to show that $T(t)$ is one-to-one and $DT(t)y$, $y \in X$, is an isomorphism. The conclusion of Theorem 4.7.1 follows from Theorem 3.8.5.

The following remark is interesting. In the proof of Theorem 4.7.1, we have shown that every orbit has its ω-limit set in the set of equilibrium points. It is a consequence of the result in Hale and Massatt [1982] that the ω-limit set of any orbit is exactly one equilibrium point. This follows because, for any equilibrium point, if zero is an eigenvalue of the linear variational equation, then it must be simple.

THEOREM 4.7.5. *Equations (7.2), (7.3) define a C^0-semigroup $T(t)$ on $Y = (H^2 \cap H_0^1) \times H_0^1$. Also, the global attractor A_f in Theorem 4.7.1 belongs to Y and is a global attractor in Y.*

PROOF. Using a proof similar to the one in the proof of Lemma 4.7.4, one has $f^e \colon H^2 \cap H_0^1 \to H_0^1$ is a local $C^{0,1}$-function and $f^e \colon H^2 \cap H_0^1 \to H^2$. Thus, $f^e \colon H^2 \cap H_0^1 \to H^1$ is compact. Since $f^e \colon H^2 \cap H_0^1 \to H_0^1$ is a local $C^{0,1}$-function, relations (7.2), (7.3) define a local $C^{0,1}$-semigroup on Y. But each solution must be defined for $t \geq 0$ since it is defined on $X = H_0^1 \times L^2$ for $t \geq 0$. The group e^{Ct} satisfies an estimate of the form (7.8) on Y from Theorem 4.6.8. Since $f^e \colon H^2 \cap H_0^1 \to H_0^1$ is compact, Theorem 4.6.5 implies $T(t)$ is a conditional α-contraction on Y.

To complete the proof of the theorem, we observe that since $f^e \colon H_0^1 \to H^1$, it is not difficult to verify that $U(t)$ satisfies property (iii) of Corollary 3.9.5 (see Ladyzenskaya [1985]). The proof is completed by applying Corollary 3.9.5.

4.7.2. A bifurcation problem.
In this section, we consider the equation

(7.15)
$$u_{tt} + 2\alpha u_t - u_{xx} = \lambda f(u), \qquad 0 < x < \pi,\ t \geq 0,$$
$$u = 0 \quad \text{at } x = 0,\ x = \pi$$

where $\alpha > 0$, $\lambda > 0$ is a parameter, $f \in C^2(R)$, and

(7.16)
$$\begin{cases} f(0) = 0, \qquad f'(0) = 1, \\ \overline{\lim}_{|x| \to \infty} f(x)/x \leq 0, \\ \operatorname{sgn} f''(x) = -\operatorname{sgn} x \quad \text{for } x \in R. \end{cases}$$

The set E of equilibrium points of (7.15) is bounded under condition (7.16). If we treat λ as a bifurcation parameter, then the bifurcation diagram is essentially the same as the one in §4.3.3 for $f(u) = u - u^3$ (see Figure 3.5). For $\lambda \in (n^2, (n+1)^2)$, there are exactly $2n+1$ equilibrium points $\phi_0 = 0$, ϕ_j^+, ϕ_j^-, $j = 1, 2, \ldots, n$, $\phi_j^{+\prime}(0) > 0$, $\phi_j^{-\prime}(0) < 0$, and $\phi_j^{\pm}$ have exactly $(j-1)$ zeros in $(0, \pi)$ and sign $f(\phi_j^{\pm}(x)) = \operatorname{sgn} \phi_j^+(x)$ for $x \in (0, \pi)$.

For the parabolic equation in §4.3.3, we observed that, if $\lambda \in (0, 1)$, then ϕ_0 is asymptotically stable and, if $\lambda \in (n^2, (n+1)^2)$, $n \geq 1$, then ϕ_1^+, ϕ_1^- are asymptotically stable, $\dim W^u(\phi_0) = n$, and $\dim W^u(\phi_j^{\pm}) = j$. It is interesting that the same results are valid for the hyperbolic equation as was observed by Webb [1973b]. Using Theorem 4.7.1, we can say more.

THEOREM 4.7.6. *For any $\lambda > 0$, $\alpha > 0$, there is a connected global attractor $A_{\lambda,\alpha}$ for (7.15). If $\lambda \in (n^2, (n+1)^2)$, then the set E of equilibrium points is $E = \{\phi_0, \phi_j^+, \phi_j^-, j = 1, 2, \ldots, n\}$, $\phi_0 = 0$, with $\dim W^u(\phi_j^{\pm}) = j - 1$, $\dim W^u(\phi_0) = n$,*

$$A_{\lambda,\alpha} = \bigcup_{\phi \in E} W^u(\phi).$$

Furthermore,

(i) *$\lambda \in (0, 1]$ implies $A_{\lambda,\alpha} = \{\phi_0\}$.*
(ii) *$\lambda \in (1, 4]$ implies $A_{\lambda,\alpha} = \operatorname{Cl} W^u(\phi_0)$; that is, there are orbits connecting ϕ_0 to ϕ_1^+ and ϕ_1^-.*
(iii) *$\lambda \in (n^2, (n+1)^2)$ implies there is an orbit connecting the set $\{\phi_0, \phi_j^{\pm}, j \geq 2\}$ to each of the equilibria ϕ_1^+, ϕ_1^-.*

PROOF. Theorem 4.7.1 shows the existence of the global attractor $A_{\lambda,\alpha}$. Chafee and Infante [1974] have discussed the bifurcation diagram and the stability of the equilibria for the parabolic problem. More specifically, for $\lambda \in (n^2, (n+1)^2)$, they have shown that the linear operator $\partial^2/\partial x^2 + \lambda f'(\phi_j^+)$ on H_0^1 has $j - 1$ positive eigenvalues with the others being negative. The proof is similar to the proof of Theorem 4.3.12. To each such eigenvalue μ and corresponding eigenfunction ψ_μ, there is a solution $e^{\varsigma t}\eta$ of the linear variational equation of (7.15) about ϕ_j^+ if and only if η is a constant multiple of ψ_μ and ς satisfies

$$\varsigma^2 + 2\alpha\varsigma - \mu = 0;$$

that is, $\varsigma = -\alpha \pm (\alpha^2 + \mu)^{1/2}$. Thus, the linear variational equation about ϕ_j^+ for (7.15) has exactly $j-1$ positive eigenvalues with the others having negative real parts. Consequently, $\dim W^u(\phi_j^+) = j-1$. The same remarks apply to ϕ_j^-. For $\phi_0 = 0$, the same argument gives $\dim W^u(\phi_0) = n$ if $\lambda \neq \lambda_n = n^2$. If $\lambda = \lambda_n$, then $\partial^2/\partial x^2 + \lambda f'(0)$ has a zero eigenvalue. The center manifold theorem and the fact that λ_n is a supercritical bifurcation to two new equilibria imply ϕ_0 is asymptotically stable along a center manifold. This proves the first part of the theorem.

To prove (i) to (iii), one needs only to observe that ϕ_0 is uniformly asymptotically stable along each center manifold and use the connectedness of $A_{\lambda,\alpha}$. This proves the theorem.

REMARK 4.7.7. Artstein and Slemrod [1982] have observed the remarks in Theorem 4.7.6 dealing with certain orbits being connected.

It would be interesting to know more about the orbit structure on $A_{\lambda,\alpha}$ for all α. A detailed study of the flow on $A_{\lambda,\alpha}$ should certainly give a better understanding of scalar hyperbolic equations. Recall that the unexpected result on transversality of stable and unstable manifolds for parabolic equations in §4.3.1 was discovered because such questions were being addressed. It is not to be expected that this transversality property will hold for all λ and α. However, a deep understanding of why nontransversal intersections can occur would be very beneficial in the study of $A_{\lambda,\alpha}$ as $\alpha \to 0$ (the limiting equation being the undamped hyperbolic equation).

One would expect that the flow on $A_{\lambda,\alpha}$ for α very large should be equivalent to the flow for the parabolic equation in §4.3.3 if $\lambda \in (n^2, (n+1)^2)$. To see this, a rescaling of time leads to the equation

$$\varepsilon u_{tt} + u_t - u_{xx} = \lambda f(u), \qquad 0 < x < 1,$$

with homogeneous Dirichlet boundary conditions and $\varepsilon = \alpha^{-2}$. Mora and Sola-Morales [1987b] and Chow and Lu [1987b] have proved that the flow on $A_{\varepsilon,\lambda}$ is equivalent to the flow on the attractor $A_{0,\lambda}$ of the parabolic equation if ε is sufficiently small and $\lambda \in (n^2, (n+1)^2)$. In §4.10.2, we make some remarks about this as well as the situation with several space variables.

4.7.3. *Nonlinear damping.* Suppose f satisfies (7.4), $g \in L^2(\Omega)$,

(7.17) $\qquad h \in C^1(R, R), \quad h(0) = 0, \qquad 0 < \alpha \leq h'(u) \leq \beta \quad \text{for } u \in R,$

where α, β are positive constants and consider the nonlinearly damped wave equation

(7.18) $\qquad\qquad u_t = v, \qquad v_t = u_{xx} + f(u) - g - h(v)$

with the Dirichlet conditions (7.2). The main result of this section is based on Lopes and Ceron [1984] and is the extension of Theorem 4.7.1 to equation (7.18), namely,

THEOREM 4.7.8. *Suppose f satisfies (7.4) and h satisfies (7.17). Then (7.2), (7.18) define a $C^{0,1}$ gradient semigroup $T(t)$ on $H_0^1 \times L^2$, the equilibrium* set E is bounded, and there is a connected global attractor $A_{h,f}$ in $H_0^1 \times L^2$. Furthermore, if each element $(\phi, 0) \in E$ is hyperbolic, then

$$A_{f,h} = \bigcup_{(\phi,0) \in E} W^u(\phi, 0).$$

Much of the same proof as used for the proof of Theorem 4.7.1 can be used here. In fact, using the same arguments as before, one obtains that $T(t)(\phi, \psi)$ is defined for $t \geq 0$, orbits of bounded sets are bounded, and E is bounded. When $h(v) = 2\alpha v$, we employed the variation of constants formula to show that $T(t)$ was an α-contraction. Another idea is needed when h is nonlinear. We show that the conditions of Theorem 2.3.6 are satisfied for the map $T(t_1)$ for some $t_1 > 0$. The appropriate estimates are obtained by using a Lyapunov function different from the one in (7.9).

Let $(\ ,\ )$ be the L^2 inner product and define

(7.19) $$E(\phi, \psi) = \frac{1}{2}|\psi|_{L^2}^2 + 2b(\phi, \psi) + \frac{1}{2}|\phi_x|_{L^2}^2,$$

where $b > 0$ is a constant such that

(7.20) $$b < \min\left(\frac{1}{4}, \frac{2\alpha}{4 + \beta^2}\right).$$

Then there are positive constants k_1, k_2 such that

(7.21) $$k_1(|\psi|_{L^2}^2 + |\phi_x|_{L^2}^2) \leq E(\phi, \psi) \leq k_2(|\psi|_{L^2}^2 + |\phi_x|_{L^2}^2).$$

Let $(u_1, v_1), (u_2, v_2)$ be two solutions of (7.2), (7.18). We know that $|u_j(t)|_{H_0^1}$, $|v_j(t)|_{L^2}$ are bounded by some constant R_1 for $t \geq 0$. If $\varsigma(t) = u_1(t) - u_2(t)$, $\eta(t) = v_1(t) - v_2(t)$, and $|\cdot|$ denotes $|\cdot|_{L^2}$, then

$$\frac{d}{dt}E(\varsigma(t), \eta(t)) = -(\eta, h(v_1) - h(v_2)) + 2b|\eta|^2 - 2b(\varsigma, h(v_1) - h(v_2))$$
$$- 2b|\varsigma_x|^2 + 2b(\varsigma, f(u_1) - f(u_2)) + (\eta, f(u_1) - f(u_2))$$
$$\leq -(\alpha - 2b)|\eta|^2 + 2b\beta|\varsigma| \cdot |\eta| - 2b|\varsigma_x|^2$$
$$+ 2b(\varsigma, f(u_1) - f(u_2)) + (\eta, f(u_1) - f(u_2))$$
$$\leq -(\alpha - 2b)|\eta|^2 + 2b\beta|\varsigma_x| \cdot |\eta| - 2b|\varsigma_x|^2$$
$$+ (2b|\varsigma| + |\eta|)|f(u_1) - f(u_2)|$$
$$\leq -\theta(|\eta|^2 + |\varsigma_x|^2) + K|f(u_1) - f(u_2)|$$
$$\leq -\frac{\theta}{k_1}E(\varsigma(t), \eta(t)) + K|f(u_1(t)) - f(u_2(t))|$$

for some constant $\theta > 0$ since b satisfies (7.20), relation (7.21) is satisfied, and $|u_j(t)|_{H_0^1}, |v_j(t)|_{L^2}$ are bounded for $t \geq 0$. This last inequality implies that there is a constant K_1 such that, for all $t \geq 0$,

(7.22) $$E(\varsigma(t), \eta(t)) \leq e^{-\theta t/k_1} E(\varsigma(0), \eta(0)) + K_1 \sup_{0 \leq s \leq t} |f(u_1(s)) - f(u_2(s))|_{L^2}.$$

From (7.21), this implies that there are constants K_2, K_3 such that

(7.23)
$$|(\varsigma(t),\eta(t))|_{H_0^1 \times L^2} \leq K_2 e^{-\theta t/k_1}|(\varsigma(0),\eta(0))|_{H_0^1 \times L^2} \\ + K_3 \sup_{0 \leq s \leq t} |f(u_1(s)) - f(u_2(s))|_{L^2}.$$

Since $u_j(t)$ is bounded in H_0^1, we have $|u_j(t,x)| \leq \sqrt{\pi}|u_j(t)|_{H_0^1}$ and there is a constant K_4 such that

(7.24) $\quad |f(u_1(s)) - f(u_2(s))|_{L^2} = \left|\int_{u_1(s,\cdot)}^{u_2(s,\cdot)} f'(\tau)\,d\tau\right|_{L^2} \leq K_4 |u_1(s) - u_2(s)|_{L^2}.$

Choose $t_1 > 0$ so that the constant $q = \exp(-\theta t/k_1) < 1$. Let $w_1^0 = (u_1(0), v_1(0))$, $w_2^0 = (u_2(0), v_2(0))$. Then (7.23), (7.24) imply that

(7.25) $\quad |T(t_1)w_1^0 - T(t_1)w_2^0|_{H_0^1 \times L^2} \leq q|w_1^0 - w_2^0|_{H_0^1 \times L^2} + \rho(w_1^0, w_2^0),$

where
$$\rho(w_1^0, w_2^0) = K_3 K_4 \sup_{0 \leq s \leq t_1} |u_1(s) - u_2(s)|_{L^2}.$$

The function $\rho(w_1^0, w_2^0)$ is a pseudometric on $H_0^1 \times L^2$. To show that ρ is precompact, suppose w_n^0 is a bounded sequence in $H_0^1 \times L^2$ and let $(u_n(t), v_n(t))$ be the solution through w_n^0 at $t = 0$. Since orbits of bounded sets are bounded, the sequence $(u_n(t), v_n(t))$ is bounded in $H_0^1 \times L^2$ uniformly for $t \geq 0$. Therefore, $\{u_n(t)\}$ has compact closure in L^2 for each $t > 0$. But $u_{nt} = v_n$ and v_n bounded in L_2 imply there is a constant K_5 such that
$$|u_n(t_1) - u_n(t_2)|_{L^2} \leq K_5|t_1 - t_2|$$
for all $t_1, t_2 \geq 0$. This implies there is a subsequence of the $u_n(t)$ convergent in L^2 and so ρ is a precompact pseudometric. Thus, $T(t_1)$ is an α-contraction by Lemma 2.3.6.

The remainder of the proof of Theorem 4.7.8 is completed as in the proof of Theorem 4.7.1.

4.7.4. Periodic forcing. In this section, we consider equation (7.18) subjected to a periodic forcing function of period p and show that the periodic map has a global compact attractor. In particular, this implies that there is a p-periodic solution of the forced equation.

Suppose f satisfies (7.4), h satisfies (7.17), and $e: R \to L^2$ is a continuous p-periodic function and consider the equation

(7.26) $\quad\quad u_t = v, \quad v_t = u_{xx} - h(v) + f(u) + e(t)$

with the Dirichlet conditions (7.2).

THEOREM 4.7.9. *If f satisfies (7.4), h satisfies (7.17), and $T(t, \sigma)$ is the solution operator of (7.26) in $H_0^1 \times L^2$, then $T(\sigma + p, \sigma)$ has a connected global attractor A_σ for every $\sigma \in R$, $A_\sigma = T(\sigma, 0)A_0$, and there is a p-periodic solution of (7.26).*

PROOF. An attempt to obtain the necessary a priori bounds on the solutions using the Lyapunov function V in (7.9) leads to difficulties because the derivative of V for the autonomous equation satisfies (7.11) and is not negative definite. To overcome this difficulty, we follow Lopes and Ceron [1984] and use a modification of the function in (7.19). A similar type of Lyapunov function was used by Ghidaghlia and Temam [1985], Babin and Vishik [1985], and Haraux [1985].

Let

(7.27) $$V(\phi, \psi) = \frac{1}{2}|\psi|^2_{L^2} + 2b(\phi, \psi) + \frac{1}{2}|\phi_x|^2_{L^2} - \int_0^\pi F(\phi(x))\,dx,$$

where $F(u) = \int_0^u f(s)\,ds$ and b satisfies (7.20). Using the hypotheses on F and the same type of estimates as before, one shows there are positive constants k_1, k_2 such that

(7.28) $$k_1(|\phi_x|^2_{L^2} + |\psi|^2_{L^2} - 1) \leq V(\phi, \psi) \leq k_2(|\phi_x|^2_{L^2} + |\psi|^2_{L^2} + c(|\phi_x|^2_{L^2})),$$

where $c\colon [0, \infty) \to [0, \infty)$ is an increasing function with $c(0) = 0$.

If $(u(t), v(t))$ is a solution of (7.26), then the argument in the previous section shows there is a $\theta > 0$ such that (here we again let $|\cdot| = |\cdot|_{L^2}$),

$$\frac{d}{dt}V(u(t), v(t)) = -(v, h(v)) + 2b(v, v) - 2b(u, h(v)) - 2b|u_x|^2$$
$$+ (v, e(t)) + 2b(u, f(u)) + 2b(u, e(t))$$
$$\leq -\theta(|u_x|^2 + |v|^2) + |v|\cdot|e(t)| + 2b(u, f(u)) + 2b|u|\cdot|e(t)|.$$

Since

$$|v|\cdot|e(t)| \leq \frac{\theta}{4}|v|^2 + \frac{1}{\theta}|e(t)|^2,$$
$$|u|\cdot|e(t)| \leq \frac{\theta}{4b}|u|^2 + \frac{b}{\theta}|e(t)|^2$$

the function V satisfies (7.28), and f satisfies (7.4), there are constants N and θ^* such that

(7.29) $$(d/dt)V(u(t), v(t)) \leq -(\theta/2)(|u_x|^2 + |v|^2) + \theta_1|e(t)|^2 + 2b(u, f(u))$$
$$\leq -\theta^* r^2(t) + N,$$

where $r^2(t) = |u_x(t)|^2_{L^2} + |v(t)|^2_{L^2}$.

If we let $a(r^2) = k_1(r^2 - 1)$, $b(r^2) = k_2(r^2 + c(r^2))$, then it follows from (7.28) that

(7.30) $$a(r^2(t)) \leq V(u(t), v(t)) \leq b(r^2(t)).$$

Relations (7.29), (7.30) imply bounded dissipativeness of $T(t, \sigma)$. In fact, suppose there are a constant r_0 and a sequence of initial conditions (u_{0n}, v_{0n}) such that $r_n^2 = |u_{0n}|^2_{H_0^1} + |v_{0n}|^2_{L^2} \leq r_0^2$ and $r_n^2(t) > 2N/\theta^*$ for $0 \leq t \leq n$, where $r_n^2(t) = |u_n(t)|^2_{H_0^1} + |v_n(t)|^2_{L^2}$ and $(u_n(t), v_n(t))$ is the solution of (7.26) through (u_{0n}, v_{0n}). Then, for $0 \leq t \leq n$, we have

$$(d/dt)V(u_n(t), v_n(t)) < -N,$$
$$V(u_n(t), v_n(t)) < -Nt + V(u_{0n}, v_{0n}) \leq -Nt + b(r_0^2).$$

In particular, for $t = n$ and $Nn > b(r_0^2) - a(2N/\theta^*)$, we have $V(u_n(n), v_n(n)) < a(2N/\theta^*)$ which is a contradiction since $a(2N/\theta^*) \leq a(r_n^2(n)) \leq V(u_n(n), v_n(n))$.

Therefore, $T(t,\sigma)$ is *bounded dissipative*; that is, for any $r > 0$, there is a $t_0 = t_0(r)$ such that

$$|(u(0), v(0))|_{H_0^1 \times L^2} \leq r \text{ implies } |(u(t), v(t))|_{H_0^1 \times L^2} \leq 2N_1 \text{ for } t \geq t_0(r).$$

In particular, this implies that orbits of bounded sets are bounded and (7.26) is point dissipative. To complete the proof of Theorem 4.7.9, we first observe that the same computations as in the previous section allow us to show that the analogue of formula (7.22) is valid; namely, there are positive constants k, α such that

$$E(T(\sigma+p,\sigma)w_1^0, T(\sigma+p,\sigma)w_2^0) \leq e^{-\alpha p} E(w_1^0, w_2^0) + \rho(w_1^0, w_2^0),$$
$$\rho(w_1^0, w_2^0) = k \sup_{\sigma \leq s \leq \sigma+p} |u_1(s) - u_2(s)|,$$

where the function E is defined in (7.19) and w_j^0, $j = 1, 2$, is the initial value of the solution $T(t,\sigma)w_j^0 = (u_j(t), v_j(t))$ at $t = \sigma$. Since E satisfies (7.21), it is an equivalent norm in $H_0^1 \times L^2$. One shows that ρ is a compact pseudometric as in the proof of Theorem 4.7.8. Thus, $T(\sigma+p,\sigma)$ is an α-contraction by Lemma 2.3.6. Corollary 3.6.2 completes the proof of the theorem.

4.8. A three dimensional damped wave equation.

In this section, we extend the results of the previous section to several space dimensions, concentrating on dimension three since this is the first case where some new ideas are needed.

4.8.1. Nonlinear damping. Suppose h satisfies (7.17), f is a C^2-function that satisfies (7.4) and, in addition,

(8.1) $$|f''(u)| \leq c(|u|^{1-\gamma} + 1) \quad \text{for } u \in R,$$

for some constants $c > 0$, $\gamma \geq 0$.

Suppose Ω is a bounded domain in R^3 with $\partial\Omega$ smooth, Δ is the Laplacian operator, $g \in L^2 = L^2(\Omega)$, and consider the equation

(8.2) $$u_{tt} + h(u_t) - \Delta u = f(u) - g \quad \text{in } \Omega, \quad u = 0 \text{ in } \partial\Omega$$

with (u, u_t) at $t = 0$ in $H_0^1 \times L^2$. Equation (8.2) is equivalent to the evolutionary system in $H_0^1 \times L^2$

(8.3) $$u_t = v, \quad v_t = \Delta u + f(u) - h(v) - g.$$

The main result of this section is

THEOREM 4.8.1. *If h satisfies (7.17) and f satisfies (7.4) and (8.1) with $\gamma > 0$, then (8.3) defines a $C^{1,1}$ gradient semigroup $T(t)$ on $H_0^1 \times L^2$ and there is a connected global attractor $A_{f,h} \subset H_0^1 \times L^2$. Furthermore, if each element of $(\phi, 0) \in E$ is hyperbolic, then*

$$A_{f,h} = \bigcup_{(\phi,0) \in E} W^u(\phi, 0).$$

REMARK 4.8.2. It will be seen from the proof that, for the existence of the attractor, one only needs f to be a C^1-function with $|f'(u)| \leq c(|u|^{2-\gamma} + 1)$,

$\gamma > 0$. The extra smoothness of f yields the result that A is the union of unstable manifolds. The additional restrictions in hypothesis (8.1) will be needed later to give a more complete discussion of the attractor in the space $(H^2 \cap H_0^1) \times H_0^1$. Also, if Ω is a domain in R^2, the same conclusion on the existence of the attractor in Theorem 4.8.1 is valid if f is a C^1-function satisfying $|f'(u)| \leq c(|u|^\tau + 1)$ for some positive constants c, τ.

REMARK 4.8.3. The proof of Theorem 4.8.1 will rely heavily upon the work of Lopes and Ceron [1984]. It generalizes the corresponding theorem for $h(v) = 2\alpha v$, $\alpha > 0$, a constant, given by Hale [1985] and Haraux [1985], which in turn generalized work of Babin and Vishik [1983], [1985].

We give a proof of Theorem 4.8.1 motivated by ideas of Lopes and Ceron [1984] in such a way that parts of it can be used when there is a nonautonomous periodic forcing function.

LEMMA 4.8.4. *Let* $f^e(\phi)(x) = f(\phi(x))$, $x \in \Omega$, *for* $\phi \in H_0^1$ *and suppose* f *satisfies* (8.1). *Then*

(i) *if* $\gamma \geq 0$, $f^e: H_0^1 \to L^2$ *is a local* $C^{1,1}$-*function*, $f^e: H^2 \cap H_0^1 \to H_0^1$ *is a local* $C^{0,1}$-*function and is compact.*

(ii) *If* $\gamma > 0$, *then* $f^e: H_0^1 \to L^2$ *is compact.*

PROOF. The Sobolev embedding theorem states that, for $\Omega \subset R^3$, the embedding $H_0^1(\Omega) \to L^q(\Omega)$ is continuous for $2 \leq q \leq 6$ and it is compact for $2 \leq q < 6 - \varepsilon$ for any $\varepsilon > 0$.

Let us first prove the first part of (i) in the lemma under the hypothesis that $\gamma = 0$. If $\phi \in H_0^1(\Omega)$, then $\phi \in L^6(\Omega)$ and

$$\int_\Omega |f^e(\phi)(x)|^2 \, dx \leq C_1^2 \int_\Omega (|\phi(x)|^3 + 1)^2 \, dx \leq C_2.$$

Thus, $f^e: H_0^1 \to L^2$. To prove it is continuous, note that, for $\phi, \psi \in H_0^1(\Omega)$.

$$|f^e(\phi(x)) - f^e(\psi(x))|^2 = \left| \int_{\psi(x)}^{\phi(x)} f'(s) \, ds \right|^2 \leq \left| \int_{\psi(x)}^{\phi(x)} |f'(s)| \, ds \right|^2$$

$$\leq C_1^2 \left| \int_{\psi(x)}^{\phi(x)} (s^2 + 1) \, ds \right|^2$$

$$\leq C_1^2 |\phi^2(x) + \psi^2(x) + |\phi(x)||\psi(x)| + 1|^2 |\phi(x) - \psi(x)|^2.$$

Since $\phi, \psi \in L^6$, we can integrate over Ω and use Hölder's inequality with $p = 6/4$, $p' = 6/2$ to obtain

$$|f^e(\phi) - f^e(\psi)|_{L^2} \leq C_3 |\phi - \psi|_{L^6}.$$

Since the embedding $H_0^1(\Omega) \to L^6(\Omega)$ is continuous, this implies f^e is locally Lipschitz continuous. To prove that f^e is a local $C^{1,1}$-function, one proceeds in an analogous manner.

To prove $f^e: H_0^1 \to L^2$ is compact for $\gamma > 0$ suppose $\{\phi_j\}_{j=1}^\infty \subset H_0^1$ is a bounded sequence. Then

$$\int_\Omega |f^e(\phi_j)(x) - f^e(\phi_k)(x)|^2\, dx = \int_\Omega \left| \int_{\phi_k(x)}^{\phi_j(x)} f'(s)\, ds \right|^2 dx$$

$$\leq C_1^2 \int_\Omega \left| \int_{\phi_k(x)}^{\phi_j(x)} (|s|^{2-\gamma} + 1)\, ds \right|^2 dx.$$

Performing this integration and then using Hölder's inequality for $p = (6 - 2\gamma)/(4 - 2\gamma)$, $p' = (6 - 2\gamma)/2$, one obtains

$$|f^e(\phi_j) - f^e(\phi_k)|_{L^2} \leq C_4 |\phi_j - \phi_k|_{L^{6-2\gamma}}.$$

Since the embedding $H_0^1 \to L^{6-2\gamma}$ is compact, there is a subsequence, which we label again as $\{\phi_j\}$, such that $\{\phi_j\}$ is a Cauchy sequence in $L^{6-2\gamma}$. Therefore $\{f^e(\phi_j)\}$ is a Cauchy sequence in L^2 and $f^e: H_0^1 \to L^2$ is compact.

The proofs of the other parts of the lemma are provided in an analogous manner.

Let $V: H_0^1 \times L^2 \to R$ be defined

(8.4)
$$V(\phi, \psi) = \int_\Omega \left[\frac{1}{2} |\nabla \phi(x)|^2 + 2b\phi(x)\psi(x) + \frac{1}{2}\psi^2(x) - F(\phi(x)) + g(x)\phi(x) \right] dx,$$

where $F(u) = \int_0^u f(s)\, ds$ and $b < \min(\lambda_1^{1/2}/4, 2\alpha(4 + \beta^2/\lambda_1))$ and λ_1 is the first eigenvalue of $-\Delta$ on H_0^1.

LEMMA 4.8.5. *Suppose f satisfies (7.4) and (8.1) with $\gamma \geq 0$ and h satisfies (7.17). Then there are constants k_1, k_2 such that*

(8.5) $\quad k_1(|\nabla\phi|_{L^2}^2 + |\psi|_{L^2}^2 - 1) \leq V(\phi, \psi) \leq k_2(|\nabla\phi|_{L^2}^4 + |\psi|_{L^2}^2 + 1).$

Also, there is a constant $k_3 > 0$ such that, for any $r > 0$, there is a $t_0 = t_0(r)$ such that any solution $(u(t), v(t))$ of (8.3) with $|\nabla u(0)|_{L^2}^2 + |v(0)|_{L^2}^2 \leq r$ satisfies

(8.6) $\quad |\nabla u(t)|_{L^2}^2 + |v(t)|_{L^2}^2 \leq k_3 \quad \text{for } t \geq t_0.$

PROOF. For any $\varepsilon > 0$, there is a constant C_ε such that $F(u) \leq \varepsilon u^2 + C_\varepsilon$ since (7.4) is satisfied (see §4.3). Since f satisfies (8.1) with $\gamma \geq 0$, $|F(u)| \leq c(u^4 + 1)$ for some constant c. Thus, with $|\cdot| = |\cdot|_{L^2}$, we have

$$V(\phi, \psi) \geq \frac{1}{2}|\nabla\phi|^2 - 2b|\phi|\cdot|\psi| + \frac{1}{2}|\psi|^2 - \varepsilon|\phi|^2 - C_\varepsilon - |g|\cdot|\phi|,$$

$$V(\phi, \psi) \leq \frac{1}{2}|\nabla\phi|^2 + 2b|\phi|\cdot|\psi| + \frac{1}{2}|\psi|^2 + c|\phi^2|^2 + |g|\cdot|\phi| + c.$$

Using the fact that H_0^1 is continuously embedded in L^4 and that $|\phi|^2 \leq \lambda_1^{-1}|\nabla\phi|^2$ and using an appropriate constant a in the formula

$$|\nabla\phi| \leq \frac{a}{2}|\nabla\phi|^2 + \frac{1}{2a},$$

one obtains (8.5).

To verify (8.6), suppose at first that $(u, u_t) \in (H^2 \cap H_0^1) \times H_0^1$ at $t = 0$. Then the solution remains in this space as long as the solution is defined. Consequently, using arguments similar to the ones in §4.7.4, we obtain a $\theta > 0$ such that

$$\begin{aligned}(d/dt)V(u(t),v(t)) &= -(v,h(v)) + 2b|v|^2 - 2b(u,h(v)) \\ &\quad - 2b|\nabla u|^2 + 2b(u,f(u)) - 2b(u,g) \\ &\leq -(\alpha - 2b)|v|^2 + 2b\beta|u|\cdot|v| - 2b|\nabla u|^2 \\ &\quad + 2b(u,f(u)) - 2b(u,g) \\ &\leq -(\alpha - 2b)|v|^2 + 2b\beta\lambda_1^{-1/2}|\nabla u|\cdot|v| \\ &\quad - 2b|\nabla u|^2 + 2b(u,f(u)) - 2b(u,g) \\ &\leq -\theta(|v|^2 + |\nabla u|^2) + 2b(u,f(u)) + 2b|u|\cdot\|g\|.\end{aligned}$$

The fact that f satisfies (7.4) implies that there are constants $\theta^* > 0$, $N > 0$ such that

$$(d/dt)V(u(t),v(t)) \leq -\theta^*(|v(t)|^2 + |\nabla u(t)|^2) + N.$$

Using the same type of argument as in the last part of the proof of Theorem 4.7.9, one obtains (8.6) for $(u(0), v(0)) \in (H^2 \cap H_0^1) \times H_0^1$. Since this set is dense in $H_0^1 \times L^2$, the same result is true for all $(u(0), v(0))$. This completes the proof of the theorem.

LEMMA 4.8.6. *Suppose f satisfies (7.4) and (8.1) with $\gamma \geq 0$ and h satisfies (7.17). Then equation (8.3) defines a $C^{1,1}$-semigroup $T(t)$ on $H_0^1 \times L^2$ and is bounded dissipative.*

PROOF. Lemma 4.8.4 implies (8.3) defines a local semigroup and Lemma 4.8.5 implies the orbits of bounded sets are bounded and the solutions are globally defined. Inequality (8.6) implies bounded dissipative.

LEMMA 4.8.7. *Suppose f satisfies (7.4) and (8.1) with $\gamma > 0$ and h satisfies (7.19). Then the semigroup $T(t)$ defined by (8.3) is an α-contraction.*

PROOF. Let
$$E(\phi,\psi) = \tfrac{1}{2}|\psi|^2 + 2b(\phi,\psi) + \tfrac{1}{2}|\nabla\phi|^2$$
with b as in (8.4). Exactly as in the proof of (7.22), one shows that the difference $(\varsigma, \eta) = (u_1 - u_2, v_1 - v_2)$ between two solutions of (8.3) satisfies

$$(8.7) \qquad E(\varsigma(t),\eta(t)) \leq e^{-\alpha t}E(\varsigma(0),\eta(0)) + K \sup_{0 \leq s \leq t}|f(u_1(s)) - f(u_2(s))|$$

for some positive constants K, α.

We now estimate $|f(\phi_1) - f(\phi_2)|_{L^2}$ in terms of $|\phi_1 - \phi_2|_{L^2}$. If $\nu = 2 - \gamma$, then $|f'(u)| \leq C(|u|^\nu + 1)$ for $u \in R$ and some constant C. One can now show that

$$\begin{aligned}|f(\phi_1) - f(\phi_2)|_{L^2} &\leq c_1 b(|\phi_1|_{L^6} + |\phi_2|_{L^6})|\phi_1 - \phi_2|_{L^r} \\ &\leq c_2 b(|\nabla\phi_1|_{L^2} + |\nabla\phi_2|_{L^2})|\phi_1 - \phi_2|_{L^r},\end{aligned}$$

where $b(s) = (s+1)^\nu$ and $r = 6/(3-\nu)$. If $a = 1 - \nu/2$, then $a > 0$ and the estimate

$$|\phi_1 - \phi_2|_{L^r} \leq c_3|\nabla\phi_1 - \nabla\phi_2|_{L^2}^{1-a}|\phi_1 - \phi_2|_{L^2}^a$$

implies that

(8.8) $\quad |f(\phi_1) - f(\phi_2)|_{L^2} \le c_4 b(|\nabla\phi_1|_{L^2} + |\nabla\phi_2|_{L^2})|\nabla\phi_1 - \nabla\phi_2|_{L^2}^{1-a}|\phi_1 - \phi_2|_{L^2}^{a}.$

Let us return to the proof that $T(t)$ is an α-contraction. Suppose $(u_1(t), v_1(t))$, $(u_2(t), v_2(t))$ are solutions of (8.3) with initial data (ϕ_1^0, ψ_1^0), (ϕ_2^0, ψ_2^0) bounded in norm by a constant r. Then the solutions are bounded by a constant $c(r)$ by Lemma 4.8.5. Then (8.8) implies there is a constant $c_5 = c_5(r)$ such that

(8.9) $\quad |f(u_1(t)) - f(u_2(t))|_{L^2} \le c_5 |u_1(t) - u_2(t)|_{L^2}^{a}.$

Fix $p > 0$ and define

$$\rho((\phi_1^0, \psi_1^0), (\phi_2^0, \psi_2^0)) = K c_5 \sup_{0 \le t \le p} |u_1(t) - u_2(t)|_{L^2}^a,$$

where K is the constant in (8.7). As in the proof of Theorem 4.7.8, one shows that ρ is a precompact pseudometric on $H_0^1 \times L^2$. From (8.7) we have

$$E(u_1 - u_2, v_1 - v_2) \le e^{-\alpha t} E(\phi_1^0 - \phi_2^0, \psi_1^0 - \psi_2^0) + \rho((\phi_1^0, \psi_1^0), (\phi_2^0, \psi_2^0)).$$

Since E is an equivalent norm in $H_0^1 \times L^2$, we have that $T(p)$ is an α-contraction from Theorem 2.3.6. This completes the proof of the lemma.

LEMMA 4.8.8. *If f satisfies (7.4) and (8.1) for $\gamma > 0$ and h satisfies (7.17), then (8.3) is gradient and every solution of (8.3) approaches the set E of equilibrium points as $t \to \infty$.*

PROOF. Consider the function

$$W(\phi, \psi) = \int_\Omega \left[\frac{1}{2}|\nabla\phi|^2 + \frac{1}{2}\psi^2(x) - F(\phi(x)) + g(x)\phi(x) \right] dx,$$

that is, V in (8.4) with $b = 0$. Then the same proof as in Lemma 4.8.5 gives

$$(d/dt) W(u(t), v(t)) = -(v(t), h(v(t)))$$

along solutions $(u(t), v(t))$ of (8.3). Since each orbit of (8.3) is precompact by Lemma 4.8.7, the ω-limit set exists. Using the properties of h, the result follows as in the proof of Theorem 4.7.1.

Theorem 4.8.1 is now a consequence of Lemmas 4.8.6, 4.8.7, 4.8.8, and Theorem 3.8.5.

4.8.2. *Nonlinear damping, periodic forcing.* Suppose f, h satisfy the conditions of the previous section, $e: R \to L^2$ is continuous and p-periodic, and consider the equation

(8.10) $\quad u_t = v, \qquad v_t = \Delta u + f(u) - h(v) + e(t) \quad \text{in } \Omega$

and $u = 0$ on $\partial\Omega$.

THEOREM 4.8.9. *If f satisfies (7.4) and (8.1) for $\gamma > 0$, h satisfies (7.17), and $T(t, \sigma)$ is the solution operator for (8.10) in $H_0^1 \times L^2$, then $T(\sigma + p, \sigma)$ has* a compact connected attractor A_σ for every $\sigma \in R$, $A_\sigma = T(\sigma, 0)A_0$, and there is a p-periodic solution of (8.10).

Following exactly the same ideas as in the proof of Lemmas 4.8.6 and 4.8.7, one shows that $T(\sigma + p, \sigma)$ is bounded dissipative and an α-contraction. The result then follows from Corollary 3.6.2.

4.8.3. *Linear damping.* In this section, we consider the equation

(8.11) $\qquad u_t = v, \qquad v_t = \Delta u - 2\alpha v + f(u) - g \quad \text{in } \Omega$

and $u = 0$ on $\partial\Omega$, where $\alpha > 0$ is a positive constant and $g \in L^2$.

THEOREM 4.8.10. *Suppose f satisfies (7.4) and (8.1) with $\gamma = 0$. Then (8.11) defines a $C^{1,1}$-semigroup on $X = H_0^1 \times L^2$ and a $C^{0,1}$-semigroup on $Y = (H^2 \cap H_0^1) \times H_0^1$. Furthermore,*
  (i) *$T(t)$ is bounded dissipative in X.*
  (ii) *$T(t)$ is bounded dissipative in Y.*
  (iii) *There is a connected global attractor $\tilde{A} \subset Y$.*
Finally, if E_Y is the set of equilibrium points of (8.11) and each point $(\phi, 0)$ in E_Y is hyperbolic, then

$$\tilde{A} = \bigcup_{(\phi,0) \in E_Y} W^u(\phi, 0).$$

Part (i) of this theorem is due to Babin and Vishik [1983]. Parts (ii) and (iii) are due to Hale and Raugel [1988a]. If $\gamma = 0$, it is not known if there is a compact attractor A in X. In the case $\gamma > 0$, this is a consequence of Theorem 4.8.1. One even can say more.

THEOREM 4.8.11. *If the conditions of Theorem 4.8.10 are satisfied with $\gamma > 0$, then, for (8.11) and $X = H_0^1 \times L^2$, $Y = (H^2 \cap H_0^1) \times H_0^1$,*
  (i) *there is a connected global attractor A in X.*
  (ii) *$A = \tilde{A}$ and is a connected global attractor in Y.*

As remarked earlier, this result has been proved by Hale [1985], Haraux [1985], and Ladyzenskaya [1987], generalizing the work of Babin and Vishik [1983], [1985]. See also Babin and Vishik [1986], Haraux [1988], and Temam [1988].

Let us prove these results. From Lemma 4.8.6, we know that (8.3) defines a $C^{1,1}$-semigroup $T(t)$ on X and $T(t)$ is bounded dissipative in X. This proves (i) of Theorem 4.8.10. Also, the fact that f^e is locally Lipschitzian from Y to H^1 allows one to show that (8.3) defines a local $C^{0,1}$-semigroup on Y. One must use some of the properties of hyperbolic equations that were used in the proof of Theorem 4.7.5. To show that $T(t)$ is a semigroup on Y, we need to know that the solutions are globally defined in Y. We obtain this property as well as much more information from the following result.

LEMMA 4.8.12. *Suppose f satisfies (7.4) and (8.1) for $\gamma \geq 0$. Then there are a positive constant c_0 and, for any $r_0 > 0$, $r_1 > 0$, two positive constants*

$c_1(r_0)$, $c_2(r_1)$ such that, for any solution (u, u_t) of (8.11) with initial data (u_0, u_1) satisfying $\|(u_0, u_1)\|_X \leq r_0$, $\|(u_0, u_1)\|_Y \leq r_1$, we have the following estimate for $t \geq 0$:

(8.12) $$\|u_{tt}(t)\|_{L^2}^2 + \|(u, u_t)\|_Y \leq c_1(r_0) + c_2(r_1)e^{-c_0 t}.$$

PROOF. It is clear that the lemma will be proved if we prove the estimate

(8.13) $$\|u_{tt}(t)\|_0^2 + \|u_t(t)\|_1^2 + \|u(t)\|_2^2 \leq c_1(r_0) + c_2(r_1)e^{-c_0 t}, \qquad t \geq 0,$$

for all solutions $u(t)$ with initial data (u_0, u_1) satisfying

$$\|u_1\|_i^2 + \|u_0\|_{i+1}^2 \leq r_i \quad \text{for } i = 0, 1.$$

In this notation, the subscript $0, 1, 2$, denotes respectively the norms in $L^2(\Omega)$, $H_0^1(\Omega)$, $H^2(\Omega) \cap H_0^1(\Omega)$. To simplify the notation, let us take $\alpha = 1/2$ so that (8.11) becomes

(8.14) $$u_t = v, \qquad v_t = \Delta u - v + f(u) - g \quad \text{in } \Omega.$$

In Lemma 4.8.4, we have already observed that $f: H_0^1 \to L^2$ is a C^1-mapping. Moreover, one can show that for $w \in H^2 \cap H_0^1$, $f'(w)$ is a continuous linear mapping from L^2 into L^2 and that

(8.15) $$\|f'(w)\|_{\mathcal{L}(L^2(\Omega), L^2(\Omega))}^2 \leq c(1 + \|w\|_1^{2-\gamma}\|w\|_2^{2-\gamma})$$

for some constant c. In fact, for any $v \in L^2$, we obtain, by using (8.1),

$$\|f'(w)v\|_0^2 \leq \|f'(w)\|_{L^\infty(\Omega)}^2 \|v\|_0^2 \leq c(1 + \|w\|_{L^\infty(\Omega)}^{2(2-\gamma)})\|v\|_0^2.$$

According to the Gagliardo-Nirenberg inequality, we have

$$\|w\|_{L^\infty(\Omega)} \leq C\|w\|_1^{1/2}\|w\|_2^{1/2}.$$

This inequality together with the previous one proves (8.15).

Assume now that $\|u_1\|_1^2 + \|u_0\|_2^2 \leq r_1$. Then $T(t)(u_0, u_1) = (u(t), u_t(t))$ belongs to $C^0([0, \infty), Y)$. We show that $u_t(t) = w(t)$ is a solution of the following linear hyperbolic equation:

(8.16) $$\begin{cases} w_{tt} + w_t - \Delta w = f'(u)u_t & \text{in } \Omega, \\ w = 0 & \text{in } \partial\Omega, \\ w(0) = u_1, \qquad w_x(0) = f(u_0) - g - u_1 - \Delta u_0. \end{cases}$$

Since $f'(u)u_t$ belongs to $C^0([0, \infty), L^2(\Omega))$ and $(w(0), w_x(0))$ belongs to X, there is a unique solution $w(t)$ of (8.16) and (w, w_t) belongs to $C([0, \infty), X)$ (see, for example, Lions and Magenes [1968, Ch. 3, §8.4] or Ladyzenskaya [1985]). One easily checks that $w = u_t$, proving the assertion above.

Our next objective is to obtain estimates on the solution w of (8.16). To do this, we again use a special Lyapunov function. Fix a constant $b > 0$ and consider the functional

$$V(\phi, \psi) = \int_\Omega \left[\frac{1}{2}|\psi(x)|^2 + 2b\phi(x)\psi(x) + \frac{1}{2}|\phi_x(x)|^2\right] dx$$

and b is small enough so that

(8.17)
$$(1 - 2b)\|\psi\|_0^2 + 2b(\phi, \psi) + 2b\|\phi\|_1^2 \geq \min(1/2, b)(\|\psi\|_0^2 + \|\phi\|_1^2),$$
$$\frac{1}{4}(\|\psi\|_0^2 + \|\phi\|_1^2) \leq V(\phi, \psi) \leq \frac{3}{4}(\|\psi\|_0^2 + \|\phi\|_1^2)$$

where (ϕ, ψ) is the L^2 inner product. If λ_1 is the first eigenvalue of $-\Delta$ with Dirichlet conditions, then b less than $\min\{1/8, \lambda_1/4, \lambda_1^{1/2}/4\}$ is sufficient.

Using estimates similar to the ones in the proof of Theorem 4.8.1, one obtains, for the solution w of (8.16), that there is a constant $k_0 > 0$ such that

(8.18)
$$\frac{d}{dt}V(w, w_t) = -(1 - 2b)\|w_t\|_0^2 - 2b\|w\|_1^2 - 2b(w_t, w)$$
$$- (f'(u)w, w_t) - 2b(f'(u)w, w)$$
$$\leq -\frac{k_0}{2}(\|w_t\|_0^2 + \|w\|_1^2) + \left(\frac{1}{2k_0} + \frac{2b^2}{k_0\lambda_1}\right)\|f'(u)w\|_0^2$$

for $t \in [0, \infty)$.

Now we must estimate $\|f'(u)w\|_0^2$. From (8.15) and the fact that $\gamma \geq 0$, we obtain

(8.19)
$$\|f'(u)w\|_0^2 \leq c\|w\|_0^2(1 + \|u\|_1^2\|u\|_2^2).$$

Since

(8.20)
$$-\Delta u = f(u) - g - u_t - u_{tt}$$

and $w = u_t$, we have

(8.21)
$$\|u\|_2^2 \leq c(\|f(u)\|_0^2 + \|g\|_0^2 + \|w\|_0^2 + \|w_t\|_0^2).$$

Moreover, by the inequality (8.1) and the fact that $H_0^1(\Omega)$ is continuously embedded into $L^6(\Omega)$, we have

(8.22)
$$\|f(u)\|_0^2 \leq c(1 + \|u\|_1^6).$$

These three latter inequalities imply that, for $0 \leq t < \infty$,

(8.23)
$$\|f'(u)w\|_0^2 \leq c\|w\|_0^2[c_3(r_0) + c_4(r_0)(\|w_t\|_0^2 + \|w\|_1^2)],$$

where $c_3(r_0)$, $c_4(r_0)$ are constants depending only on r_0.

Using (8.17), (8.18), (8.23), we have, for $0 \leq t < \infty$, that there are a constant $k_2 > 0$ and constants $c_5(r_0)$, $c_6(r_0)$ so that

(8.24)
$$\frac{d}{dt}V(w, w_t) \leq -k_2 V(w, w_t) + c_5(r_0)\|w\|_0^2 + c_6(r_0)\|w\|_0^2 V(w, w_t).$$

To obtain an estimate for $V(w, w_t)$, we show that

(8.25)
$$\int_0^\infty \|w\|_0^2 \, dt = \int_0^\infty \|u_t\|_0^2 \, dt \leq c_7(r_0)$$

for some constant $c_7(r_0)$. In fact, using the Lyapunov function $\tilde{V}(\phi, \psi)$ in (8.4) for $b = 0$, one easily sees that, for the solution u of (8.14),

$$\tilde{V}(u, u_t) = \tilde{V}(u_0, u_1) - \int_0^t \|u_t\|_0^2 \, dt$$

which implies (8.25).

Using (8.24), (8.25), a Gronwall inequality, and (8.17), one shows there are constants $c_8(r_0)$, $c_9(r_0)$ such that

(8.26) $\quad \|w_t\|_0^2 + \|w\|_1^2 \leq c_8(r_0)(\|w_t(0)\|_0^2 + \|w(0)\|_1^2)e^{-k_2 t} + c_9(r_0).$

Using the fact that $w = u_t$ and relations (8.20), (8.21), (8.22), one obtains the estimate (8.12) for $t \in [0, \infty)$. This completes the proof of Lemma 8.12.

Now suppose that $(u_0, u_1) \in Y$. We have already observed that the solution $T(t)(u_0, u_1) = (u, u_t)$ is defined on an interval $[0, \tau)$ for some $\tau > 0$. If $\|(u_0, u_1)\|_Y \leq r_1$, then there is a constant r_0 such that $\|(u_0, u_1)\|_X \leq r_0$. The estimate (8.12) implies that (u, u_t) is defined on $[0, \infty)$ and there are a constant $C(r_1)$ and a constant $t_0(r_1)$ such that $\|(u, u_t)\|_Y \leq C(r_1)$ for $t \geq t_0(r_1)$; that is, $T(t)$ is bounded dissipative in Y. This proves part (ii) of Theorem 4.8.12.

To prove that there is a global attractor for $T(t)$ in Y, it is sufficient from Theorem 3.4.7 to show that $T(t)$ is an α-contraction on Y. The equation (8.11) can be written in X or Y as an abstract evolutionary equation

(8.27) $\quad dw/dt = Cw + \tilde{f}^e(w) - \tilde{g}$

where

(8.28) $\quad w = \begin{bmatrix} u \\ v \end{bmatrix}, \quad C = \begin{bmatrix} 0 & I \\ \Delta & -2\alpha I \end{bmatrix}, \quad \tilde{f}^e(w) = \begin{bmatrix} 0 \\ f^e(u) \end{bmatrix}, \quad \tilde{g} = \begin{bmatrix} 0 \\ g \end{bmatrix}$

and $f^e(\phi(x)) = f(\phi(x))$, $x \in \Omega$. The variation of constants formula for (8.27) is

(8.29) $\quad w(t) = e^{Ct} w(0) + \int_0^t e^{C(t-s)} [\tilde{f}^e(w(s)) - g] \, ds, \quad t \geq 0.$

If we let $w(0) = (\phi, \psi)$, and define

(8.30) $\quad U(t)(\phi, \psi) = \int_0^t e^{C(t-s)} [\tilde{f}^e(w(s)) - g] \, ds, \quad t \geq 0,$

then

(8.31) $\quad T(t) = e^{Ct} + U(t).$

From §4.6, there are positive constants $k > 0$, $\alpha > 0$, such that

(8.32) $\quad |e^{Ct}|_X, |e^{Ct}|_Y \leq k e^{-\alpha t}, \quad t \geq 0.$

Let us now prove that $U(t)$ is completely continuous in Y. Let us remember that $f(0) = 0$. If B is a bounded set in Y, then $\{T(t)B, 0 \leq t < \infty\}$ is bounded in Y from part (ii) of Theorem 4.8.10. Since f is a C^2-function satisfying (8.1) and $H^2(\Omega)$ is continuously embedded in $L^\infty(\Omega)$, it follows that $\{f(T(s)(u_0, u_1)): 0 \leq s \leq t, (u_0, u_1) \in B\}$ is a bounded set in $H^2(\Omega) \cap H_0^1(\Omega)$ and therefore a compact set of $H_0^1(\Omega)$. From Theorem 4.6.1, it follows that $U(t)$ is completely continuous.

Therefore, relations (8.31), (8.32) imply that $T(t)$ is an α-contraction. Now suppose that $(\phi, \psi) \in Y$. Then $\gamma_Y^+(\phi, \psi)$ is precompact in Y and in particular is precompact in X. Therefore, (ϕ, ψ) has an ω-limit set $\omega_X(\phi, \psi)$ in X. The proof of Lemma 4.8.8 implies that $\omega_X(\phi, \psi) \subset E_X$ = set of equilibrium points in X. This implies that $E_Y \subset E_X$ and $\omega_Y(\phi, \psi) \subset E_Y$. The same argument applies to show that the α-limit set of a bounded negative orbit $\gamma_Y^-(\phi, \psi)$ is in E_X and therefore in E_Y. This is enough to conclude the representation of $\tilde{A}$ as stated in the theorem. This completes the proof of Theorem 4.8.10.

To prove Theorem 4.8.11, we observe first that part (i) is a special case of Theorem 4.8.1. For the proof of the second part, we need the following lemma of Haraux [1985]. A special case was proved by Babin and Vishik [1983].

LEMMA 4.8.13. *Let $A \subset X$ be the global attractor in* (i) *of Theorem 4.8.1. Then A belongs to a bounded set of Y.*

PROOF. Only a sketch of the proof will be given. For any $q \in L^\infty(R, L^2(\Omega))$, let

$$(8.33) \qquad (\mathcal{T}q)(t) = \int_{-\infty}^0 e^{-Cs} \begin{bmatrix} 0 \\ q(t+s) \end{bmatrix} ds,$$

where C is defined in (8.28). If $w(t) = (u(t), u_t(t))$ is a bounded solution of (8.11) for $t \in R$, then

$$(8.34) \qquad w = \mathcal{T}(f^e(u(\cdot)) - g)$$

where $f^e(\phi)(x) = f(\phi(x))$. The derivative $W(t) = w_t(t) = (u_t(t), u_{tt}(t))$ will be a solution of the equation

$$(8.35) \qquad W = \mathcal{T}(f^{e\prime}(u(\cdot))).$$

On the other hand,

$$(8.36) \qquad -\Delta u = -g + f(u) - 2\alpha u_t - u_{tt}.$$

For any $\phi \in H^{-1}(\Omega)$ let $K\phi \in \mathcal{L}(H^{-1}(\Omega), H_0^1(\Omega))$ satisfy $-\Delta K\phi = \phi$. From (8.35),

$$(8.37) \qquad u(t) = -Kg + Kf^e(u(t)) - 2\alpha K u_t - K u_{tt}.$$

Suppose now that we know a priori that $u(t, \cdot) \in L^\infty(R, C(\overline{\Omega}))$. Then, $f^e(u) \in L^\infty(R, H_0^1(\Omega))$. From (8.34), we obtain $w \in C_b(R, Y)$, the continuous bounded functions from R to Y. This implies the conclusion in the lemma.

Thus, the problem is to obtain the a priori bound $u \in L^\infty(R, C(\overline{\Omega}))$. To obtain this bound, let Λ be the selfadjoint operator defined by

$$D(\Lambda) = H^2(\Omega) \cap H_0^1(\Omega), \qquad \Lambda u = -\Delta u \quad \text{for } u \in D(\Lambda).$$

For each $\sigma \geq 0$, let D_σ be the domain of the fractional power Λ^σ. As a consequence of interpolation theory (see Lions and Magenes [1968]), one obtains

LEMMA 4.8.14.
(a) *For any* $p \in [0,1]$, $p \neq 1/2$, $K(H^{-1+p}) \subset D_{(1+p)/2}$;
(b) *for any* $\sigma \geq 0$, $D_\sigma \subset H^{2\sigma}(\Omega)$;
(c) *for any* $\sigma \geq 0$, *if* $q \in L^\infty(R, D_\sigma)$, *then* $\mathcal{T}q \in L^\infty(R, D_{1+\sigma} \times D_\sigma)$.

The idea for proving that $u(t,\cdot) \in L^\infty(R, C(\overline{\Omega}))$ is to use a "bootstrap" argument with (8.35), (8.37), Lemma 4.8.14, and the fact that for any $q \in [1,2]$, we have

$$(8.38) \qquad L^q(\Omega) \hookrightarrow H^{-s}(\Omega) \quad \text{if } s = 3\left(\frac{1}{q} - \frac{1}{2}\right).$$

Since $w \in L^\infty(R, X)$, it follows that $u \in L^k(\Omega)$ for $k \geq 6$. Fix $k \geq 6$ and let $r = 2 - \gamma$. Let $h(t) = f^{e'}(u(t))u_t(t)$. Since $u \in L^k$, $w \in L^\infty(R, X)$, we have $h \in L^\infty(R, L^q(\Omega))$ for $1/q = r/k + 1/2$. Thus, (8.38) implies that $h \in L^\infty(R, H^{-s}(\Omega))$ with $s = 3r/k$. If $s \neq 1/2$, then Lemma 4.8.14 implies that $u \in L^\infty(R, H^{2-s}(\Omega))$.

If $r < 1$, take $k = 6$ and then $s < 1/2$. Then $H^{2-s}(\Omega) \hookrightarrow C(\overline{\Omega})$ and the a priori bound $u \in L^\infty(R, C(\overline{\Omega}))$ is obtained in one step.

If $r = 1$, then $u \in L^\infty(R, H^{3/2-\alpha}(\Omega))$ for any $\alpha > 0$. Therefore, $h \in L^\infty(R, L^{2-\delta}(\Omega))$ for any $\delta \in (0,1)$. For δ small enough, we conclude that $h \in L^\infty(R, H^{-p}(\Omega))$ with $p \in (0, 1/2)$. Then Lemma 4.8.14 implies that $Kh \in L^\infty(R, D(\Lambda^{1-p/2}))$ and thus $(Ku_t, Ku_{tt}) = (\mathcal{T}Kh) \in L^\infty(R, D_{1+\sigma} \times D_2)$ with $\sigma = 1 - p/2 > 3/4$. This implies that $Ku_{tt} \in L^\infty(R, C(\overline{\Omega}))$. Now one applies (8.37) to obtain $u \in L^\infty(R, C(\overline{\Omega}))$.

If $r > 1$, define a sequence inductively by the formula

$$(8.39) \qquad \frac{1}{k_{n+1}} = \frac{1}{2} - \frac{1}{3}\left(2 - \frac{3r}{k_n}\right) = -\frac{1}{6} + \frac{r}{k_n}$$

with $k_1 = 6$. We stop the sequence at the last index n_0 such that $k_n \leq 6r$. Such an n_0 always exists.

If $k_{n_0} < 6r$, then k_{n_0+1} defined by (8.39) is finite, positive, and $k_{n_0+1} > 6r$. Thus, after $n_0 + 1$ steps as above, one can use Lemma 4.8.14 to see that $u \in H^{3/2+\eta}(\Omega)$ with $\eta > 0$. Since $H^{3/2+\eta}(\Omega) \hookrightarrow C(\overline{\Omega})$, we obtain the desired a priori bound.

If $K_{n_0} = 6r$, one arrives at the same conclusion using an argument as in the case $r = 1$. This proves the theorem.

Using Lemma 4.8.13 and Theorem 4.8.10, one completes the proof of Theorem 4.8.11 since A being invariant in Y implies $A \subset \tilde{A}$. Obviously, $\tilde{A} \subset A$.

REMARK 4.8.15. It is certainly an interesting problem to try to understand whether or not there is a compact attractor A in $X = H_0^1 \times L^2$ if $\gamma = 0$. The solution operator is bounded dissipative in X and there is a compact attractor in Y. For (8.11), Ball [1973a] (and for (8.3) Ceron [1984]) has shown that the ω-limit set of any point in X exists and belongs to the set E of equilibrium points. Since E is bounded, this implies point dissipativeness in X. Thus, if there are situations where there does not exist a compact attractor in X, it would be very surprising and extremely interesting.

REMARK 4.8.16. Since e^{Ct} is a group on X, it follows that $T(t)$ is also a group on the attractor A in Theorem 4.8.11. The results in Hale and Scheurle [1985] in §3.10 imply that the flow restricted to the local unstable sets $W_{\text{loc}}^u(\phi,\psi)$ is as smooth in t as the function $\tilde{f}^e$, even up to analyticity. Since $T(t)$ is a group on A_f and these sets are finite dimensional, it follows that $T(t)|A_f$ is as smooth in t as $\tilde{f}^e$. See also Ghidaglia and Temam [1987].

4.8.4. *Linear damping, periodic forcing.* Suppose $e\colon R \to L^2(\Omega)$ is a p-periodic C^2-function and consider the equation

(8.40) $\qquad u_t = v, \qquad v_t = \Delta u - 2\alpha v + f(u) + e(t) \quad \text{in } \Omega$

with $u = 0$ on $\partial\Omega$ and $\alpha > 0$ a positive constant.

THEOREM 4.8.17. *Suppose f satisfies (7.4) and (8.1) with $\gamma > 0$ and let $T(t,\sigma)$ be the solution operator of (8.40). Then $T(t,\sigma)$ is defined on $X = H_0^1 \times L^2$ and $Y = (H^2 \cap H_0^1) \times H_0^1$ and*

(i) *$T(\sigma+p,\sigma)$ has a connected global attractor A_σ in X for every $\sigma \in R$, $A_\sigma = T(\sigma,0)A_0$.*

(ii) *$A_\sigma \in Y$ and A_σ is a global attractor for $T(\sigma+p,\sigma)$ in Y.*

Finally, there is a p-periodic solution of (8.40) in Y.

Following the same ideas as in §4.8.2, one shows that $T(\sigma+p,\sigma)$ is bounded dissipative and is given by

$$T(\sigma+p,\sigma) = e^{Cp} + U(p,\sigma)$$

where

$$U(t,\sigma)(\phi,\psi) = \int_\sigma^{t+\sigma} e^{C(t-s)}[\tilde{f}^e(u(s,\sigma,\phi,\psi)) + \tilde{e}(s)]\,ds,$$

where $u(s,\sigma,\phi,\psi), v(s,\sigma,\phi,\psi)$ is the solution of (8.40) with initial data ϕ,ψ at $s = \sigma$. One now uses the same type of argument as in the previous section to show that $U(t,\sigma)$ is completely continuous. One completes the proof as in the previous section.

4.9. Remarks on other applications.

4.9.1. *Retarded equations with infinite delays.* In this section, we consider equations of the type considered in §4.1, but with the delay being infinite rather than finite. The choice of the space X of initial functions plays an important role in whether or not the solution operator is an α-contraction. The spaces are of the fading memory type considered in Hale and Kato [1978] and, in some respects, are more restrictive than the ones in Coleman and Mizel [1968] and Schumacher [1978].

We do not describe the most general situation for X, but simply remark that X includes the space

$$C_\gamma = \{\phi\colon (-\infty,0] \to \mathbf{R}^n, \text{ continuous, } e^{\gamma\theta}\phi(\theta) \to \text{ a limit as } \theta \to \infty\},$$
$$|\phi|_{C_\gamma} = \sup_{\theta\in(-\infty,0]} e^{\gamma\theta}|\phi(\theta)|$$

with $\gamma > 0$. It also includes the space B isomorphic to $L_p(\mu) \times \mathbf{R}^n$ given by
$$B = \{\phi \colon (-\infty, 0] \to \mathbf{R}^n, \text{ measurable, } |\phi|_B < \infty\},$$
$$|\phi|_B = \{|\phi(0)|^p + \int_{-\infty}^0 g(\theta)|\phi(\theta)|^p \, d\theta\}^{1/p}$$
where $g \colon (-\infty, 0] \to [0, \infty)$ is continuous and satisfies $g(t + s) \leq g(t)g(s)$.

If $f \colon X \to \mathbf{R}^n$ is a C^r-function, $r \geq 1$, then one can consider the equation

(9.1) $$\dot{x} = f(x_t)$$

and define $T(t)\phi = x_t(\cdot, \phi)$, where $x(t, \phi)$ is the solution of (9.1) through ϕ at time 0.

Suppose $\phi \in C_\gamma$. The solution $T(t)\phi$ of (9.1) through ϕ can be written as
$$T(t) = S(t)\phi + U(t)\phi,$$
$$[S(t)\phi](\theta) = \begin{cases} \phi(t + \theta) - \phi(0), & t + \theta < 0, \\ 0, & t + \theta \geq 0, \end{cases}$$
$$[U(t)\phi](\theta) = \begin{cases} \phi(0), & t + \theta < 0, \\ \phi(0) + \int_0^t f(T(s)\phi) \, ds, & t + \theta \geq 0. \end{cases}$$

If $\phi \in C_\gamma$, $t > 0$, then
$$|S(t)\phi|_{C_\gamma} = \sup_{\theta \in (-\infty, 0]} |[S(t)\phi](\theta)|e^{\gamma\theta} = \sup_{\theta \in (-\infty, -t]} |[S(t)\phi](\theta)|e^{\gamma\theta}$$
$$= e^{-\gamma t} \sup_{\theta \in (-\infty, 0]} |[S(t)\phi](\theta - t)|e^{\gamma\theta} = e^{-\gamma t} \sup_{\theta \in (-\infty, 0]} |\phi(\theta) - \phi(0)|e^{\gamma\theta}$$
$$= e^{-\gamma t}|\phi - \phi(0)|_{C_\gamma}.$$

Also, one easily shows that $U(t) \colon C_\gamma \to C_\gamma$ is conditionally completely continuous. This implies that $T(t)$ is a conditional α-contraction. It is an α-contraction if T is a bounded map.

Using the same type of argument, one shows that T satisfies the same property in the space B. More general spaces have been considered in Hale and Kato [1978].

The preceding theory applies to this case. In particular, if $T(t)$ is point dissipative and orbits of bounded sets are bounded, then Theorem 3.4.6 implies there is a global attractor for $T(t)$.

One also can consider the periodic system

(9.2) $$\dot{x}(t) = f(t, x_t)$$

with $f \colon R \times B \to R^n$, $f(t, \phi) = f(t + \omega, \phi)$. Using ideas similar to the ones in §4.5.3 of exploiting the properties of solutions in two spaces, Massatt [1981] has shown that system (9.2) being point dissipative in B is sufficient to imply that (9.2) has an ω-periodic solution.

For a discussion of the relationship between various types of stability in fading memory spaces, see Murakami and Naito [1987].

4.9.2. Strongly damped quasilinear evolution equations. We begin this section with abstract results due to Fitzgibbon [1981] and Massatt [1983], which in turn were motivated by Webb [1980]. Suppose X is a Banach space, $\alpha > 0$ is a constant, A is a sectorial operator on X, and f is a nonlinear operator defined on a subset of $X \times X$ satisfying certain regularity and growth properties which will be specified later. We consider the abstract equation

$$(9.3) \qquad u_{tt} + \alpha A u_t + A u = f(u, u_t),$$

where $\alpha > 0$ is a real constant.

A common example where this type of equation arises is in the modelling of longitudinal vibrations in a homogeneous bar in which there are viscous effects. The equation considered is

$$(9.4) \qquad u_{tt} - \alpha \Delta u_t - \Delta u = f(u, \nabla u, u_t, \nabla u_t)$$

on a smooth bounded domain with, say, Dirichlet boundary conditions. The term $a\Delta u_t$ indicates that the stress is proportional not only to the strain, as with Hooke's law, but also to the strain rate as in a linearized Kelvin material.

Another special case of (9.3) is

$$(9.5) \qquad u_{tt} + au_{xxxx} - \left(\beta + K \int_0^1 \left(\frac{\partial u(\xi,t)}{\partial \xi}\right)^2 d\xi\right) u_{xx}$$

$$+ \alpha u_{xxxxt} - \sigma \left[\int_0^1 \left(\frac{\partial u}{\partial \xi}\right)\left(\frac{\partial^2 u}{\partial \xi \partial t}\right) d\xi\right] u_{xx} + \delta u_t = 0$$

on $0 \leq x \leq 1$, subject to the boundary conditions

$$u(0,t) = u(1,t) = u_{xx}(0,t) = u_{xx}(1,t) = 0.$$

This equation arises as a model for the transverse motion of an extensible beam whose ends are held a fixed distance apart. The boundary conditions correspond to the case where the ends of the beam are hinged.

If $\alpha > 0$ and all other constants are nonnegative, the existence of a global attractor can be proved. This will also be true even if δ, β are negative if they are not too much so with respect to α.

The equation

$$u_{tt} - \alpha u_{xxt} - u_{xx} + \beta u_t = -\sin u + \gamma, \qquad 0 < x < l,$$

where α, β, γ are positive constants occurs in the theory of a Josephson transmission line (see, for example, Scott, Chu, and Reible [1976] and Lomdahl, Soerensen and Christiansen [1982]). With Neumann boundary conditions at $x = 0$ and $x = l$, this equation is a special case of (9.3).

It is usually more convenient to consider (9.3) as a system of first order ordinary differential equations in a Banach space. If $v = u_t$, then (9.3) is equivalent to

$$\begin{pmatrix} u \\ v \end{pmatrix}_t = \begin{pmatrix} 0 & I \\ -A & -\alpha A \end{pmatrix} \begin{pmatrix} u \\ v \end{pmatrix} + \begin{pmatrix} 0 \\ f(u,v) \end{pmatrix}$$

or

(9.6)
$$z_t = -Bz + F(t,z),$$
$$B = \begin{pmatrix} 0 & -I \\ A & \alpha A \end{pmatrix}, \quad z = \begin{pmatrix} u \\ v \end{pmatrix}, \quad F(z) = \begin{pmatrix} 0 \\ f(u,v) \end{pmatrix}.$$

Let X^β be the fractional power spaces of X associated with the sectorial operator A. One can prove (see, for example, Fitzgibbon [1981] or Massatt [1983])

LEMMA 4.9.1. *The operator B in (9.6) generates an analytic semigroup e^{-Bt} on $X^\beta \times X^\sigma$ for any $0 \leq \sigma \leq \beta < 1$. Furthermore, if $\operatorname{Re}\sigma(A) \geq 0$, then there is a $\delta > 0$ and $C > 0$ such that*

(9.7)
$$|e^{-Bt}| \leq Ce^{-\delta t}, \quad t \geq 0.$$

If we now assume that $f(u,v)$ is locally Lipschitzian as a map from $X^\beta \times X^\sigma \to X$, then one proves in the standard way that (9.3) defines a local $C^{0,1}$-semigroup. If the solutions are globally defined, then (9.3) defines a $C^{0,1}$-semigroup $T(t)$ on $X^\beta \times X^\sigma$. Furthermore, if the resolvent of A is compact, then e^{-Bt} is compact for $t > 0$. As a consequence of the variation of constants formula and Theorem 4.6.4, we can state

THEOREM 4.9.2. *If A has compact resolvent, then the semigroup $T(t)$ of (9.3) on $X^\beta \times X^\sigma$ satisfies*
$$T(t) = e^{-Bt} + U(t)$$
where $U(t)$ is conditionally compact and e^{-Bt} satisfies (9.7). Thus, in an appropriate norm, $T(t)$ is an α-contraction on $X^\beta \times X^\sigma$.

If the conditions of Theorem 4.9.2 are satisfied, $T(t)$ is point dissipative and is a bounded map, then Theorem 3.4.6 implies there is a global attractor A for (9.6).

For specific equations, one can exploit the fact that (9.3) defines a semigroup in more than one space and use the results on dissipativeness in two spaces in §3.9 to obtain regularity properties of the attractor A and also show that A is an attractor in more than one space. We remark that the theory is directly applicable to the equation (9.4) considered by Webb [1980] where the function f depends only on $u \in R$, is a C^1-function with $f'(u) \leq c_0$ for some constant $c_0 \geq 0$, $f(0) = 0$, and $\overline{\lim}_{|u| \to \infty} f(u)/u \leq 0$. Webb [1980] proves that this system is a gradient system. The additional properties of the attractor mentioned above follow from Massatt [1983]. This paper of Massatt also contains a discussion of the existence of ω-periodic solutions of (9.3) when $f = f(t, u, u_t)$ is ω-periodic in t.

4.9.3. *A beam equation.* In the previous section, we considered a general equation for which a special case was the beam equation (9.5) with dissipation—the dissipation coming from the viscous term αu_{xxxt}. This was an especially simple case since the linear part of the equation generated a compact, analytic semigroup. If there is no viscous term, the situation is completely different. In this section, we consider no viscous term and obtain dissipation by introducing a damping term. To simplify notation, we take $\sigma = 0$ in (9.5); that is, we consider the damped equation

$$(9.8) \qquad u_{tt} + h(u_t) + au_{xxxx} = \left(\beta + K \int_0^1 u_x^2(\xi, t)\, d\xi\right) u_{xx}, \qquad 0 < x < 1,$$

where $0 < \gamma \le h'(\sigma) \le \delta$ for all $\sigma \in R$, where γ, δ are constants. Also, suppose the ends of the beam are clamped

$$(9.9) \qquad u(0,t) = u(1,t) = u_x(0,t) = u_x(1,t) = 0.$$

The constants a, K are positive and $\beta \in R$.

The initial value problem for (9.8) will be specified in $X = (H^2 \cap H_0^1) \times L^2$.

THEOREM 4.9.3. *There is a connected global attractor A for (9.8), (9.9) in X. Furthermore, (9.8), (9.9) is a gradient system.*

Let us give an outline of the proof. It is natural to choose the Lyapunov function as

$$V(u, u_t) = \frac{1}{2}|u_t|^2 + \frac{a}{2}|u_{xx}|^2 + \frac{\beta}{2}|u_x|^2 + \frac{K}{4}|u_x|^2,$$

where $|\cdot|$ is the L^2-norm. For smooth initial data,

$$(d/dt)V(u, u_t) = -\int_0^1 u_t h(u_t)\, dt \le 0.$$

This implies that orbits of bounded sets are bounded. If an orbit is precompact, then it shows also that the ω-limit is in the set of equilibrium points. This set is bounded and therefore the semigroup generated by the equation would be point dissipative.

To complete the proof, it is sufficient to show that the semigroup is an α-contraction. At first glance, this is not to be expected since the nonlinear terms in the equation considered as a mapping from $H^2 \cap H_0^1 \to L^2$ are not compact. However, due to the special form of the nonlinearity, this difficulty can be overcome as was pointed out by Lopes and Ceron [1984]. The ideas follow the ones in §4.7.3, but much more care is needed in choosing the appropriate modification of the Lyapunov function. Rather than do the general case of nonlinear damping $h(v)$, let us assume $h(v) = 2\alpha v$ where $\alpha > 0$ is a constant.

If $h(v) = 2\alpha v$ and $(u_1(t), v_1(t))$, $(u_2(t), v_2(t))$ are two solutions of (9.8) and (9.9) and $z(t) = u_1(t) - u_2(t)$, then z satisfies the equation

$$z_{tt} + 2\alpha z_t + az_{xxxx} = \left(\beta + K\int_0^1 u_{1x}^2(\xi, t)\, d\xi\right) u_{1xx}$$
$$\qquad - \left(\beta + K\int_0^1 u_{2x}^2(\xi, t)\, d\xi\right) u_{2xx}.$$

Therefore, integrating by parts, we obtain

$$(9.10) \quad \begin{aligned} z_{tt} + 2\alpha z_t + az_{xxxx} &- \left(\beta + K\int_0^1 u_{1x}^2(\xi, t)\, d\xi\right) z_{xx} \\ &= -K\left[\int_0^1 (u_1 - u_2)(u_{1xx} + u_{2xx})\, d\xi\right] u_{2xx}. \end{aligned}$$

Since u_{1xx}, u_{2xx} are bounded in L^2, this implies that

$$K\left|\left[\int_0^1 (u_1 - u_2)(u_{1xx} + u_{2xx})\, dx\right] u_{2xx}\right|_{L^2} \leq k|u_1 - u_2|_{L^2}$$

for some constant k.

We can rewrite (9.10) as a system

$$z_t = w,$$

(9.11)
$$w_t = -az_{xxxx} - 2\alpha w + \left(\beta + K\int_0^1 u_{1x}^2\right) z_{xx}$$
$$- K\left[\int_0^1 (u_1 - u_2)(u_{1xx} + u_{2xx})\right] u_{2xx}.$$

Let

$$E(z, w) = \int_0^1 \left(\frac{1}{2}w^2 + 2bzw + \frac{a}{2}z_{xx}^2\right) dx,$$

$$V(z, w) = E(z, w) - \int_0^1 \frac{1}{2}\left(\beta + K\int_0^1 u_{1x}^2\right) z_x^2\, dx.$$

By following the proof of Theorem 4.8.1, one can choose $b > 0$ so that $[E(z, w)]^{1/2}$ is an equivalent norm in $(H^2 \cap H_0^1) \times L^2$ and also so that there is a $\theta > 0$ such that

$$\frac{dV(z, w)}{dt} \leq -\theta\left[V(z, w) + \int_0^1 \frac{1}{2}\left(\beta + K\int_0^1 u_{1x}^2\right) z_x^2\, dx\right] + h(t).$$

The function $h(t)$ has an upper bound

$$|h(t)| \leq k_1|u_1(t) - u_2(t)|_{L^2}$$

since (z, w) is bounded in $(H^2 \cap H_0^1) \times L^2$. Also, the function

$$\int_0^1 \frac{1}{2}\left(\beta + K\int_0^1 u_{1x}^2\right) z_x^2\, dx = \int_0^1 \frac{1}{2}\left(\beta + K\int_0^1 u_{1x}^2\right) zz_{xx}\, dx$$
$$\leq k_2|z|_{L^2} = k_2|u_1(t) - u_2(t)|_{L^2}.$$

Integrating the above differential inequality, we obtain

$$V(z(t), w(t)) \leq e^{-\theta t}V(z(0), w(0)) + k_3 \sup_{0\leq s\leq t} |u_1(s) - u_2(s)|_{L^2}$$

and

$$E(z(t), w(t)) \leq e^{-\theta t}E(z(0), w(0)) + k_4 \sup_{0\leq s\leq t} |u_1(s) - u_2(s)|_{L^2}.$$

Since $[E(z, w)]^{1/2}$ is an equivalent norm in $(H^2 \cap H_0^1) \times L^2$ it follows that there are positive constants k_1, δ_1 such that

$$|(z, z_t)|_{(H^2 \cap H_1^0) \times L^2} \leq k_1 e^{-\delta_1 t}|(z_0, z_{0t})|_{(H^2 \cap H_0^1) \times L^2}$$
$$+ k_1 \sup_{0\leq s\leq t} |u_1(s) - u_2(s)|_{L^2}^{1/2}.$$

The function

$$\rho((u_1, v_1), (u_2, v_2)) = \sup_{0\leq s\leq t} |u_1(s) - u_2(s)|_{L^2}^{1/2}$$

is a compact pseudometric for each t. This shows that the semigroup generated by (9.8) and (9.9) is an α-contraction (see Lemma 2.3.6).

Lopes and Ceron [1984] have also considered the case of several space variables.

Ball [1973a,b,c] has considered the case where $h(v) = \alpha v$ in (9.8) and proved that every solution approaches an equilibrium point. The concept of a weakly invariant set was employed.

Since the attractor A for (9.8), (9.9) is connected, it follows that each stable equilibrium point is connected by an orbit in X to some other equilibrium point. Artstein and Slemrod [1982] have proved this latter result in the weak topology.

Let us now consider the forced beam equation

$$(9.12) \qquad u_{tt} + h(u_t) + au_{xxxx} = \left(\beta + K \int_0^1 u_x^2(\xi,t)\,d\xi\right) u_{xx} + e(t)$$

where $e(t)$ is a C^1-function which is p-periodic. The results of Lopes and Ceron [1984] together with Theorem 2.6.4 imply also the following result.

THEOREM 4.9.4. *For the p-periodically forced beam equation (9.12) with boundary conditions (9.9), the solution operator $T(t,\sigma)$ in $(H^2 \cap H_0^1) \times L^2$ is such that $T(\sigma + p, \sigma)$ has a connected global attractor A_σ for every $\sigma \in R$, $A_\sigma = T(\sigma,0)A_0$, and there is a p-periodic solution of the equatiion (9.12).*

4.9.4. Other hyperbolic systems. In the theory of transmission lines, the voltage $v(t,x)$ and current $i(t,x)$ can satisfy the differential equations

$$L\frac{\partial i}{\partial t} = -\frac{\partial v}{\partial x} - Ri,$$

$$C\frac{\partial v}{\partial t} = -\frac{\partial i}{\partial x} - Gv$$

for $0 \leq x \leq l$, $t \geq 0$, with boundary conditions

$$v(t,0) + R_0 i(t,0) = E(t),$$

$$i(t,l) = C_1 \frac{\partial}{\partial t} v(t,l) + g(v(l,t)).$$

The function $E(t)$ is periodic and C^1, the function g is nonlinear and L, C, R_0, C_1 are positive constants and R, G are nonnegative constants. If $R = G = 0$, this system can be reduced to NFDE of the type discussed in §4.5 (see, for example, Hale [1977] and the references therein). If R, G are different from zero, this cannot be done.

It is convenient to change variables so that the boundary conditions at $x = 0$ are homogeneous. If $I = i$, $V = v - E$, then

$$(9.13) \qquad \begin{aligned} \frac{\partial I}{\partial t} &= -\frac{1}{L}\frac{\partial V}{\partial x} - \frac{R}{L}I, \\ \frac{\partial V}{\partial t} &= -\frac{1}{C}\frac{\partial I}{\partial t} - \frac{G}{C}V + h_1(t), \end{aligned}$$

with the boundary conditions

$$(9.14) \qquad \begin{aligned} V(t,0) + R_0 I(t,0) &= 0, \\ C_1(\partial/\partial t)V(t,l) + g(V(t,l) + E(t)) + h_2(t) &= I(t,l), \end{aligned}$$

where $h_1 = -GE/C - \dot{E}$ and $h_2 = C_1 \dot{E}$.

This problem can be viewed as an abstract evolutionary equation

(9.15) $$\dot{u} = Au + f(t, u)$$

in the real Hilbert space $X = L^2(0,l) \times L^2(0,l) \times R$ with norm $|(\phi, \psi, b)|^2 = \int_0^l \phi^2 + \int_0^l \psi^2 + b^2$ and

$A: D(A) \subset X \to X$,

$$A(\phi, \psi, b) = \left(-\frac{1}{L}\psi' - \frac{R}{L}\phi, -\frac{1}{C}\phi' - \frac{G}{C}\psi, \frac{1}{C_1}\phi(l)\right),$$

$D(A) = \{(\phi, \psi, b) \in X : \phi, \psi \in H^1(0,l), \psi(0) + R_0\phi(0) = 0, b = \psi(l)\}$,

$$f(\cdot, u) = \left(0, h_1, -\frac{1}{C_1}g(b+E) - \frac{1}{C_1}h_2\right).$$

It is not difficult to show that A is the infinitesimal generator of a C^0-semigroup e^{At} on X. Also, if g is a C^1-function, then $f(t,u)$ is continuous in t, u and Lipschitzian in u for (t, u) in bounded sets. One then obtains local existence and uniqueness of solutions of (9.15) through the initial value (σ, ϕ, ψ, b). Let $T(t, \sigma)(\phi, \psi, b)$ be this solution. It is now possible to follow the arguments in Lopes and Ceron [1984] to show that

$$T(t, \sigma) = e^{A(t-\sigma)} + U(t, \sigma), \qquad t \geq \sigma,$$

where $U(t, \sigma)$ is compact if $T(t, \sigma)$ is a bounded map.

Ribeiro, Neves, and Lopes [1986] have shown that the essential spectral radius $r_e(e^{At})$ of e^{At} is given by $e^{-\sigma_0 t}$ where

$$\sigma_0 = \frac{RG + LG}{2LC} - \frac{Z}{2l}\ln\frac{Z - R_0}{Z + R_0}, \qquad Z = (L/C)^{1/2}.$$

Consequently, if $\sigma_0 > 0$, then $T(t, \sigma)$ is a conditional α-contraction and the theory of dissipative processes in Chapter 3 applies to (9.13).

Under some additional conditions on the coefficients and the nonlinear function g, Lopes and Ceron [1984] have shown that $T(t, \sigma)$ is bounded dissipative. Thus, there is a global attractor A_σ for $T(\omega + \sigma, \sigma)$ for each $\sigma \in R$ and an ω-periodic solution of (9.15).

Lopes [1984], Ribeiro, Neves, and Lopes [1986], and Lopes [1988] have considered hyperbolic equations more general than (9.13) and discussed the essential spectral radius of the semigroup generated by the linear part. Lin and Neves [1986] have extended the class of equations (9.14) in such a way that it can include even RFDE's. Without going into detail, we simply write the form of the equation

$$\frac{\partial}{\partial t}\begin{bmatrix}u(t,x)\\v(t,x)\end{bmatrix} + K(x)\frac{\partial}{\partial x}\begin{bmatrix}u(t,x)\\v(t,x)\end{bmatrix} + C(x)\begin{bmatrix}u(t,x)\\v(t,x)\end{bmatrix} = 0, \qquad 0 < x < 1,$$

$(d/dt)[v(1,t) - Du(1,t)] = Fu(\cdot, t) + Gv(\cdot, t),$

$u(0,t) = Ev(0,t),$

u, v are vectors, K, C, D are matrices, and F, G are functionals on $(0, 1)$.

5. Kuramoto-Sivashinsky equation. In this section we consider the ...oto-Sivashinsky equation in one space dimension

$$u_t + \nu u_{xxxx} + u_{xx} + \tfrac{1}{2}(u_x)^2 = 0 \quad \text{in } (0,1),$$
$$u_x = u_{xxx} = 0 \quad \text{at } x = 0, \ x = 1.$$

... physical motivation for this equation and references, see Nicolaenko, ..., and Temam [1985]. The results presented below are based on this latter paper but the presentation follows more closely some notes of Sakamoto [1987].

The equation (9.16) as it stands may not generate a dissipative semiflow since $|u(\cdot,t)|_{L^2(0,1)}$ could be unbounded as $t \to \infty$. In fact, if $\bar{u}(t) = \int_0^1 u(x,t)\,dx$, then

$$\frac{d}{dt}\bar{u}(t) + \frac{1}{2}\int_0^1 |u_x(t,x)|^2\,dx = 0.$$

This relation implies that $\bar{u}(t)$ is unbounded as $t \to \infty$ unless $u(t,\cdot)$ approaches a constant function as $t \to \infty$. Since $|\bar{u}(t)| \leq (\text{constant}) \cdot |u(\cdot,t)|_{L^2(0,1)}$, it follows that the L^2-norm of $u(\cdot,t)$ may be unbounded as $t \to \infty$. This motivates the consideration of the difference $u(x,t) - \bar{u}(t)$ rather than $u(t,x)$. This difference satisfies the equation

$$(9.18) \qquad u_t + \nu u_{xxxx} + u_{xx} + \frac{1}{2}(u_x)^2 = \frac{1}{2}\int_0^1 (u_x)^2\,dx$$

with the boundary conditions (9.17).

Let X be the closed subspace in $L^2(0,1)$ consisting of all functions which are orthogonal to $\sin n\pi x$, $n = 1, 2, \ldots$. If $D(A) = H^4(0,1) \cap X$, define

$$A\colon D(A) \to X,$$
$$A\phi = \nu\phi_{xxxx} + \phi_{xx}.$$

Then one can prove that A satisfies the following properties (see, for example, Sakamoto [1986]):

(i) A is selfadjoint with compact resolvent;
(ii) $-A$ generates a compact analytic semigroup on X.

For $\phi \in X^{1/2} = H^2 \cap X$, let

$$f\colon X^{1/2} \to X,$$
$$f(\phi) = \frac{1}{2}\phi_x^2 - \frac{1}{2}\int_0^1 \phi_x^2\,dx.$$

Then f is an analytic function.

Equation (9.18) now can be written as an abstract evolutionary equation in X:

$$(9.19) \qquad du/dt + Au + f(u) = 0.$$

THEOREM 4.9.5. *Equation (9.19) generates a C^1-semigroup $T(t)$ on $\dot{X}^{1/2}$ $= X^{1/2} \cap \{\phi \in L^2(0,1) \colon \int_0^1 \phi = 0\}$ and $T(t)$ is compact for $t > 0$. Furthermore, $T(t)$ is point dissipative and there is a global attractor A for $T(t)$. For any $\phi \in A$, the map $t \mapsto T(t)\phi$ is analytic.*

The local existence and uniqueness of the solution of the initial value problem is obtained in the usual way by using the variation of constants formula and the contraction mapping principle. For the global existence of solutions, one must use various energy type estimates. They follow a standard type of procedure and can be found in Sakamoto [1986]. The fact that the resulting semigroup is compact follows from Theorem 4.6.4.

The most difficult part of the proof of theorem is to show that $T(t)$ is point dissipative. This is a consequence of the following important and nontrivial result of Nicolaenko, Scheurer, and Temam [1985]. There is a constant c such that

$$\varlimsup_{t\to\infty} |u_x(t)|_{L^2(0,1)} \leq c$$

for each solution u of (9.18). The existence of the global attractor follows from Theorem 3.4.6. The regularity of the map $t \mapsto T(t)\phi$ is a consequence of Hale and Scheurle [1985] taking into account that the semigroup for the linear part of the equation is an analytic semigroup.

4.9.6. A nonlinear diffusion problem. The material in this section is a summary of Aronson, Crandall, and Peletier [1982]. Consider the equation

(9.20)
$$\begin{aligned} u_t &= (u^m)_{xx} + f(u) \quad \text{in } Q, \\ u(\pm L, t) &= 0 \quad \text{in } \mathbf{R}^+, \end{aligned}$$

where $m > 1$ is a parameter, f is locally Lipschitz continuous, $Q = \Omega \times \mathbf{R}^+$, and $\Omega = (-L, L)$. For any $T > 0$, let $Q_T = \Omega \times (0, T]$. For a given $u_0 \in L^\infty(\Omega)$, a function u on $[0, T]$ is said to be a *solution through u_0* if

(i) $\quad u \in C([0,T], L^1(\Omega)) \cap L^\infty(Q_T),$

(ii) $\quad \displaystyle\int_\Omega u(t)\phi(t)\,dx - \iint_{Q_t}(u\phi_t + u^m \phi_{xx})\,dx\,dt$
$$= \int_\Omega u_0 \phi(0)\,dx + \iint_Q f(u(t))\phi\,dx\,dt$$

for all $\phi \in C^2(\overline{Q}_T)$ such that $\phi \geq 0$, $\phi = 0$ at $x = \pm L$, and $0 \leq t \leq T$.

THEOREM 4.9.6. *Suppose $f(0) = f(1) = 0$ and $u_0 \in L^\infty(\Omega)$, $0 \leq u_0 \leq 1$. Then, for every $T > 0$, there is a unique solution u of (9.20) through u_0, $0 \leq u \leq 1$, and u is continuous in u_0. More precisely, if $u, \hat{u}$ are solutions on $[0, T]$ and if K is the Lipschitz constant for f on $[-M, M]$ where $M = \max(|u|_{L^\infty(Q_T)}, |\hat{u}|_{L^\infty(Q_T)})$, then*

$$|u(t) - \hat{u}(t)|_{L^1(\Omega)} \leq e^{Kt} |u_0 - \hat{u}_0|_{L^1(\Omega)}.$$

Furthermore, for each $\tau > 0$, there is a constant M_τ, independent of u_0, such that

(i) $u^m(t)_x \in L^\infty(\Omega)$ for $t > \tau$,
(ii) $|u^m(t)_x|_{L^\infty(\Omega)} \leq M_\tau$ and essential variation $u^m(t)_x \leq M_\tau$ for $t \geq \tau$.

Now, let X_τ be the metric space consisting of those $w \in L^\infty(\Omega)$ such that $0 \leq w \leq 1$, $(w^m)_x \in L^\infty(\Omega)$, $|(w^m)_x|_{L^\infty(\Omega)} \leq M_\tau$ and essential variation $(w^m)_x \leq M_\tau$, where M_τ is as in Theorem 4.9.6, equipped with the metric

$$d(u,v) = |u - v|_{L^1(\Omega)} + |(u^m - v^m)_x|_{L^2(\Omega)}.$$

The space X_τ is a compact metric space.

If we let $T(t)u_0 = u(t, u_0)$ and $B_1^\infty = \{u_0 \in L^\infty(\Omega) : 0 \leq u_0 \leq 1\}$ with metric $d_{B_1^\infty}(u,v) = |u - v|_{L^1(\Omega)}$, then $T(t) : B_1^\infty \to B_1^\infty$ for $t \geq 0$ is a C^0-semigroup and, for any $\tau > 0$, $T(t)B_1^\infty \subseteq X_\tau$ for $t \geq \tau$; in particular $T(t)B_1^\infty$ is compact for $t \geq \tau$. Let

$$X = \{u \in B_1^\infty : (u^m)_x \in L^2(\Omega)\}$$

be equipped with the metric in X_τ.

THEOREM 4.9.7. *Under the hypotheses of Theorem 4.9.6, there is a connected global attractor A_f of $T(t)$ in $X \subset B_1^\infty$ and*

$$d_X(T(t)B_1^\infty, A_f) \to 0 \quad \text{as } t \to \infty.$$

Also, $A_f = \text{Cl } W^u(E_f)$, where E_f is the set of equilibrium points of (9.20), *that is, the solutions of*

$$(u^m)_{xx} + f(u) = 0 \quad \text{in } \Omega,$$
$$u = 0 \quad \text{in } \partial\Omega.$$

4.9.7. Age-dependent populations. In this section, we rephrase some results of Webb [1985] on age-dependent populations so that they fit into the theory of dissipative processes in Chapter 3.

Let R^n denote an n-dimensional linear vector space with norm $|x| = \sum_{i=1}^n |x_i|$, $x = (x_1, \ldots, x_n)^T \in R^n$. Let $L^1 = L^1(0, \infty, R^n)$ be the Banach space of equivalence classes of Lebesgue measurable functions from $[0, \infty)$ to R^n which agree almost everywhere on $[0, \infty)$ with norm

$$|\phi|_{L^1} = \int_0^\infty |\phi(a)| \, da.$$

Let $T > 0$ and let $L_T = C([0,T], L^1)$. Elements of L_T can be identified with functions in $L^1((0,\infty) \times (0,T); R^n)$. We use the symbol l to denote both of these elements in that

$$l(t)(a) = l(a,t), \quad 0 \leq t \leq T, \quad \text{a.e. in } a > 0.$$

If $l(\cdot, t) = (l_1(\cdot, t), \ldots, l_m(\cdot, t))^T$, then $l_i(\cdot, t)$ denotes the density with respect to age at time t of the ith subclass of a population divided into n subclasses. Let $P(t)$ represent the total population at time t,

(9.21) $$P(t) = \int_0^\infty l(a,t) \, da, \quad t \geq 0.$$

The average rate of change in the total population size in the time interval $(t, t+h)$ is

$$h^{-1}[P(t+h) - P(t)] = h^{-1}\int_0^h l(a, t+h)\, da + \int_0^\infty h^{-1}[l(a+h, t+h) - l(a, t)]\, da.$$

As $h \to 0$, the term on the left converges to the instantaneous rate of change of the total population at time t, the first term on the right-hand side converges to the instantaneous birth rate at time t, and the second term on the right-hand side converges to instantaneous rate of change of the total population at time t due to causes other than births.

If we let $F: L^1 \to R^n$ designate the birth rate at time t and let

$$\int_0^\infty G(l(\cdot, t))(a)\, da, \qquad G: L^1 \to L^1$$

designate the rate of change of the total population due to other causes then, letting $h \to 0^+$ in the above relation, we have

(9.22) $$\frac{d}{dt}P(t) = F(l(\cdot, t)) + \int_0^\infty G(l(\cdot, t))(a)\, da.$$

If all functions are sufficiently smooth this would say that

(9.23) $$l(0, t) = F(l(\cdot, t)),$$
(9.24) $$Dl(\cdot, t) = G(l(\cdot, t))$$

where

(9.25) $$Dl(a, t) = \partial l(a, t)/\partial a + \partial l(a, t)/\partial t.$$

Relation (9.24) is referred to as the *balance law* for the population. Equations (9.23) and (9.24) together with an initial age distribution $\phi \in L^1$ at $t = 0$ represent the dynamics of the population.

If the functions are not smooth, the weak form of (9.23) and (9.24) must be used:

(9.26) $$\lim_{h \to 0^+} h^{-1} \int_0^h (l(a, t+h) - F(l(\cdot, t)))\, da = 0,$$

(9.27) $$\lim_{h \to 0^+} \int_0^\infty (h^{-1}[l(a+h, t+h) - l(a, t)] - G(l(\cdot, t))(a))\, da = 0$$

for $0 \le t \le T$.

Let us proceed formally to derive an equivalent integral equation for the initial value problem for (9.26) and (9.27) using (9.23), (9.24), and (9.25).

The characteristics for the differential operator (9.25) are $a - t = c$, where c is a constant. If we let $w_c(t) = l(t+c, t)$, then the rate of change in $l(a, t)$ along characteristics is (from (9.24))

$$(d/dt)w_c(t) = G(l(\cdot, t))(t+c).$$

Integrating this relationship, one obtains

$$w_c(t) = \begin{cases} w_c(t-a) + \int_{t-a}^{t} G(l(\cdot,s))(s+c)\,ds, & a < t, \\ w_c(0) + \int_0^t G(l(\cdot,s))(s+c)\,ds, & a \geq t. \end{cases}$$

Replacing c by $a - t$ and using (9.23), one obtains

(9.28) $\quad l(a,t) = \begin{cases} F(l(\cdot, t-a)) + \int_{t-a}^{t} G(l(\cdot,s))(s+a-t)\,ds, & a \in (0,t), \\ \phi(a-t) + \int_0^t G(l(\cdot,s))(s+a-t)\,ds, & a \in (t,\infty), \end{cases}$

where it is understood that the equality holds almost everywhere.

To proceed further, we need to impose some additional conditions on F, G. The conditions are the ones imposed by Webb [1985]. In the enumeration of these conditions, c will denote a nondecreasing continuous function from $R_+ \to R_+$. If X, Y, Z are subsets of Banach spaces, then a function $f\colon X \to Z$ will be said to be *locally Lipschitz continuous* if

$$|f(x) - f(\bar{x})|_Z \leq c(r)|x - \bar{x}|_X \quad \text{for } |x|_X, |\bar{x}|_X \leq r.$$

A function $f\colon X \times Y \to Z$, $(x,y) \to f(x,y)$ is said to be *locally Lipschitz in x uniformly with respect to y* if

$$|f(x,y) - f(\bar{x},y)|_Z \leq c(r)|x - \bar{x}|_X \quad \text{for } |x|_X, |\bar{x}|_X \leq r.$$

A function $f\colon X \times Y \to Z$, $(x,y) \to f(x,y)$ is *Lipschitz in x for each $y \in Y$* if

$$|f(x,y) - f(\bar{x},y)|_Z \leq c(|y|_Y)|x - \bar{x}|_X \quad \text{for all } x, \bar{x} \in X,$$

and it is said to be *uniformly bounded for each $y \in Y$* if

$$|f(x,y)|_Z \leq c(|y|_Y).$$

Let $R_+^n = \{x \in R^n\colon x_i \geq 0,\ i = 1, 2, \ldots, n\}$, $L_+^1 = \{\phi \in L^1\colon \phi(a) \in R_+^n$ for almost all $a > 0\}$.

The hypotheses to be imposed on F, G are

(9.29) $\quad F\colon L^1 \to R^n, G\colon L^1 \to L^1$ are locally Lipschitz.

(9.30) $\quad G(\phi) + c_3(r)\phi \in L_+^1$ if $r > 0$, $\phi \in L_+^1$, $|\phi| \leq r$, where $c_3\colon [0,\infty) \to [0,\infty)$ is an increasing function.

(9.31) $\quad F\colon L_+^1 \to R_+^n$.

(9.32) $\quad G(\phi)(a) = -\mu(a,\phi)\phi$, where $\mu(a,\phi)$ is an $n \times n$ matrix function which is Lipschitz in a for each $\phi \in L^1$, uniformly bounded for each $\phi \in L^1$, and locally Lipschitz in ϕ uniformly in a.

(9.33) $\quad$ There is a constant $\nu > 0$ such that $\sum_{i=1}^n \int_t^\infty G(\phi)_i(a)\,da \leq -\nu \sum_{i=1}^n \int_t^\infty \phi_i(a)\,da$ for each $\phi \in L_+^1$, $t \geq 0$.

(9.34) $F(\phi) = \int_0^\infty \beta(a,\phi)\phi(a)\,da$, where $\beta(a,\phi)$ is an $n \times n$ matrix which is Lipschitz in a for each $\phi \in L^1$, uniformly bounded for each $\phi \in L^1$, and $|\beta(a,\phi) - \beta(a,\overline{\phi})| \leq c(r)\sum_{i=1}^n ||\phi_i|_{L^1} - |\overline{\phi}_i|_{L^1}|$ for all $\phi, \overline{\phi} \in L^1, |\phi|, |\overline{\phi}| \leq r, a \geq 0$.

The following important theorems are proved in Webb [1985].

THEOREM 4.9.8. *Under assumptions* (9.29) *and* (9.30) *and the additional hypothesis that the solution of* (9.28) *is defined for* $t \geq 0$, *the family of mappings* $T(t): L_+^1 \to L_+^1$, $t \geq 0$, *defined by* $T(t)\phi = l(\cdot,t)$ *where* $l(\cdot,t)$ *is the unique solution of* (9.28), *is a* C^0-*semigroup. Also, the infinitesimal generator* B *of* $T(t)$ *is given by* $B = -A$, *where*

$$A\phi = \phi' - G(\phi) \quad \text{for } \phi \in D(A),$$
$$D(A) = \{\phi \in L_+^1 : \phi' \in AC([0,\infty), R^n), \phi' \in L^1, \phi(0) = F(\phi)\}.$$

Webb [1985] also gives conditions under which the solution operator $T(t)$ is an α-contraction. From (9.28), there is a natural decomposition of the operator $T(t)$ corresponding to that part of the population with age $\geq t$ (the members of the population which were existent at time zero) and the population with age $< t$ (the members of the population born after time zero). Define the operators $S(t), U(t): L_+^1 \to L_+^1$ by the relations

(9.35) $$(S(t)\phi)(a) = \begin{cases} 0, & \text{a.e. } a \in (0,t), \\ (T(t)\phi)(a), & \text{a.e. } a \in (t,\infty), \end{cases}$$

(9.36) $$(U(t)\phi)(a) = \begin{cases} (T(t)\phi)(a), & \text{a.e. } a \in (0,t), \\ 0, & \text{a.e. } a \in (t,\infty). \end{cases}$$

THEOREM 4.9.9. *If* F, G *satisfy the conditions* (9.29)–(9.34), *then* $T(t) = S(t) + U(t)$ *in Theorem* 4.9.8 *where* $S(t), U(t)$ *are given in* (9.35) *and* (9.36) *and satisfy*

(9.37) $$|S(t)\phi|_{L^1} \leq e^{-\nu t}|\phi|_{L^1}, \quad t \geq 0,$$
$$U(t) \text{ is compact for } t > 0.$$

This result is not stated explicitly in Webb [1985] but follows from Proposition 3.14, p. 102, and the proof of Proposition 3.16, p. 106 of Webb [1985]. Lemma 2.2.4 implies that $T(t)$ in Theorem 4.9.8 is asymptotically smooth. We state this as

COROLLARY 4.9.10. *Under the hypotheses of Theorem* 4.9.9, *the semigroup* $T(t)$ *is asymptotically smooth.*

EXAMPLE 4.9.11. Let $R^n = R$, $\beta \in C^1(R, [0,\infty))$, $\gamma \in C^1(R_+, R_+)$, γ bounded, $P: L^1 \to R$, $P\phi = \int_0^\infty \phi(a)\,da$ and $F: L^1 \to R$,

(9.38) $$F\phi = \int_0^\infty \beta(P\phi)\gamma(a)\phi(a)\,da.$$

Let $\mu \in C^1(R,[0,\infty))$ and define $G: L^1 \to L^1$ by

(9.39) $$G(\phi)(a) = -\mu(P\phi)\phi(a).$$

Also, suppose that there is an $\omega \in R$ such that

(9.40) $$\beta(P) - \mu(P) \leq \omega.$$

(This last condition is imposed in order to obtain global existence of solutions.) Then (9.28) defines a C^0-semigroup $T(t)$ with infinitesimal generator B given by
(9.41)
$$B\phi = -\phi'(a) - \mu(P\phi)\phi(a) \quad \text{a.e.,}$$

$$D(B) = \{\phi \in L^1_+, \phi \in AC([0,\infty), R^n), \phi' \in L^1, \phi(0) = \int_0^\infty \beta(P\phi)\gamma(a)\phi(a)\,da\}.$$

For a proof, see Webb [1985]. Special cases of interest in the applications are when there is a constant $\alpha > 0$ such that
  (i) $\gamma(a) = e^{-\alpha a}$,
  (ii) $\gamma(a) = 1 - e^{-\alpha a}$,
  (iii) $\gamma(a) = ae^{-\alpha a}$.

The reproductive capacity is concentrated on the youngest ages in case (i), on the intermediate ages in case (iii), and reproduction is possible at all ages in case (ii).

THEOREM 4.9.12. *For Example 4.9.11 above, if there is a constant $\nu > 0$ such that*

(9.42) $$\mu(s) \geq \nu > 0 \quad \text{for all } s \geq 0,$$

then the corresponding semigroup $T(t)$ satisfies $T(t) = S(t) + U(t)$ where $S(t)$, $U(t)$ satisfy (9.37). Thus, $T(t)$ is asymptotically smooth.

We remark as before that the following are consequences of Theorem 4.9.12:

(i) Asymptotic stability of an equilibrium point is equivalent to uniform asymptotic stability.

(ii) More generally, asymptotic stability of an isolated invariant set is equivalent to uniform asymptotic stability.

(iii) If orbits of bounded sets are bounded and $T(t)$ is point dissipative, then there is a global attractor A with $\omega(B) \subset A$ for all bounded sets B.

REMARK 4.9.13. Webb [1985] gives a matrix version of Example 4.9.11 which also satisfies the condition of Theorem 4.9.12.

REMARK 4.9.14. Webb [1985, p. 156] gives conditions that are almost enough for the hypotheses in (iii).

REMARK 4.9.15. In Webb [1985, Theorem 4.5, p. 158], one can use the above results to show that the global attractor is $\{0\}$. Thus, the conclusion can be made much stronger by saying that $\text{dist}(T(t)B, \{0\}) \to 0$ as $t \to \infty$ where B is an arbitrary bounded set. This same remark applies to the other examples considered there.

4.10. Dependence on parameters and approximation of the attractor.

In the previous sections, we have given many classes of evolutionary problems for which the corresponding semigroup has a global attractor. In some cases, we also have discussed two other important questions concerning the dependence of the attractor and the flow on the attractor on parameters in the problem.

Let us recall that the attractor A_λ is upper semicontinuous in λ at λ_0 if $\delta(A_\lambda, A_{\lambda_0}) \to 0$ as $\lambda \to \lambda_0$, where, for any two subsets A, B of X,

$$\delta(A, B) = \sup_{x \in A} \inf_{y \in B} d(x, y)$$

and $d(x, y)$ is the distance function. The attractor A_λ is lower semicontinuous at λ_0 if $d(A_{\lambda_0}, A_\lambda) \to 0$ as $\lambda \to \lambda_0$. It is continuous at λ_0 if it is upper and lower semicontinuous. In §3.5, a general result was presented for the upper semicontinuity of the attractor A_λ in λ when the semigroup $T_\lambda(t)$ has the property that $T_\lambda(t)x$ is continuous in (t, x, λ) with the continuity in λ being uniform with respect to t, x in bounded sets. The upper semicontinuity in λ under these strong hypotheses on $T_\lambda(t)$ is an easy consequence of the strong stability properties of the attractor. In classical ordinary differential equations, it would have been referred to as stability under constantly acting disturbances.

Without some further hypotheses on the flow restricted to the attractor A_{λ_0}, there will be no lower semicontinuity of the sets A_λ at λ_0. Examples are easily constructed even in one-dimensional ODE's where the attractor for A_{λ_0} is a line segment with the flow as indicated with one hyperbolic equilibrium point and one not hyperbolic $\cdot \to \cdot$ and the attractor for A_λ is a point if $\lambda \neq \lambda_0$. This drastic change in the size of the attractor for this example is caused by the fact that one of the equilibria is not hyperbolic. If the system is Morse-Smale, the attractor A_λ is continuous at $\lambda = 0$.

In a specific problem, it may be very difficult to prove that $T_\lambda(t)x$ has the strong continuity property mentioned above and, in others, this continuity property may not hold. Even in the latter case it still may be possible to assert that A_λ is upper semicontinuous or lower semicontinuous. This will be possible because A_λ is *compact* and has strong stability properties. The restriction of the discussion to the compact attractors and not to a comparison of the semigroups on arbitrary bounded sets plays a crucial role.

In this section, we give a summary of some recent results that have been obtained on the upper semicontinuity of the attractor when the parameters are physically interesting but their effect on the semigroup is not easily ascertained. The topics considered are the effects of diffusion coefficients and boundary conditions in reaction diffusion systems, some singular perturbation problems, and some numerical procedures. We also present one result on lower semicontinuity for gradient systems.

In this presentation, we shall primarily concentrate on local attractors. Let us recall the definition. A set A_λ is said to be a *local attractor* for a semigroup $T_\lambda(t)$ on X if A_λ is compact, invariant, and there is a neighborhood U of A

such that $\omega(U) \subset A_\lambda$, and $\text{dist}(T(t)U, A_\lambda) \to 0$. It is very easy to show that a compact invariant set A_λ is a local attractor if and only if it is uniformly asymptotically stable. For a given λ_0 and a given neighborhood U of A_{λ_0}, we say A_λ is upper semicontinuous at λ_0 if $\delta(A_\lambda \cap U, A_{\lambda_0}) \to 0$ as $\lambda \to \lambda_0$; that is, we are only concerned with that part of the attractor A_λ at λ which belongs to U. Global results can be obtained but require a detailed analysis of the effects of the parameters on very large orbits, a subject that we do not wish to discuss.

4.10.1. *Reaction diffusion equations.* Consider the system of equations

(10.1) $\qquad \partial u/\partial t = D\Delta u + f(u) \quad \text{in } \Omega,$

(10.2) $\qquad (1-\theta)D\partial u/\partial n + \theta E(x)u = \quad \text{in } \partial\Omega,$

where $u \in R^N$, $\Omega \subset R^n$ is a bounded open set with $\partial\Omega$ smooth, $D = \text{diag}(d_1, \ldots, d_N)$, $E = \text{diag}(e_1, \ldots, e_N)$ with each $d_j > 0$ and each $e_j \colon \partial\Omega \to R$ continuous, $j = 1, 2, \ldots, N$, and $\theta \in [0, 1]$ is constant. The function $f \colon R^N \to R^N$ is a $C^{1,1}$ function, that is, continuous and has a Lipschitz continuous first derivative.

Let $X = L^2(\Omega, R^N)$ and define the operator

(10.3) $\qquad \begin{aligned} &A = A_{D,\theta} \colon D(A) \to X, \qquad A\phi = -D\Delta\varphi, \\ &D(A) = \{u \in H^2(\Omega, R^N) \colon u \text{ satisfies the boundary conditions (10.2)}\}. \end{aligned}$

The operator A is sectorial and one can define the fractional powers A^α of A for $\alpha \geq 0$ and the space $X^\alpha = D(A^\alpha)$ with the graph norm. The space X^α depends on D, θ; that is, $X^\alpha = X^\alpha_{D,\theta}$. When studying the effects of the boundary conditions and diffusion coefficients on the attractor, one must have more precise information about the dependence of the spaces $X^\alpha_{D,\theta}$ on D and θ. The proof of the following lemma is technical and is obtained by appropriately adapting classical methods in elliptic regularity theory and Sobolev inequalities. The complete proof can be found in Hale and Rocha [1987b].

LEMMA 4.10.1.
  (i) *If* $n = 1$, *then* $X^{1/2}_{D,\theta} = H^1(\Omega, R^N)$ *for all* $\theta \in [0, 1)$ *and* $D > 0$.
  (ii) *If* $n > 1$ *and* $\alpha > (5n-6)/4n$, *then* $X^\alpha_{D,\theta} \subset H^1(\Omega, R^N) \cap L^\infty(\Omega, R^N)$ *for all* $\theta \in [0, 1)$, $D > 0$ *and, for any* $d_0 > 0$, *there is a constant* $k(d_0)$ *such that, for any* $D \geq d_0 I_N$, $I_N = $ *identity*, $\theta \in [0,1)$ *and any* $u \in X^\alpha_{D,\theta}$, *we have*

$$|u|_{L^\infty} \leq k(d_0)|u|_{X^\alpha_{D,\theta}}.$$

Let us first consider the case of Neumann boundary conditions since the notation and ideas are very simple. For large diffusion coefficients one can prove (see Hale [1986])

THEOREM 4.10.2. *Consider* (10.1), (10.2) *with* $\theta = 0$, *that is, Neumann boundary conditions, and suppose that the ODE*

(10.4) $\qquad dz/dt = f(z)$

has a local (or global) attractor A. *If* $\alpha > (5n-6)/4n$, *then there exists a* $d_2^0 > 0$ *such that* A *considered as a subset of* X^α *is a local (or global) attractor for the*

PDE (10.1), (10.2) with $\theta = 0$ if $\lambda D_2 \geq d_2^0 I_N$, where λ is the first nonzero eigenvalue of $-\Delta$ on X.

The attractor A for (10.4) can be embedded in a natural way into X^α as the set $\tilde{A} = \{\varphi \in X^\alpha \colon$ there exists a $z \in A$ such that $\phi(x) = z$ for $x \in \Omega\}$. It will be seen from the indication of the proof below that, if $u(t,x)$ is a solution of (10.1), (10.2) with $\theta = 0$ and $\bar{u}(t) = |\Omega|^{-1} \int_\Omega u(t,x)\,dx$, then there is a function $g(t) \to 0$ exponentially as $t \to \infty$ such that $\bar{u}$ satisfies the differential equation

$$d\bar{u}/dt = f(\bar{u}) + g(t)$$

and

$$|u(t,\cdot) - \bar{u}(t)|_{X^\alpha} \to 0 \quad \text{exponentially as } t \to \infty.$$

For the case in which the ODE (10.4) has an invariant rectangle, this latter result about the average $\bar{u}$ was proved by Conway, Hoff, and Smoller [1978]. Such a condition puts many restrictions on the vector field f. As we will see from the sketch of the proof below, one could have (10.1), (10.2) be a system of diffusive functional differential equations and obtain the same result; that is, the function f in (10.1) depends on $u(t+s, x)$, $s \in [-r, 0]$ for some given constant $r > 0$. Invariant rectangles for such equations have little meaning.

The idea for the proof of Theorem 4.10.2 is very simple. Let

$$u = z + v,$$
$$z = |\Omega|^{-1} \int_\Omega u(\cdot, x)\,dx, \qquad \int_\Omega v(\cdot, x)\,dx = 0$$

and obtain the new equation for z, v:

(10.5)
$$dz/dt = |\Omega|^{-1} \int_\Omega f(z + v(\cdot, x))\,dx,$$
$$\partial v/\partial t = D\Delta v + f(z+v) - |\Omega|^{-1} \int_\Omega f(z+v(\cdot,x))\,dx \quad \text{in } \Omega,$$
$$\partial v/\partial n = 0 \quad \text{in } \partial\Omega.$$

The set $A \times \{0\}$ is an invariant set for (10.5). The semigroup $\exp(D\Delta t)$ restricted to functions of average zero decays exponentially with rate $-d\lambda$ where $d = \min(d_1, \ldots, d_N)$. Since the vector field in the v equation in (10.5) vanishes for $v = 0$, the solution v will decay exponentially as $t \to \infty$ at a rate proportional to $-d\lambda$ if it is known that solutions stay close to the invariant set $A \times \{0\}$. The stability of the attractor for the ODE implies that solutions of (10.5) will stay close to $A \times \{0\}$. The proof that $A \times \{0\}$ is an attractor for (10.5) is completed in this way.

In Hale and Sakamoto [1987], appropriate modification of these ideas have led to a much more general result. One can suppose that some of the diffusion coefficients are large relative to others and obtain information about the PDE's by considering a related set of PDE's coupled with some ODE's. To be slightly more precise, let us consider a coupled system

(10.6)
$$\partial v/\partial t = D_1 \Delta v + f(v, w),$$
$$\partial w/\partial t = D_2 \Delta w + g(v, w) \quad \text{in } \Omega,$$
$$\partial v/\partial n = 0, \quad \partial w/\partial n = 0 \quad \text{in } \partial\Omega,$$

where $v \in R^M$, $w \in R^N$.

Associated with this equation, we consider the "shadow" system (the term comes from Nishuira and Fujii [1985])

(10.7)
$$\partial v/\partial t = D_1 \Delta v + f(v, z) \quad \text{in } \Omega,$$
$$dz/dt = |\Omega|^{-1} \int_\Omega f(v(\cdot, x), z) \, dx,$$
$$\partial v/\partial n = 0 \quad \text{in } \partial\Omega.$$

If we suppose that $D_1 \geq d_1^0 I_M$, $d_1^0 > 0$, fixed and that the system (10.7) has a local attractor $A_{D_1} \subset K$, a compact set, and there is a $\delta > 0$ such that $\omega(N(\delta, A_{D_1})) \subset A_{D_1}$ for all $D_1 \geq d_1^0 I_M$, then one can show that (10.6) has a local attractor A_{D_1, D_2} if $D_2 \geq d_2^0 I_N$, where d_2^0 is sufficiently large and $\delta(A_{D_1, D_2}, \tilde{A}_{D_1}) \to 0$ as $D_2 \to \infty$ where $\tilde{A}_{D_1}$ is a natural embedding of A_{D_1} in the function space in which (10.6) is considered, that is, z is the spatial average of w.

If we now consider the case of mixed boundary conditions in (10.1), (10.2), then analogous results can be obtained (see Hale and Rocha [1987a], [1987b]). At first, suppose $\theta \in [0, \theta_0]$, $0 < \theta_0 < 1$. If all diffusion coefficients are sufficiently large, then the associated shadow system of ODE's is

(10.8)
$$dz/dt = -\bar{\theta}\varsigma z + f(z),$$
$$\varsigma = \int_{\partial\Omega} E(x) \, dx, \quad \bar{\theta} = \theta(1-\theta)^{-1}.$$

Let us now assume that (10.8) has a local attractor A_θ. For the operator $-D\Delta$ on X, there are N eigenvalues $\lambda_j(D, \theta)$ and normalized eigenfunctions $\phi_j(D, \theta)$ such that, if $D \geq dI_N$, then

$$B_{D,\theta} = \text{diag}(\lambda_1(D, \theta), \ldots, \lambda_N(D, \theta)) \to \bar{\theta}\varsigma,$$
$$\Phi_{D,\theta} = (\phi_1(D, \theta), \ldots, \phi_N(D, \theta)) \to |\Omega|^{-1/2} I_N$$

as $d \to \infty$. The remaining eigenvalues of $-D\Delta$ on X are approaching ∞ as $d \to \infty$ at a rate proportional to d.

Let us decompose the space X^α as

$$X^\alpha = X_1^\alpha \oplus X_2^\alpha,$$
$$u = \Phi_{D,\theta} z + v, \quad z \in R^N, \quad \int_\Omega \Phi_{D,\theta} v = 0.$$

If this decomposition is applied to (10.1), (10.2), then

$$\frac{dz}{dt} = -B_{D,\theta} z + \int_\Omega \Phi_{D,\theta} f(\Phi_{D,\theta} z + v) \, dx,$$
$$\frac{\partial v}{\partial t} = -D\Delta v + f(\Phi_{D,\theta} z + v) - \Phi_{D,\theta} \int_\Omega \Phi_{D,\theta} f(\Phi_{D,\theta} z + v) \, dx$$

with v satisfying the boundary conditions.

Using the ideas above for Neumann boundary conditions and the theory of integral manifolds, Hale and Rocha [1987a] show that there is a $d_0 > 0$ such that, for $D \geq d_0 I_N$, this latter equation has a local integral manifold $M_{D,\theta} = \{(z,v): v = h(z,D,\theta), |z| \leq \delta\}$ with $h(z,D,\theta) \to 0$ as $d \to \infty$ if $D \geq dI_N$ and the flow on this manifold is determined by the ODE

$$dz/dt = B_{D,\theta} z + \int_\Omega \Phi_{D,\theta} f(\Phi_{D,\theta} z + h(z,D,\theta))\, dx.$$

If $D \geq dI_N$, we know that $B_{D,\theta} \to \bar{\theta}\varsigma$ and $\Phi_{D,\theta} \to |\Omega|^{-1/2}$ as $d \to \infty$. Thus, the flow of the above ODE is close to the flow defined by

$$dz/dt = -\bar{\theta}\varsigma z + |\Omega|^{1/2} f(|\Omega|^{-1/2} z).$$

If we rescale z to $|\Omega|^{1/2} z$, we obtain (10.8). As a consequence, the ODE describing the flow on the integral manifold has a local attractor which will correspond to a local attractor $A_{D,\theta}$ for the complete equations (10.1), (10.2).

To compare the local attractor $A_{D,\theta}$ with the local attractor A_θ for (10.8), embed A_θ into X_α by defining

$$\tilde{A}_{D,\theta} = \{\phi \in X^\alpha : \phi = \Phi_{D,\theta} z,\ z \in A_\theta\}.$$

Using the above remarks about the ODE's, Hale and Rocha [1987a] prove the following

THEOREM 4.10.3. *Suppose (10.8) has a local attractor A_θ and $\tilde{A}_{D,\theta}$ is defined as above. Then there is a $d_0 > 0$ such that (10.1), (10.2) has a local attractor $A_{D,\theta}$ for $D \geq d_0 I_N$, $\theta \in [0,\theta_0]$, and $\delta(A_{D,\theta}, \tilde{A}_{D,\theta}) \to 0$ as $d_0 \to \infty$. If the flow of (10.8) restricted to the attractor is stable, the flow of (10.1), (10.2) restricted to $A_{D,\theta}$ is equivalent to the flow of (10.8) restricted to A_θ.*

Hale and Rocha [1987b] discuss also the case where $\theta \in [0,1]$ and under appropriate hypotheses, they obtain the existence of a local attractor $A_{D,\theta}$ for $D \geq d_2^0 I_N$, d_2^0 sufficiently large. For $\theta \geq \theta_1$, $0 < \theta_1 < 1$, $A_{D,\theta}$ is a singleton which is shown to converge to a solution of the Dirichlet problem. These results use Lemma 4.10.1 in a significant way.

Application of these results can be seen in the cited references.

For the hyperbolic equation,

$$u_{tt} + 2au_t = d\Delta u + f(u) \quad \text{in } \Omega,$$
$$\partial u/\partial n = 0 \quad \text{in } \partial\Omega,$$

with f globally Lipschitzian with Lipschitz constant L, let λ be the first positive eigenvalue of $-\Delta$ on Ω and suppose $d\lambda - L > L^2/4a^2$. Sola-Morales and Valencia [1986] have shown that the spatial average $\bar{u}(t)$ of a solution satisfies

$$|u(t,\cdot) - \bar{u}(t)|_{H'(\Omega)} \to 0 \quad \text{as } t \to \infty$$

and $\bar{u}(t)$ satisfies the ordinary differential equation

$$\bar{u}'' + 2a\bar{u}' = f(u) + g(t),$$

where $g(t) \to 0$ as $t \to \infty$. The proof is similar to the proof of Theorem 4.10.2. Estimates on $u - \bar{u}$ are obtained from the skewed energy function

$$\Phi(v, v_t) = \frac{1}{2} \int_\Omega (d|\text{grad } v|^2 + v_t^2 + 2a^2 v^2 + 2avv_t)\, dx.$$

4.10.2. Singular perturbations. In this section, we mention some results on the upper semicontinuity of the attractor for some equations for which there is a singularly perturbed term. One result will concern a neutral functional differential equation (NFDE) and the other concerns an hyperbolic PDE.

Let us first consider the NFDE. The results stated below represent some unpublished work obtained in collaboration with X.-B. Lin. In the notation of §4.5, let us consider the NFDE

(10.9) $$(d/dt)[x(t) - \varepsilon M x_t] = f(x_t),$$

where $M: C \to R^n$ is a bounded linear operator, f is Lipschitz continuous in every bounded set, and ε is a small parameter.

THEOREM 4.10.4. *Suppose the RFDE*

(10.10) $$(d/dt)x(t) = f(x_t)$$

has a local attractor $A_0 \subset C$. Then there is an $\varepsilon_0 > 0$ such that (10.9) has a local attractor A_ε which is upper semicontinuous in ε at $\varepsilon = 0$.

PROOF. We give the main idea of the proof since the ideas are used in other types of problems and also bring out the role of the stability of the attractor A_0 for (10.10) as well as the manner in which (10.9) approximates (10.10).

We need the following lemma.

LEMMA 4.10.5. *Suppose the constants K, K_1 are related by the inequality*

$$|f(\phi_1) - f(\phi_2)| \leq K_1 |\phi_1 - \phi_2| \quad \text{for } \phi_1, \phi_2 \in N(K, A_0),$$

where $N(K, A_0)$ is the K-neighborhood of A_0. Suppose $t_1 > 0$, $\varepsilon > 0$ are fixed and $x^\varepsilon(t)$, $x^0(t)$ are solutions respectively of (10.9), (10.10) defined for $0 \leq t \leq t_1$ and having $x_t^\varepsilon, x_t^0 \in N(K, A_0)$, for $0 \leq t \leq t_1$ and $x_0^0 = x_0^\varepsilon$, $0 \leq \varepsilon \leq \varepsilon_0$. Then there is a continuous function $\delta(\varepsilon)$, $0 \leq \varepsilon \leq \varepsilon_0$, $\delta(0) = 0$, such that

$$|x_t^\varepsilon - x_t^0| \leq \delta(\varepsilon), \qquad 0 \leq t \leq t_1.$$

PROOF. The equation $\dot{x}(t) = 0$ considered as an RFDE in C generates a semigroup $\overline{T}(t)$ with $(\overline{T}(t)\phi)(\theta) = \phi(t+\theta)$, $t+\theta \leq 0$, $(\overline{T}(t)\phi)(\theta) = \phi(0)$, $t+\theta \geq 0$. If $X_0: [-r, 0] \to R^n$ is the $n \times n$ matrix function $X_0(\theta) = 0$, $\theta < 0$, and $X_0(0) = I_n$, then $\text{Var}_{0-\leq s \leq t}\overline{T}(t-s)X_0 \leq K_2$ for $0 \leq t < t_1$. Using the variation of constants formula for (10.9) (see Hale [1977, p. 302]), one obtains

$$x_t^\varepsilon = \overline{T}(t)\phi + \int_0^t \overline{T}(t-s)X_0 f(x_s^\varepsilon)\, ds - \varepsilon \int_0^{t^+} d_s[\overline{T}(t-s)X_0][Mx_s^\varepsilon - M\phi],$$

$$x_t^0 = \overline{T}(t)\phi + \int_0^t \overline{T}(t-s)X_0 f(x_s^0)\, ds.$$

Using the hypotheses in the lemma, one obtains

$$|x_t^\varepsilon - x_t^0| \leq \int_0^t K_1 K_2 |x_s^\varepsilon - x_s^0|\, ds + 2\varepsilon K K_2 \|M\|.$$

By the Gronwall inequality, we have $|x_t^\varepsilon - x_t^0| \leq K_3 \varepsilon$ for $0 \leq t \leq t_1$ and some constant K_3. This proves the lemma.

Returning to the proof of Theorem 4.10.4, let us now use the fact that A_0 is an attractor for (10.10). Let $T_0(t)$ be the semigroup generated by (10.10). There is a δ_2-neighborhood $N(\delta_2, A_0)$ of A_0 such that, for any $0 < \delta_1 < \delta_2$, there is a $t_0 > 0$ such that $T_0(t) N(\delta_2, A_0) \subset N(\delta_1, A_0)$ for $t \geq t_0$. We can now apply Lemma 4.10.5 to ensure that there is an $\varepsilon_0 > 0$ such that $T_\varepsilon(t) N(\delta_1/2, A_0) \subset N(\delta_1, A_0)$ for $0 \leq t \leq t_0$ and that there is a $\tau > 0$ such that $T_\varepsilon(t) N(\delta_1/2, A_0) \subset N(\delta_2/2, A_0)$ for $0 \leq t \leq t_0 + \tau$ for $0 \leq \varepsilon \leq \varepsilon_0$. Using the semigroup property of the semigroups, this is enough to show that, for any $\eta > 0$, there is a $t_1 > 0$, $\varepsilon_1 > 0$, such that $T_\varepsilon(t) N(\delta_1/2, A_0) \subset N(\eta, A_0)$ for $t \geq t_1$, $0 \leq \varepsilon \leq \varepsilon_1$. The details are omitted.

We now use the fact that $T_\varepsilon(t)$ is an α-contraction from Theorem 4.5.3. Since the orbit $T_\varepsilon(t) N(\delta_1/2, A_0)$, $t \geq 0$, is bounded, its ω-limit set is compact and attracts $N(\delta_1/2, A_0)$ from Lemma 3.2.1. Let A_ε be this ω-limit set. Then A_ε is a local attractor for $T_\varepsilon(t)$ and $A_\varepsilon \subset N(\eta, A_0)$. Since $\eta > 0$ is arbitrary, this implies that A_ε is upper semicontinuous and the proof of the theorem is complete.

In the comparison of NFDE with RFDE, restriction of the flows to the attractors is perhaps the most that can be expected. The semigroup $T_0(t)$ for the RFDE is compact for $t \geq r$, whereas the semigroup $T_\varepsilon(t)$ for the NFDE in general defines a group with no smoothing properties. Therefore, it is impossible to compare $T_\varepsilon(t)$ and $T_0(t)$ on the whole space. On the other hand, Theorem 4.10.4 shows that the local attractors are upper semicontinuous. The flows $T_\varepsilon(t)|A_\varepsilon$ and $T_0(t)|A_0$ are as smooth as the vector field f (see Theorem 4.5.4) and therefore it is feasible to compare these flows. One would not expect them to be the same without more restrictions on $T_0(t)|A_0$. If this latter flow is stable, one would expect the flow $T_\varepsilon(t)|A_\varepsilon$ to be equivalent to the one defined by $T_0(t)|A_0$, but this has not been considered.

In the analysis of this latter problem, Theorem 4.10.4 will probably play an important role since it can be used to show that the NFDE is a regular nonautonomous perturbation of the RFDE. In fact, since A_0 is compact and the sets A_ε, $0 \leq |\varepsilon| \leq \varepsilon_0$, are upper semicontinuous in ε, there is a bounded set B in C such that $A_\varepsilon \subset B$ for $0 \leq |\varepsilon| \leq \varepsilon_0$. Thus, there is a constant $k > 0$ such that $|f(\phi)| \leq k$ for ϕ in $\bigcup_{0 \leq |\varepsilon| \leq \varepsilon_0} A_\varepsilon$. If $x(t)$ is a solution of (10.9) for t in R, x_t in A_ε for $t \in R$, then $\dot{x}(t)$ is continuous and

$$\dot{x}(t) - \varepsilon M \dot{x}_t = f(x_t).$$

Since $M: C \to R^n$ is bounded, we may assume ε_0 so small that this relation implies there is a constant k_1 such that

$$|\dot{x}(t)| \leq k_1 \sup_{s \in R} |f(x_s)| \leq k_1 k$$

for $0 \leq |\varepsilon| \leq \varepsilon_0$. This implies that there is a bounded set U in $C^1([-r,0], R^n)$;

$$U = \{\phi \in B: |\dot\phi| \leq k_1 k\}$$

such that $A_\varepsilon \subset U$ for $0 \leq |\varepsilon| \leq \varepsilon_0$. The set U is in a compact set in C. The solution of the NFDE on A_ε can be written as

$$\dot x(t) = f(x_t) + ch(t)$$

where $h: R \to R^n$ satisfies $|h(t)| \leq |M| \cdot k_1 k$, $t \in R$. This shows the manner in which the NFDE is a regular perturbation of the RFDE. This latter equation certainly makes it plausible that the flow for the NFDE can be shown to be topologically equivalent to the RFDE if the latter is stable.

Let us now give an example of Hale and Raugel [1988] for the hyperbolic equation

(10.11) $\qquad \varepsilon u_{tt} + u_t - \Delta u = -f(u) - g \quad \text{in } \Omega, \qquad u = 0 \quad \text{in } \partial\Omega$

where $\varepsilon > 0$ is a small parameter, Ω is a bounded smooth domain in R^n, $n = 1, 2, 3$, g is a given function in $L^2(\Omega_0)$, and the initial data is taken in $X = H_0^1(\Omega) \times L^2(\Omega)$. Suppose that f belongs to $C^2(R, R)$ and satisfies the relation

(10.12) $\qquad \limsup_{|u| \to \infty}[-f(u)/u] \leq 0.$

Assume also that, for $n \geq 2$, there is a positive constant $c > 0$ such that

(10.13) $\qquad |f''(u)| \leq C(|u|^\gamma + 1) \quad \text{for } u \in R$

where

$$0 \leq \gamma < \infty \quad \text{if } n = 2,$$
$$0 \leq \gamma < 1 \quad \text{if } n = 3.$$

Along with equation (10.11), we consider the limiting parabolic equation when $\varepsilon = 0$,

(10.14) $\qquad u_t - \Delta u = -f(u) - g \quad \text{in } \Omega$

with Dirichlet boundary conditions.

Under the above hypotheses on f, there is a global attractor $\tilde A_0$ of (10.14) in H_0^1. In addition, $\tilde A_0$ is in $H_0^1(\Omega) \cap H^2(\Omega)$. The proof follows along the same lines as in §4.3 for the case $n = 1$ making use of the regularity theory for parabolic systems. Also, we have shown in §4.8 that (10.11) has a global attractor A_ε in X. How is A_ε related to $\tilde A_0$ for ε small? To make a comparison, we define

$$A_0 = \{(\phi, \psi): \phi \in \tilde A_0, \psi = \Delta\phi - f(\phi) - g, \phi \in \tilde A_0\}.$$

The set A_0 is a natural embedding of the attractor $\tilde A_0 \subset H_0^1(\Omega)$ into X. One can now prove (see Hale and Raugel [1988a])

THEOREM 4.10.6. *Under the above hypotheses, the sets $\{A_\varepsilon, A_0\}$ are upper semicontinuous at $\varepsilon = 0$; that is, $\lim_{\varepsilon \to 0} \delta(A_\varepsilon, A) = 0$.*

The outline of the proof, which will not be given in detail, is as follows. Let $Y = (H^2(\Omega) \cap H_0^1(\Omega)) \times H_0^1(\Omega)$. We have already remarked in §4.8 that the global attractor $A_\varepsilon \subset X$. If $(u^\varepsilon, u_t^\varepsilon)$ denotes the solution of (10.11) for $0 < \varepsilon \leq \varepsilon_0$, then, using the Lyapunov function

$$V_\varepsilon(\phi, \psi) = \int_\Omega \left[\frac{1}{2}\varepsilon|\psi(x)|^2 + \frac{1}{2}|\nabla\phi(x)|^2 + 2\varepsilon b\phi(x)\psi(x) + F(\phi(x)) + g(x)\phi(x) \right] dx$$

and an argument similar to the proof of Lemma 4.8.5, one shows that there is a positive constant c and, for any $r_0 > 0$, a positive constant $t_0 = t_0(r_0)$ such that, for $0 < \varepsilon \leq \varepsilon_0$,

(10.15) $\qquad \varepsilon \|u_t(t)\|_0^2 + \|u(t)\|_1^2 \leq c^2 \quad \text{for } t \geq t_0(r_0)$

provided that the initial data (u_0, u_1) satisfies $\varepsilon \|u_1\|_0^2 + \|u_0\|_1^2 \leq r_0^2$. A subscript 0, 1, or 2 on the norm designates respectively the norm in L^2, H_0^1, $H^2 \cap H_0^1$.

Using the Lyapunov function

$$V_\varepsilon(\phi, \psi) = \int_\Omega \left[\frac{1}{2}\varepsilon|\psi(x)|^2 + 2\varepsilon b\phi(x)\psi(x) + \frac{1}{2}|\nabla\phi(x)|^2 \right] dx$$

and arguments similar to the proof of Lemma 4.8.12, one can show that there are a constant $c_0 > 0$ and constants $c_1(r_0), c_2(r_1)$ such that, for $0 < \varepsilon \leq \varepsilon_0$, the estimate

(10.16) $\qquad \varepsilon \|u_{tt}^\varepsilon\|_0^2 + \|u_t^\varepsilon\|_1^2 + \|u^\varepsilon\|_2^2 \leq c_1(r_0) + \frac{c_2(r_1)}{\varepsilon} e^{-c_0 t}, \quad t \geq 0$

holds provided that $\varepsilon \|u_1\|_i^2 + \|u_0\|_{i+1}^2 \leq r_i^2$ for $i = 0, 1$.

Now suppose that $(u(t), u_t(t)) \in A_\varepsilon$ for $t \in R$. We know from (10.15) that

$$\varepsilon \|u_t(t)\|_0^2 + \|u(t)\|_1^2 \leq c^2 \quad \text{for } t \in R.$$

We also know from Lemma 4.8.13 that A_ε belongs to a bounded set in Y. Thus, there is an $r_1 = r_1(\varepsilon)$ such that $\varepsilon \|\psi\|_1^2 + \|\phi\|_2^2 \leq r_1^2$ for $(\phi, \psi) \in A_\varepsilon$. Let us now apply (10.16) with this r_1 and $r_0 = c$. Since A_ε is invariant, it follows that

$$\varepsilon \|u_{tt}^\varepsilon(\tau)\|_0^2 + \|u_t^\varepsilon(\tau)\|_1^2 + \|u^\varepsilon(\tau)\|_2^2 \leq c_1(r_0) \quad \text{for } \tau \in R.$$

We have therefore proved the following

LEMMA 4.10.7. *There is a bounded set $B \subset Y$ and a positive constant c such that, for $0 < \varepsilon \leq \varepsilon_0$,*

$$A_\varepsilon \subset B, \qquad \varepsilon^{1/2} \|u_{tt}^\varepsilon\|_0 \leq c, \qquad t \in R,$$

for any solution (u, u_t) of (10.11) with initial data in A_ε.

This lemma has several important implications. If $(u^\varepsilon, u_t^\varepsilon)$ is a solution in A_ε and $h_\varepsilon(t) = \varepsilon u_{tt}$, then $\|h_\varepsilon(t)\|_0 \to 0$ as $\varepsilon \to 0$ uniformly in $t \in R$. Therefore, the solution of the hyperbolic equation is a solution of the perturbed parabolic equation

$$u_t = \Delta u + f(u) - g + h_\varepsilon(t) \quad \text{in } \Omega \tag{10.17}$$

with homogeneous Dirichlet conditions. The singularly perturbed problem (10.11) has been reduced to a regular perturbation problem (10.17). The fact that $A_\varepsilon \subset B$ for $0 < \varepsilon \leq \varepsilon_0$ with B bounded in Y implies that A_ε belongs to a special type of compact set in X. These two facts allow us to prove the upper semicontinuity at $\varepsilon = 0$.

In fact, suppose that ε_n is a sequence approaching zero as $n \to \infty$ and let $u^n = u^{\varepsilon_n}$. Then one can assume that $u^n \to u$ in $C(J, H_0^1(\Omega))$ for any compact interval $J \subset R$. Using the above estimates, one observes that

$$u_t^n \to \Delta u^n - f(u) - g \quad \text{in } C(J, H^{-1}(\Omega)).$$

Observing that $u_t^n \in C_b(R, H^{-1}(\Omega))$ where C_b denotes bounded continuous, and $|u_t^n|_1 \leq k$, a constant, for $t \in R$, one obtains that $u_t^n \to u_t$ in $C(J, L^2(\Omega))$ and $u_t \in C_b(R, L^2(\Omega))$. Thus, $\Delta u \in C_b(R, L^2(\Omega))$. Using the regularity theory for the Laplacian, we have $u \in L^\infty(R, H^2(\Omega))$. Therefore, we have proved that $u \in L^\infty(R, H^2(\Omega)) \cap W^{1,\infty}(R, L^2(\Omega))$. As a consequence, $(u, u_t) \in A_0$ for any $t \in R$. Since $(u^n, u_t^n) \to (u, u_t)$ in $C(J, H_0^1(\Omega) \times L^2(\Omega))$, it follows that the initial data converge in X and belong to A_0. This shows that the family of sets $\{A_\varepsilon, \varepsilon > 0, A_0\}$ is upper semicontinuous at $\varepsilon = 0$. Otherwise, there would be a sequence $\varepsilon_n \to 0$ as $n \to \infty$, a $\nu > 0$, and a solution in A_{ε_n} with

$$\delta((u^n(0), u_t^n(0)), A_0) \geq \nu.$$

This contradicts the properties above concerning the convergence of solutions.

For the case of one space variable, a related problem has been considered by Mora [1987] and more general results are obtained. More specifically, consider the equation

$$\varepsilon u_{tt} + u_t = u_{xx} + f(u) \quad \text{in } (0, \pi),$$
$$u = 0, \quad \text{at } x = 0, \pi$$

where $f(u)$ satisfies (10.12). Then there is an integer n_0 and a function $\beta(n)$, n an integer, $\beta(n) \to \infty$ as $n \to \infty$, such that, for any $n \geq n_0$, and $\varepsilon^{-1} > \beta(n)$, the global attractor A_ε of the above equation lies in a C^1-local invariant submanifold of $H_0^1(0, \pi) \times L^2(0, \pi)$ of dimension n. Such an invariant manifold often is referred to as an *inertial manifold* (see §4.10.5).

This result of Mora implies that the flow on the attractor can be described by an ODE of dimension n provided that $n \geq n_0$ and the singular term ε is sufficiently small. More recently, Mora and Sola-Morales [1987] and Chow and Lu [1987b] have shown that the vector field $h(\cdot,\varepsilon)$ varies continuously in ε in the C^1-topology even as $\varepsilon \to 0$. For $\varepsilon = 0$, the vector field describes the flow on an integral manifold containing the global attractor A_0 of the parabolic equation

$$u_t = u_{xx} + f(u) \quad \text{in } (0,\pi),$$
$$u = 0 \quad \text{at } x = 0,\pi.$$

In particular, the global attractors A_ε are upper semicontinuous at $\varepsilon = 0$. If the flow on A_0 is stable, then the flow on A_ε is equivalent to the one on A_0 if ε is sufficiently small.

For the case of (10.1), (10.2) with $n > 1$, there do not exist such precise results at this time. In §10.4, we return to some aspects of this problem.

For the case $n = 1$, another question immediately comes to mind. Does the attractor A_ε lie on a C^1-local invariant manifold for each $\varepsilon > 0$? In general, this is not the case. For ε large, the exponential approach to the global attractor is governed by an infinite number of eigenvalues with real part $(2\varepsilon)^{-1}$ and this fact prevents the embedding of A_ε into a C^1-local invariant manifold. For the details, see Mora and Sola-Morales [1986] or Hale and Sell [1988].

4.10.3. *Approximation of attractors.* In previous sections, we have given many illustrations of evolutionary equations for which there exist global attractors. Due to the complexity of some of the equations, the location of the attractor and the flow on the attractor must be determined by some numerical approximation to the equation through Galerkin approximations, splines, or discretizations in time. Let the original semigroup be $T(t)$ and the approximate semigroups be $T_h(t)$, where h is a small parameter. Many important problems present themselves, two of which are

(i) If $T(t)$ has an attractor A, does $T_h(t)$ have an attractor A_h for h small and does $\delta(A_h, A) \to 0$ as $h \to 0$?

(ii) If $T_h(t)$ has an attractor A_h for h small, does $T(t)$ have an attractor A and does $\delta(A_h, A) \to 0$ as $h \to 0$?

One also would like to be able to show that $\delta(A, A_h) \to 0$ and relate the corresponding flows on the attractor. It is somewhat surprising that very little research has been done on any of these problems.

Obviously, problem (ii) is the most important. Once the approximate equations have become the center of attention, information is obtained only from these equations. Therefore, predicting the behavior of the exact system is of major concern. Problem (ii) is also very difficult and few results are available. In addition, the results that have been obtained require stringent hypotheses. We mention the papers of Constantin, Foiaş, and Temam [1984] and Kloeden [1984], [1986] for stable equilibrium solutions of the Navier-Stokes equations.

It is possible to obtain abstract results on the upper semicontinuity of the attractors for general semigroups if some hypotheses are satisfied. There are two types (see Theorem 2.5.4). One is to assume that the approximate attractors A_h all belong to a bounded set U, uniformly attract this set, and the semigroups

$T_h(t)$ approximate $T(t)$ pointwise. In addition, $T(t)$ must be asymptotically smooth. This was used by Hale, Lin, and Raugel [1985]. One could also assume that A_0 is an attractor for $T(t)$ and $T_h(t)y - T(t)y \to 0$ as $h \to 0$ uniformly for $t \in [0, \tau]$ and a bounded neighborhood U of A_0. This method was used by Kloeden and Lorenz [1986] and Lin and Raugel [1986]. For boundary value problems, see Schmitt, Thompson, and Walter [1978].

Let us now discuss problem (i) above; that is, if $T(t)$ has an attractor A, does $T_h(t)$ have an attractor and does $\delta(A_h, A) \to 0$ as $h \to \infty$. Abstract results generalizing the ideas used in the proof of Theorem 4.10.3 have been given by Hale, Lin, and Raugel [1985]. Applications are then given to spectral projection methods for sectorial evolutionary equations and Galerkin approximations for parabolic equations as well as discretizations in time. A linear damped wave equation is also discussed.

The above discussion has been concerned only with the convergence of A_h to A as $h \to 0$ considered in the sense of sets. Some results are also known about the relationship between the dynamics on A_h and the dynamics on A. For the case of parabolic equations which are Morse-Smale on A, Lin and Raugel [1986] have shown that the flow on A_h is equivalent to the flow on A. Using the Conley index, Khalsa [1985] has discussed the dynamics of a discrete approximation for a parabolic equation in one space variable with a cubic nonlinearity. Numerical computations using Galerkin approximations have been done for a similar example by Rutkowski [1983] and Mora [1984].

4.10.4. *Lower semicontinuity of the attractor.* In this section, we present a special case of a result of Hale and Raugel [1988b] on lower semicontinuity of the attractor for gradient systems.

Before stating the precise result, let us discuss, in an intuitive way, the expected behavior of gradient systems. In §3.8, we observed that hyperbolicity of the equilibria and transversal intersection of stable and unstable manifolds imply the flow on the global attractor is stable under smooth perturbations of the semigroup. This implies in particular that a family of perturbations depending on a parameter ε has the property that the attractor is continuous in ε.

Hyperbolicity of equilibria is a local property, the verification of which consists of discussing the eigenvalues of some linear operator. Transversality is a global property for which no general procedure is available for verification. Therefore, it is natural to determine as much information as possible about the flow when transversality may not hold. We will see in this case that the attractors still are continuous in the parameter. Intuitively, this seems as if it should be true because the attractor is the union of the unstable manifolds of the equilibria and these unstable manifolds vary continuously with the parameter.

Our objective is to state a precise result in the spirit of the above discussion. We also want the result to apply to situations where the dependence of the semigroup on parameters is not continuous, but it is continuous when the discussion is restricted to the attractor as, for example, in problems of the type discussed in this chapter.

Let X be a Banach space and, for $0 \leq \varepsilon \leq \varepsilon_0^*$, let $T_\varepsilon(t)$, $t \geq 0$, be a family of semigroups on X. We make the following hypotheses on $T_\varepsilon(t)$ for $\varepsilon \in [0, \varepsilon_0^*]$:

(H1)$_0$ $T_0(t)$ is a C^1-gradient system, asymptotically smooth with orbits of bounded sets bounded.

(H2)$_0$ The set E_0 of equilibrium points of $T_0(t)$ is bounded in X.

(H3)$_0$ Each element of E_0 is hyperbolic.

Since $T_0(t)$ satisfies (H1)$_0$, (H2)$_0$, the global attractor A_0 exists. Furthermore, (H3)$_0$ implies that $E_0 = \{\phi_1, \ldots, \phi_N\}$ is a finite set and

$$A_0 = \bigcup_{\phi_j \in E_0} W^u(\phi_j)$$

where $W^u(\phi_j)$ is the unstable manifold of ϕ_j.

(H4)$_\varepsilon$ For $\varepsilon \neq 0$, $T_\varepsilon(t)$ is a C^1-semigroup and there is a neighborhood U_0 of A_0 which is independent of ε such that $T_\varepsilon(t)$ has a local attractor A_ε which attracts U_0 (of course, A_ε can be the global attractor for $T_\varepsilon(t)$).

(H5)$_\varepsilon$ If E_ε is the set of equilibrium points of $T_\varepsilon(t)$, then there exists a neighborhood W_0 of E_0 such that $W_0 \cap E_\varepsilon = \{\phi_{1,\varepsilon}, \ldots, \phi_{N,\varepsilon}\}$ where each $\phi_{j,\varepsilon}$, $1 \leq j \leq N$, is hyperbolic and $\phi_{j,\varepsilon} \to \phi_j$ as $\varepsilon \to 0$.

(H6)$_\varepsilon$ $\delta_X(W^u_{\text{loc}}(\phi_j), W^u_{\text{loc},\varepsilon}(\phi_{j,\varepsilon})) \to 0$ as $\varepsilon \to 0$.

(H7)$_\varepsilon$ For any $\eta > 0$, $t_0 > 0$, $\tau_0 > 0$, there are a $\delta^* > 0$ and an $\varepsilon_0^* > 0$ such that

$$\|T_\varepsilon(t)y_\varepsilon - T_0(t)x\|_X < \eta \quad \text{for } t_0 \leq t \leq \tau_0,$$

provided that

$$x \in A_0, \quad y_\varepsilon \in A_\varepsilon, \quad \text{and} \quad \|y_\varepsilon - x\|_X \leq \delta^*.$$

THEOREM 4.10.8. *Under the hypotheses* (H1)$_0$ *to* (H7)$_\varepsilon$, *the family of sets* $\{A_\varepsilon, 0 \leq \varepsilon \leq \varepsilon_0\}$ *is lower semicontinuous at* $\varepsilon = 0$.

We obtain a useful corollary for a family of gradient systems if we replace (H4)$_\varepsilon$ and (H7)$_\varepsilon$ by the stronger hypotheses:

(H.4bis)$_\varepsilon$ For $\varepsilon \neq 0$, $T_\varepsilon(t)$, $t \geq 0$, is a C^1-semigroup which is asymptotically smooth.

(H.7bis)$_\varepsilon$ $T_\varepsilon(t)x$ is continuous in (t, x, ε) and continuous in ε uniformly with respect to (t, x) in bounded sets of $R \times X$.

With these new hypotheses, we know from Theorem 3.8.8 that the family of sets $\{A_\varepsilon, 0 \leq \varepsilon \leq \varepsilon_0\}$ is upper semicontinuous at $\varepsilon = 0$. Combining this fact with Theorem 4.10.8, we have

COROLLARY 4.10.9. *Under the hypothesis* (H1)$_0$ *to* (H.7bis)$_\varepsilon$ *the family of sets* $\{A_\varepsilon, 0 \leq \varepsilon \leq \varepsilon_0^*\}$ *is continuous at* $\varepsilon = 0$.

A result of this type has been stated also by Babin and Vishik [1986].

REMARK 4.10.10. The hypotheses in Theorem 4.10.8 are weak enough to allow one to prove the lower semicontinuity of the attractor when the semigroup $T_\varepsilon(t)$ does not converge to $T_0(t)$ as $\varepsilon \to 0$ in a uniform way. For specific applications to numerical methods, see Hale and Raugel [1988a]. The results are also applicable to the singular hyperbolic equation (10.10). In fact, one can prove (see Hale and Raugel [1988b])

THEOREM 4.10.11. *With the notation of Theorem 4.10.6, the global attractors A_ε for the hyperbolic equation together with the global attractor A_0 for the parabolic equation are continuous at $\varepsilon = 0$ provided the equilibrium points are hyperbolic.*

PROOF OF THEOREM 4.10.8. We apply the decomposition of the attractor A_0 for $T_0(t)$ that is given in Theorem 3.8.7. Let $E_0 = \{\phi_1, \ldots, \phi_N\}$, $v^1 > v^2 > \cdots > v^M$ be the distinct points of the set $\{V(\phi_1), \ldots, V(\phi_N)\}$,

$$E_0^j = \{x \in E_0 : V(x) = v^j\},$$

$$A_0^k = \bigcup \{W^u(x) : x \in \bigcup_{j=k}^{M} E_0^j\},$$

$$U_0^k = \{x \in X : V(x) < v^k\}.$$

Theorem 3.8.7 asserts that A_0^k is a compact attractor for $T_0(t)$ restricted to U_0^{k-1}. In particular, for any compact set in U_0^{k-1}, $\delta(T_0(t)K, A_0^k) \to 0$ as $t \to \infty$. Therefore, for any $\eta > 0$, there is a $\tau > 0$ such that

$$T_0(t)(U_0^{k-1} \setminus N_\eta(E_0^{k-1})) \subset N_\eta(A_0^k) \quad \text{for } t > \tau,$$

where $N_\eta(\cdot)$ designates an η-neighborhood. Thus, for any $x \in U_0^{k-1} \setminus N_\eta(E_0^{k-1})$, there is an $\tilde{x} \in \bigcup_{j=k}^{M} E_0^j$ such that $\delta(T_0(\tau)x, W^{u,0}(\tilde{x})) < \eta$.

This is the reason that one suspects Theorem 4.10.8 to be true since $(H7)_\varepsilon$ is satisfied. To make the argument precise, one must proceed with some care.

If ϕ_j is an hyperbolic equilibrium point of $T_0(t)$, then there is an $r > 0$ such that $N_r(\phi_j) \cap W^u(\phi_j) \subset W^u_{\text{loc}}(\phi_j)$ and, if

$$\Gamma_j^r = W^u_{\text{loc}}(\phi_j) \cap \partial N_r(\phi_j),$$

then

(10.18) $$W^u(\phi_j) = W^u_{\text{loc}}(\phi_j) \cup \left(\bigcup_{t \geq 0} T(t)\Gamma_j^r\right).$$

We also can choose r so that $N_{2r}(\phi_j) \cap N_{2r}(\phi_k) = \emptyset$ for $j \neq k$, $1 \leq j, k \leq N$.

The sets $\{\Gamma_j^r, j = 1, 2, \ldots, N\}$ are compact. From the definition of a gradient system, for any integer k, $1 \leq k \leq N - 1$, there is a $t_k > 0$ such that, for any $\phi_{jk} \in E_0^k$,

(10.19) $$V(T_0(t)\Gamma_{jk}^r) \leq v^{k+1} + d/2 \quad \text{for } t \geq t_k,$$

where $d = \min\{v^{i-1} - v^i, i = 2, 3, \ldots, M\}$. Set $t_0 = \max\{t_k : 1 \leq k \leq M - 1\}$.

Now suppose $\eta > 0$ is a positive number. In order to prove the theorem, it is sufficient to show that, for any $x \in A_0$,

(10.20) $$\inf_{y \in A_\varepsilon} |T_0(t)x - y| \leq \eta, \quad t \geq 0.$$

If x is an equilibrium point, then $(H5)_\varepsilon$ implies that (10.20) is satisfied for $0 < \varepsilon \leq \bar{\varepsilon}_0$. If $x \in A_0 \setminus E_0$, then there is a $\phi_i \in E_0$ such that $x \in W^u(\phi_i)$ and there is an s_i such that $x = T_0(s_i)\phi_i^*$ where $\phi_i^* \neq \phi_i$, $\phi_i^* \in W^u_{\text{loc}}(\phi_i) \cap N_r(\phi_i)$.

By $(H6)_\varepsilon$, there is a real number $\bar{\varepsilon}_1 < \bar{\varepsilon}_0$ such that, for $0 < \varepsilon \leq \bar{\varepsilon}_1$, we have, for $1 \leq j \leq N$,

$$\delta_X(W^u_{\text{loc}}(\phi_j), W^u_{\text{loc},\varepsilon}(\phi_{j\varepsilon})) \leq \eta. \tag{10.21}$$

Therefore, it is sufficient to show that, for $1 \leq j \leq N$, and for any ϕ_j^* in Γ_j^r, we have

$$\inf_{y \in A_\varepsilon} \|T(t)\phi_j^* - y\|_X \leq \eta \quad \text{for } t \geq 0 \tag{10.22}$$

and $0 < \varepsilon \leq \varepsilon_0$, for ε_0 sufficiently small.

Without loss in generality, it is sufficient to give the proof of (10.22) only for the case in which $\phi_j = \phi_{j1}$ belongs to E_1 and $\phi_{j1}^* \in \Gamma_{j1}^r$. The proof will be by induction consisting of at most M steps.

Since A_0 is compact, there exists a constant ν_d such that, for any $x_1 \in A_0$, $x_2 \in A_0$, $\|x_1 - x_2\|_X < \nu_d$, we have

$$\|V(x_1) - V(x_2)\| < d/4. \tag{10.23}$$

Define

$$B_k = \{x \in A_0 : V(x) \leq v^k - d/4\}$$

and fix a constant $t_0^* > 0$.

We next construct two sequences of real numbers

$$\eta_{M-1} > \eta_{M-2} > \cdots > \eta_0,$$
$$\varepsilon_{M-1} > \varepsilon_{M-2} > \cdots > \varepsilon_0,$$

in the following manner. Let

$$\eta_{M-1} = \min\{\eta/2, \eta_{d/2}, r/2\}$$

and choose $\varepsilon_{M-1} \leq \varepsilon_0^*$ so that $\|\phi_j - \phi_{j,\varepsilon}\|_X \leq \eta_{M-1}$ for $1 \leq j \leq N$.

Suppose $k \geq 1$ and assume that $\eta_{M-1} > \cdots > \eta_k$ and $\varepsilon_{M-1} > \cdots > \varepsilon_k$ have been chosen. There exists a $\tau_k > 0$ such that, for $t \geq \tau_k$,

$$T_0(t)B_k \subset N_{\eta_k}(A_0^{k+1}). \tag{10.24}$$

From the representation of A_0^{k+1}, we deduce that, for any $\phi \in B_k$, there is a $\psi \in \bigcup_{j=k+1}^M E^j$ such that

$$\delta_X(T_0(\tau_k)\phi, W^u(\psi)) \leq \eta_k.$$

Let t_0 be the number defined after formula (10.19). Since $T(t)$ is a C^0-semigroup, there is an $\eta_{k-1}^* > 0$ such that

$$\|T_0(t)x - T_0(t)x'\|_X < \eta_k \quad \text{for } 0 \leq t \leq (t_0 + \tau_k) + 2t_0^* \tag{10.25}$$

as soon as $x \in A_0$, $x' \in A_0$, and $\|x - x'\|_X \leq 2\eta_{k-1}^*$. By $(H7)_\varepsilon$, there are positive real numbers $\eta_{k-1} \leq \min(\eta_{k-1}^*, \eta_k/2)$ and $\varepsilon_{k-1}^* < \varepsilon_k$ such that, for $0 < \varepsilon \leq \varepsilon_{k-1}^*$, we have

$$\|(T_\varepsilon(t)y_\varepsilon - T_0(t)x\|_X < \eta_k \quad \text{for } t_0^* \leq t \leq (t_0 + \tau_k) + 2t_0^*, \tag{10.26}$$

provided that $x \in A_0$, $y_\varepsilon \in A_\varepsilon$, and $\|x - y_\varepsilon\|_X \leq 2\eta_{k-1}$. By (H6)$_\varepsilon$, we can choose $0 < \varepsilon_{k-1} < \varepsilon_{k-1}^*$ so that, for $0 < \varepsilon \leq \varepsilon_{k-1}$,

(10.27) $\qquad \delta_X(W_{\text{loc}}^u(\phi_j), W_{\text{loc},\varepsilon}^u(\phi_{j,\varepsilon})) \leq \eta_{k-1} \quad \text{for } 1 \leq j \leq N.$

This completes the construction of the sequence η_k, ε_k.

We now prove by induction that, for $0 < \varepsilon \leq \varepsilon_0$, (10.22) holds for $\phi_{j1} \in E^1$ and ϕ_{j1}^* in Γ_{j1}^r.

Step 1. By (10.19), $T(t_1)\Gamma_{j1}^r \subset B_1$ and $T_0(t_1 + t)\Gamma_{j1}^r \subset N_{\eta_1}(A^2)$ for $t \geq \tau_1$. There is a $\tilde{\phi}_{j1}^*$ in $W_{\text{loc}}^u(\phi_{j1})$ such that $T(t_0)\tilde{\phi}_{j1}^* = \phi_{j1}^*$. By (H6)$_\varepsilon$, (H7)$_\varepsilon$, and the definitions of ε_0, η_0, there is a $\tilde{\phi}_{j1,\varepsilon}^*$ in $W_{\text{loc},\varepsilon}^u(\phi_{j1,\varepsilon})$ such that

$$\|\tilde{\phi}_{j1}^* - \tilde{\phi}_{j1,\varepsilon}^*\|_X \leq \eta_0,$$
$$\|T_0(t_0+t)\tilde{\phi}_{j1}^* - T_\varepsilon(t_0+t)\tilde{\phi}_{j1,\varepsilon}^*\|_X \leq \eta_1 \quad \text{for } 0 \leq t \leq (t_1 + \tau_1) + t_0^*.$$

Since $T(t_0)\tilde{\phi}_{j1}^* = \phi_{j1}^*$, we have in particular that, for $s_1 = t_1 + \tau_1$,

(10.28) $\qquad \inf_{y \in A_\varepsilon} \|T_0(t)\phi_{j1}^* - y\|_X \leq \eta_1 \quad \text{for } 0 \leq t \leq s_1 + t_0^*$

and $T_0(s_1 + s)\phi_{j1}^* \in N_{\eta_1}(A^2)$ for $s \geq 0$.

Step k. Let us assume that there exists an $s_{k-1} > 0$ such that

(10.29) $\qquad T_0(s_{k-1} + s)\phi_{j1}^* \in N_{\eta_{k-1}}(A^k) \quad \text{for } s \geq 0,$

(10.30) $\qquad \inf_{y \in A_\varepsilon} \|T_0(t)\phi_{j1}^* - y\|_X \leq \eta_{k-1} \quad \text{for } 0 \leq t \leq s_{k-1} + t_0^*.$

From (10.30) and (10.26), we have

(10.31) $\qquad \inf_{y \in A_\varepsilon} \|T_0(t)\phi_{j1}^* - y\|_X \leq \eta_k \quad \text{for } 0 \leq t \leq s_{k-1} + (t_0 + \tau_k) + 3t_0^*.$

From (10.29), there exists $\phi_{jk}^s \in \bigcup_{j=k}^M E^j$ and $\tilde{\phi}_{jk}^s \in W^u(\phi_{jk}^s)$ such that

(10.32) $\qquad \|T_0(s_{k-1} + s)\phi_{j1}^* - \tilde{\phi}_{jk}^s\|_X \leq \eta_{k-1}.$

Let us first put $s = 0$. If $\tilde{\phi}_{jk}^0$ belongs to $W_{\text{loc}}^u(\phi_{jk}^0) \cap \{y \in X : \|y - \phi_{jk}^0\|_X < r\}$, then we have

(10.33) $\qquad \|T_0(s_{k-1})\phi_{j1}^* - \phi_{jk}^0\|_X \leq \eta_{k-1} + r \leq 3r/2$

and there exists a real number $\sigma_0 > 0$ such that $T_0(s_{k-1} + s)\phi_{j1}^* \in N_{2r}(\phi_{jk}^0)$ for $0 \leq s \leq \sigma_0$. Moreover, if, for an s in $(0, \sigma_0]$, the element $\tilde{\phi}_{jk}^s$ given in (10.32) belongs to $W_{\text{loc}}^u(\phi_{jk}^s) \cap N_r(\phi_{jk}^s)$, then $\phi_{jk}^s = \phi_{jk}^0$ since $N_{2r}(\phi_j) \cap N_{2r}(\phi_k) = \emptyset$ for $j \neq k$. This property means that we have only two possibilities: either

(a) $\tilde{\phi}_{jk}^s$ belongs to $W_{\text{loc}}^u(\phi_{jk}^0) \cap N_r(\phi_{jk}^0)$ for any $s \geq 0$; that is, the ω-limit set $\omega(\phi_{j1}^*)$ is ϕ_{jk}^0. Then, the induction hypothesis (10.30), relation (10.31), and (H6)$_\varepsilon$ (see (10.27)) imply that, for $0 < \varepsilon \leq \varepsilon_0$, we have

(10.34) $\qquad \inf_{y_\varepsilon \in A_\varepsilon} \|T_0(t)\phi_{j1}^* - y_\varepsilon\|_X \leq 2\eta_{k-1} \quad \text{for } t \geq 0;$

or

(b) there is a first time $\sigma \geq 0$ such that the element $\tilde{\phi}^\sigma_{jk}$ given in (10.32) does not belong to $W^u_{\text{loc}}(\phi^\sigma_{jk}) \cap N_r(\phi^\sigma_{jk})$ and, if $\sigma > 0$, $\tilde{\phi}^s_{jk} \in W^u_{\text{loc}}(\phi^0_{jk}) \cap N_r(\phi^0_{jk})$ for $0 \leq s < \sigma$.

In case (b), $\tilde{\phi}^\sigma_{jk} = T_0(s^\sigma_{jk})\phi^{*\sigma}_{jk}$ where $\phi^{*\sigma}_{jk} \in \Gamma^r(\phi^\sigma_{jk}) = \{y \in X : \|y - \phi^\sigma_{jk}\| = r\}$ and $s^\sigma_{jk} \geq 0$. Thus, for $s \geq t_k$, $V(T_0(s)\tilde{\phi}^\sigma_{jk}) \leq v^{k-1} + d/2$. From property (10.32) and (10.25), we obtain

(10.35) $\qquad \|T_0(s_{k-1} + \sigma + s)\phi^*_{j1} - T_0(s)\tilde{\phi}^\sigma_{jk}\|_X \leq \eta_k \quad$ for $0 \leq s \leq t_k$.

As $\eta_k \leq \eta_d$, we conclude from (10.35) that $T_0(s_{k-1} + \sigma + s)\phi^*_{j1}$ belongs to B_k for $s \geq t_k$ and, therefore, $T_0(s_k + s)\phi^*_{j1} \in N_{\eta_k}(A^{k+1}_0)$ for $s \geq 0$, where $s_k = s_{k-1} + \sigma + t_k + \tau_k$.

It now remains to prove (10.30) with $k - 1$ replaced by k; that is,

(10.36) $\qquad \inf_{y_\varepsilon \in A_\varepsilon} \|T(t)\phi^*_{j1} - y_\varepsilon\|_X \leq \eta_k \quad$ for $0 \leq t \leq s_k + t^*_0$.

If $\sigma \leq 2t^*_0$, (10.36) is the same as (10.31).

Suppose now that $\sigma > 2t^*_0$. By (H6)$_\varepsilon$ (see (10.27)), for $0 \leq s < \sigma$, there exist $\tilde{\phi}^s_{jk,\varepsilon} \in W^u_{\text{loc},\varepsilon}(\phi^0_{jk,\varepsilon})$ such that

(10.37) $\qquad \|\tilde{\phi}^s_{jk,\varepsilon} - \tilde{\phi}^s_{jk}\|_X \leq \eta_{k-1}$.

The estimates (10.31) and (10.37) imply, for $0 \leq s \leq \sigma$,

(10.38) $\qquad \|\tilde{\phi}^s_{jk,\varepsilon} - T_0(s_{k-1} + s)\phi^*_{j1}\|_X \leq 2\eta_{k-1} < \eta_k$.

As a consequence of (10.26) and (10.38) for $s = \sigma - t^*_0$, we have

(10.39) $\|T_0(s_{k-1} + \sigma - t^*_0 + t)\phi^*_{j1} - T_\varepsilon(t)\tilde{\phi}^{\sigma-t_0}_{jk,\varepsilon}\| \leq \eta_k \quad$ for $t^*_0 \leq t \leq (t_0 + \tau_k) + 2t^*_0$.

(10.36) follows from the induction hypothesis (10.30) and (10.38), (10.39).

Step M. Finally, arguing as above $M - 2$ times at most, one shows that either there exists a $k_0 \leq M - 1$ such that $\omega(\phi^*_{j1}) \subset E^{k_0}$ and then (10.34) holds for $k = k_0$, or there exists $s_{M-1} > 0$ such that $T_0(s_{M-1} + s)\phi^*_{j1} \in N_{\eta_{M-1}}(E^M)$ for $s \geq 0$ and the inequality (10.34) holds for $k = M - 1$. Since $W^u_{\text{loc}}(\phi_{jM}) = \phi_{jM}$ for $\phi_{jM} \in E^M$, the hypothesis (H6)$_\varepsilon$ (see (10.27)) implies that

$$\inf_{y_\varepsilon \in A_\varepsilon} \|T(s_{M-1} + s)\phi^*_{j1} - y_\varepsilon\|_X \leq 2\eta_{M-1} \quad \text{for } s \geq 0$$

and the estimate (10.22) is proved. This completes the proof of the theorem.

4.10.5. Remarks on inertial manifolds. Throughout this book, we have emphasized the importance of restricting the flow of a dissipative system to the global attractor. In several of the examples, we also have seen that the flow on the global attractor is equivalent to the flow defined by an ordinary differential equation on a finite-dimensional manifold. It is natural to try to classify those dissipative systems for which the flow on the global attractor is equivalent to such an ordinary differential equation. This subject belongs to the theory of inertial manifolds.

Suppose we are given an evolutionary equation in a Banach space X which defines a semigroup $T(t)$ on X and has a global attractor A. We say that M is an *inertial manifold* of the equation if M is a finite-dimensional submanifold of X which contains the attractor A and is positively invariant under $T(t)$; that is, $T(t)M \subset M$. If the semigroup $T(t)$ is one-to-one on M, then the flow restricted to M (and, thus, restricted to A) is determined by some ordinary differential equation.

Suppose that $T(t)$ depends on a parameter, say $T(t) = T_\lambda(t)$, λ in a Banach space E, and, for each λ, there is an inertial manifold M_λ with $\dim M_\lambda = n$, constant for $\lambda \in \Lambda$, and a corresponding n-dimensional ordinary differential equation

$$\dot{x} = f_\lambda(x)$$

which determines the flow on A_λ. Suppose also that $f_\lambda(x)$ is continuous in λ, x together with its first derivative in x. If the flow defined by f_{λ_0} is (structurally) stable, then we know that each of the flows $T_\lambda(t)|A_\lambda$ is equivalent to the flow $T_{\lambda_0}(t)A_{\lambda_0}$. This fact alone is sufficient justification to devote some effort to the problem of the existence of inertial manifolds.

At the present time, this is an active area of research in differential equations and a complete discussion is inappropriate here. We will give some of the prevailing ideas and refer the reader to the papers of Foiaş, Sell, and Temam [1987], Mallet-Paret and Sell [1987], Chow and Lu [1987b], and the book of Temam [1988] for extensive references.

One of the basic approaches for the existence of inertial manifolds has been to apply the classical methods of center manifold theory. This method makes the theory of inertial manifolds similar to the theory of global center manifolds. To show existence of such global center manifolds, one must be able to achieve a type of normal hyperbolicity in the sense that the flow in X toward the inertial manifold M must be stronger than the dispersion of the flow on M. Since the attractor A may be large, the dispersion of the flow on A may be large. Since very little is known about A, obtaining the strong attractivity property imposes many restrictions on the original evolutionary equation. These difficulties have been overcome for some partial differential equations in one space variable and some special domains in two space variables by exploiting the large gaps in the eigenvalues of the operator consisting of the highest order derivatives. The inertial manifold is obtained as a graph over part of the finite-dimensional subspace of X generated by the first n eigenfunctions with n sufficiently large. An example has been discussed in §4.10.2. There is an inertial manifold in Theorem 4.10.2 where the dimension is identical to the dimension of the original system of equations. In the discussion of shadow systems in (10.7), one can obtain an inertial manifold in one space dimension due to the gaps in the eigenvalues. In the case of several space variables it is not known if there is an inertial manifold. Other examples are in the references above.

In many cases, no inertial manifold will exist. Examples have been given by Mallet-Paret and Sell [1987]. There also is an example in §4.10.2 of Mora and

Sola-Morales [1986]. Conditions for the existence of an inertial manifold should be related to general properties of the flow on the attractor and relationships between the stable and unstable manifolds of nonwandering sets. It is difficult to make this precise in a general situation. However, for gradient systems, a compatibility condition and a stable intersection property of the stable and unstable manifolds of equilibria assures the existence of an inertial manifold, and one can describe the minimal dimension of the manifold in terms of the spectral properties of the linearization near the equilibria (see Hale and Sell [1988]). Much more research is needed in this area.

APPENDIX

Stable and Unstable Manifolds.

In this section, we state the basic properties of the stable and unstable manifolds of an hyperbolic equilibrium point of an abstract evolutionary equation. Indications of the proofs also will be given. To simplify the presentation and in order not to obscure the fundamental ideas, we concentrate on ordinary differential equations in finite dimensions and then give references for the appropriate modifications in infinite dimensional cases.

Consider the system of differential equations

(A.1) $$\dot{x} = Ax + f(x),$$

where $x \in R^n$, A is an $n \times n$ constant matrix whose eigenvalues have nonzero real parts, $f: R^n \to R^n$ is a Lipschitz continuous function satisfying

(A.2) $$\begin{aligned} &f(0) = 0, \\ &|f(x) - f(y)| \leq \eta(\sigma)(x - y) \quad \text{if } |x|, |y| \leq \sigma, \end{aligned}$$

where $\eta: [0, \infty) \to [0, \infty)$ is a continuous function with $\eta(0) = 0$.

For any $x_0 \in R^n$, let $\phi(t, x_0)$ be the solution of (A.1) through x_0. The unstable set $W^u(0)$ and the stable set $W^s(0)$ of 0 are defined as

$$W^u(0) = \{x_0 \in R^n : \phi(t, x_0) \text{ is defined for } t \leq 0 \text{ and } \phi(t, x_0) \to 0 \text{ as } t \to -\infty\},$$

$$W^s(0) = \{x_0 \in R^n : \phi(t, x_0) \text{ is defined for } t \geq 0 \text{ and } \phi(t, x_0) \to 0 \text{ as } t \to +\infty\}.$$

For a given neighborhood U of 0, we can also define

$$W^u(0, U) = \{x_0 \in W^u(0): \phi(t, x_0) \in U, \ t \leq 0\},$$
$$W^s(0, U) = \{x_0 \in W^s(0): \phi(t, x_0) \in U, \ t \geq 0\}.$$

These latter sets also are called local unstable and stable sets and are designated by $W^u_{loc}(0), W^s_{loc}(0)$.

Since the eigenvalues of A have nonzero real parts, there is a projection operator $P: R^n \to R^n$ such that PR^n and QR^n, $Q = I - P$, are invariant under A and the spectrum $\sigma(AP)$ of AP has positive real parts and $\sigma(AQ)$ has negative real parts.

A basic lemma is the following.

LEMMA A.1. *If $\phi(t, x_0)$, $t \leq 0$, is a bounded solution of (A.1), then $\phi(t, x_0)$ satisfies the integral equation*

(A.3) $$y(t) = e^{At} P x_0 + \int_0^t e^{A(t-s)} P f(y(s)) \, ds + \int_{-\infty}^t e^{A(t-s)} Q f(y(s)) \, ds.$$

If $\phi(t, x_0)$, $t \geq 0$, is a bounded solution of (A.1), then $\phi(t, x_0)$ satisfies the integral equation

(A.4) $$z(t) = e^{At} Q x_0 + \int_0^t e^{A(t-s)} Q f(z(s)) \, ds - \int_t^\infty e^{A(t-s)} P f(z(s)) \, ds.$$

Conversely, if $y(t)$, $t \leq 0$ [or $t \geq 0$], is a bounded solution of (A.3) [or (A.4)], then $y(t)$ satisfies (A.1).

PROOF: Let $y(t) = \phi(t, x_0)$, $t \leq 0$, be a bounded solution of (A.1). Then, for any τ in $(-\infty, 0]$,

(A.5) $$Q y(t) = e^{A(t-\tau)} Q y(\tau) + \int_\tau^t e^{A(t-s)} Q f(y(s)) \, ds.$$

There are positive constants k, α such that

(A.6) $$|e^{A(t-\tau)} Q| \leq k e^{-\alpha(t-\tau)}, \qquad t \geq \tau.$$

If we let $\tau \to -\infty$ in (A.5) using the fact that $y(s)$ is bounded in s, we obtain

$$Q y(t) = \int_{-\infty}^t e^{A(t-s)} Q f(y(s)) \, ds.$$

Since

$$P y(t) = e^{At} P x_0 + \int_0^t e^{A(t-s)} P f(y(s)) \, ds,$$

we see that $y(t)$ must satisfy (A.3). The proof for the case when $x_0 \in W^s(0)$ is similar and therefore omitted.

The converse statement is proved by direct computation.

We say that $W^u(0, U)$ is a *Lipschitz graph* over PR^n if there is a neighborhood V of 0 in PR^n such that $W^u(0, U) = \{y \in R^n : y = g(x), x \in V,$ where g is a Lipschitz continuous function$\}$. The set $W^u(0, U)$ is said to be *tangent to* PR^n at 0 if $|Qx|/|Px| \to 0$ as $x \to 0$ in $W^u(0, U)$. Similar definitions hold for $W^s(0, U)$.

A classical theorem on local unstable and stable sets which can be traced initially to the work of Lyapunov and Poincaré is contained in the following result.

THEOREM A.2. *Suppose f satisfies (A.2) and $\operatorname{Re} \sigma(A) \neq 0$. There is a neighborhood U of 0 in R^n such that $W^u(0, U)$ [or $W^s(0, U)$] is a Lipschitz graph over PR^n [or QR^n] which is tangent to PR^n [or QR^n] at 0.*

PROOF: The proof is a standard application of the contraction mapping principle (for example, see Hale [1969] or [1980]) and gives exponential decay rates of the solutions to the origin. Suppose k, α are chosen so that (A.6) is satisfied and also so that

(A.7) $$|e^{At} P| \leq k e^{\alpha t}, \qquad t \leq 0.$$

Choose $\delta > 0$ so that $4k\eta(\delta) < \alpha$, $8k^2\eta(\delta) < \alpha$. For $x_0 \in PR^n$ with $|x_0| \leq \delta/2k$, define $S(x_0, \delta)$ as the set of continuous functions $x:(-\infty, 0] \to R^n$ such that $|x| = \sup_{-\infty \leq t \leq 0} |x(t)| \leq \delta$ and $Px(0) = x_0$. The set $S(x_0, \delta)$ is a closed bounded subset of the Banach space of all continuous functions taking $(-\infty, 0]$ into R^n with the uniform topology. For any x in $S(x_0, \delta)$, define

$$(A.8) \quad (Tx)(t) = e^{At}x_0 + \int_0^t e^{A(t-s)} Pf(x(s))\,ds + \int_{-\infty}^t e^{A(t-s)} Qf(x(s))\,ds$$

for $t \leq 0$.

It is easy to show that $T: S(x_0, \delta) \to S(x_0, \delta)$ is a contraction mapping with contraction constant x_0 and therefore has a unique fixed point $x^*(\cdot, x_0)$. This fixed point satisfies (A.3) and thus is a solution of (A.1) from Lemma A.1.

The function $x^*(\cdot, x_0)$ is continuous in x_0. Also,

$$|x^*(t, x_0)| \leq ke^{\alpha t}|x_0| + k\eta(\delta) \int_t^0 e^{\alpha(t-s)}|x^*(s, x_0)|\,ds$$
$$+ k\eta(\delta) \int_{-\infty}^t e^{-\alpha(t-s)}|x^*(s, x_0)|\,ds.$$

From this inequality, one can prove that (see, for example, Lemma 6.2, p. 110 of Hale [1980]) that

$$(A.9) \quad |x^*(t, x_0)| \leq 2ke^{\alpha t/2}|x_0|, \quad t \leq 0.$$

This estimate shows that $x^*(\cdot, 0) = 0$ and also that $x^*(0, x_0) \in W^u(0)$.

The same type of computations show that

$$(A.10) \quad |x^*(t, x_0) - x^*(t, \bar{x}_0)| \leq 2ke^{\alpha t/2}|x_0 - \bar{x}_0|, \quad t \leq 0.$$

In particular, $x^*(\cdot, x_0)$ is Lipschitzian in x_0.

One next observes that

$$|x^*(0, x_0) - x^*(0, \bar{x}_0)| \geq |x_0 - \bar{x}_0| - \int_{-\infty}^0 k\eta(\delta)e^{\alpha s}|x^*(s, x_0) - x^*(s, \bar{x}_0)|\,ds$$
$$\geq |x_0 - \bar{x}_0|\left[1 - \frac{4k^2\eta(\delta)}{3\alpha}\right] \geq \frac{1}{2}|x_0 - \bar{x}_0|.$$

Thus, the mapping $x_0 \mapsto x^*(0, x_0)$ is one-to-one with a continuous inverse.

These estimates together with the fact that $\phi(t, x_0)$, $t \leq 0$, $x_0 \in W^u(0)$ must satisfy (A.3) imply that $W^u(0, U)$ for some U satisfies the conclusions of the theorem. The same type of argument applied to (A.4) will yield a complete proof of the theorem.

It is more difficult to obtain more regularity of the manifolds $W^u_{\text{loc}}(0), W^s_{\text{loc}}(0)$. If we assume that the vector field is C^k, then a standard application of the contraction mapping principle to (A.8) will not show that the above manifolds are C^k. However, with considerable effort, one can show that they are $C^{k-1,1}$; that is, they are represented by a function which is C^{k-1} with the $k-1$ first derivatives being Lipschitz (see, for example, Carr [1981], Sijbrand [1985]). One must then use some other method to show that the manifolds actually are C^k. A convenient way is to use the following lemma of Henry [1983] based on a remark in Hirsch, Pugh, and Shub [1977, p. 35].

LEMMA A.3. *Let X, Y be Banach spaces, $Q \subset X$ an open set, and $g: Q \to Y$ locally Lipschitzian. Then g is continuously differentiable if and only if, for each $x_0 \in Q$,*

(A.11) $$|g(x+h) - g(x) - g(x_0 + h) + g(x_0)| = o(|h|_X)$$

as $(x, h) \to (x_0, 0)$.

PROOF. It is easy to see that (A.11) holds if g is a C^1-function. Without loss in generality, we take Q to be a ball and g to be Lipschitzian in Q. If the derivative g' of g exists at each point of Q and (A.11) is satisfied, then g' is continuous. Thus, it is enough to prove that g' exists at each point of Q.

Case 1. Let us first suppose that $X = Y = R$. Since g is absolutely continuous, it is differentiable almost everywhere. For any $x_0 \in Q$, $\varepsilon > 0$, there is a $\delta > 0$ such that

$$|g(x+h) - g(x) - g(x_0 + h) + g(x_0)| \le \varepsilon|h| \quad \text{if} \quad |x - x_0| + |h| < \delta.$$

There is an x^* in $(x_0 - \delta, x_0 + \delta)$ such that $g'(x^*)$ exists. Thus, for $h \ne 0$ sufficiently small,

$$\left| \frac{g(x_0 + h) - g(x_0)}{h} - g'(x^*) \right| \le 2\varepsilon,$$

$$0 \le \{\varlimsup_{h \to 0} - \varliminf_{h \to 0}\} \frac{g(x_0 + h) - g(x_0)}{h} \le 4\varepsilon.$$

Since ε is arbitrary, this implies that $g'(x_0)$ exists.

Case 2. Now suppose that $X = R$ and Y is a Banach space with Y^* being the dual space. If $\eta \in Y^*$, then Case 1 implies that ηg is a C^1-function and $|(\eta g)'(x)| \le |\eta| \cdot \text{Lip } g$. If $D(x): \eta \to (\eta g)'(x)$, then $D(x) \in Y^{**}$ is continuous in x from (A.11).

For $x_0 \in Q$, $\eta \in Y^*$, $|\eta| \le 1$, we have

$$\eta\left(\frac{g(x_0 + h) - g(x_0)}{h}\right) - D(x_0)\eta = \frac{1}{h}\int_{x_0}^{x_0 + h} (D(x) - D(x_0))\eta \to 0$$

as $h \to 0$ uniformly for $|\eta| \le 1$. Let $\tau: Y \to Y^*$ be the canonical inclusion. Then $\tau[g(x_0 + h) - g(x_0)]/h \to D(x_0)$ as $h \to 0$. Since τ is an isometry, this implies that $[g(x_0 + h) - g(x_0)]/h \to$ a limit in Y as $h \to 0$; that is, $g(x_0)$ exists.

Case 3. Finally, let X, Y be arbitrary Banach spaces. From Case 2, for any $x \in Q$, $h \in X$, the map $t \mapsto g(x + th)$ taking R to Y is C^1 if t is small. Thus, the Gateaux derivative $dg(x, h) = dg(x + th)/dt|_{t=0}$ exists. Condition (A.11) implies that $dg(x, h) \to dg(x_0, h)$ in Y as $x \to x_0 \in Q$, uniformly for $|\eta| \le 1$. This implies that $h \mapsto dg(x, h)$ is linear and continuous. Thus, $dg(x, h)$ is the Fréchet derivative at h of g and the proof is complete.

THEOREM A.4. *Suppose $\text{Re } \sigma(A) \ne 0$, f satisfies (A.2) and is also a C^1-function. Then the sets $W^u(0, U)$, $W^s(0, U)$ of Theorem A.2 are C^1-manifolds.*

PROOF. Let $x^*(t, x_0)$ be the fixed point of the map T in (A.8) which was used to define $W^u_{loc}(0) = W^u(0, U)$. Define $y(x, x_0, h)(t) = x^*(t, x+h) - x^*(t, x) - x^*(t, x_0 + h) + x^*(t, x_0)$. From Lemma (A.3), it is sufficient to show that

(A.12) $$\overline{\lim_{(x,h) \to (x_0,0)}} \frac{1}{h} y^*(x, x_0, h)(t) = 0$$

uniformly for $-\infty < t \leq 0$.

From the definition of $x^*(t, x_0)$ and the fact that f is a C^1-function, we have, for $t \leq 0$,

$$y^*(x, x_0, h)(t)$$
$$= \int_0^t e^{A(t-s)} P f_x(x^*(s, x_0)) y^*(x, x_0, h)(s)\, ds$$
$$+ \int_{-\infty}^t e^{A(t-s)} Q f_x(x^*(s, x_0)) y^*(x, x_0, h)(s)\, ds$$
$$+ \int_0^t e^{A(t-s)} P[f_x(x^*(s, x)) - f_x(x^*(s, x_0))][x^*(s, x_0 + h) - x^*(s, x_0)]\, ds$$
$$+ \int_{-\infty}^t e^{A(t-s)} Q[f_x(x^*(s, x)) - f_x(x^*(s, x_0))][x^*(s, x_0 + h) - x^*(s, x_0)]\, ds$$
$$+ o(h)$$

as $|h| \to 0$. Using the estimates (A.6), (A.7), (A.2), and (A.10), we have

$$|y^*(x, x_0, h)(t)| \leq k\eta(\delta) \int_t^0 e^{\alpha(t-s)} |y^*(x, x_0, h)(s)|\, ds$$
$$+ k\eta(\delta) \int_{-\infty}^t e^{-\alpha(t-s)} |y^*(x, x_0, h)(s)|\, ds$$
$$+ \frac{8k^2}{\alpha} |h| e^{\alpha t/2} \sup_{s \leq 0} |f_x(x^*(s, x)) - f_x(x^*(s, x_0))|$$
$$+ o(h)$$

as $h \to 0$.

One can now show that there is a constant $K > 0$ such that (see, for example, Lemma 6.2, p. 110 of Hale [1980])

$$\frac{1}{|h|} |y^*(x, x_0, h)(t)| \leq K \sup_{s \leq 0} |f_x(x^*(s, x)) - f_x(x^*(s, x_0))|$$

for $t \leq 0$. Since the function $x^*(\cdot, x)$ is continuous in x and f is a C^1-function, we have (A.12) is satisfied and the theorem is proved.

Let us now suppose that $f = f(x, \lambda)$ depends on a parameter λ varying in an open subset V of a Banach space $\Lambda, 0 \in V$. Also, suppose that $f(x, \lambda)$ is continuous in (x, λ) and there is a continuous function $\eta: [0, \infty) \times \Lambda \to [0, \infty), \eta(0, 0) = 0$, such that

$$|f(x, \lambda) - f(y, \lambda)| \leq \eta(\sigma, \lambda) |x - y| \quad \text{if } |x|, |y| \leq \sigma, \qquad \lambda \in \Lambda$$
$$f(0, 0) = 0.$$

The equilibrium points of the equation are the solutions of $x = -A^{-1}f(x,\lambda)$. An application of the uniform contraction principle shows that there is a unique fixed point ϕ_λ of the map $-A^{-1}f(x,\lambda)$ for (x,λ) in a neighborhood of zero. Furthermore, $\phi_0 = 0$ and ϕ_λ is continuous in λ. If $x \mapsto x + \phi_\lambda$, then one can use the proof of Theorem A.2 and the uniform contraction principle to see that there are neighborhoods U of zero in R^n and V_1 of zero in Λ such that the local unstable manifold $W_\lambda^u(\phi_\lambda, U)$ of ϕ_λ and the local stable manifold $W_\lambda^s(\phi_\lambda, U)$ of $\phi_\lambda, \lambda \in V_1$, are continuous in λ.

If we suppose in addition that $f(x,\lambda)$ is a C^1-function in (x,λ), then the Implicit Function Theorem implies that ϕ_λ is C^1 in λ. From Theorem A.4, we also know that $W_\lambda^u(\phi_\lambda, U), W_\lambda^s(\phi_\lambda, U)$ are C^1-manifolds for each $\lambda \in V_1$. To show that these manifolds are C^1 in λ, let $x^*(t, x_0, \lambda)$ be the fixed point of the operator $T = T_\lambda$ in (A.8) [we have made the transformation above: $x \mapsto \phi_\lambda + x$]. The same type of argument used in the proof of Theorem A.2 shows thatf $x^*(t, x_0, \lambda)$ is Lipschitz continuous in λ. One now may apply Lemma A.3 and the argument in the proof of Theorem A.4 to obtain the C^1-dependence on λ.

We have proved

THEOREM A.5. *Suppose* $\operatorname{Re}\sigma(A) \neq 0$, $f = f_\lambda$ *depends on a parameter* λ *in an open subset* V *in a Banach space* Λ, $\sigma \in V$, *and* $f_\lambda(x)$ *is a C^1-function of λ, x with* $f(0,0) = 0$, $D_x f_\lambda(0,0) = 0$. *Then the manifolds* $W_\lambda^u(\phi_x, U), W_\lambda^s(\phi_\lambda, U)$ *are represented by a C^1-function of x, λ.*

One can extend Theorems A.4 and A.5 to the following

THEOREM A.6. *Suppose* $\operatorname{Re}\sigma(A) \neq 0$, f *satisfies* (A.2) *and is a C^k-function*, $k \geq 1$. *Then the sets* $W^u(0, U)$, $W^s(0, U)$ *of Theorem A.2 are C^k-manifolds. In addition, if $f = f_\lambda$ depends on a parameter λ in an open subset V of a Banach space* $\Lambda, 0 \in V$ *and* $f_\lambda(x)$ *is a C^k-function of x, λ with* $f_0(0) = 0, D_x f_0(0) = 0$, *then the manifolds* $W_\lambda^u(\phi_\lambda, U)$, $W_\lambda^s(\phi_\lambda, U)$ *are represented by a C^k-function of x, λ.*

We do not prove this theorem. A complete proof can be found in Hirsch, Pugh,and Shub [1977] where they used the graph transform (in contrast to fixed points of T in Formula (A.8)) to show the existence of the unstable and stable sets and they used a fiber contraction theorem to obtain the smoothness. Another proof is contained in Henry [1983] Chow and Lu [1987a] where they use (A.8) for existence, Lemma A.3, and a C^r-section theorem similar to the one in Hirsch, Pugh, and Shub [1977, p. 31] for the smoothness. Another proof has been given by Vanderbauwhede and van Gils [1987] using (A.8) and a fixed point theorem in weighted Banach spaces. Recently, Chow and Lu [1987b] have given a proof using (A.8), weighted Banach spaces and the contraction theorem. We should remark that the results in Hirsch, Pugh, and Shub [1977] deal with the more general problem of stable and unstable sets (as well as the persistence under perturbations) of normally hyperbolic invariant sets.

For infinite dimension problems, there are analogues of Theorem A.6 for some situations. These include all of the examples discussed in this book except the nonlinear diffusion problem of §4.9.6. The analogue of Theorem A.2 can be found in Ball [1973] and the presentation in Henry [1977], as well as in Vanderbauwhede and van Gils [1987], Chow and Lu [1987a], [1987b], is given for infinite dimensional problems.

We state the result in the infinite dimensional case. Let $X \subset Y$ be Banach spaces with the embedding being continuous. Let $S(t): Y \to Y, t \geq 0$, and the spaces X, Y satisfy the following properties:

(H_1) $S(t)$ is a strongly continuous linear semigroup.

(H_2) There is a decomposition $Y = Y_1 \oplus Y_2$ with continuous projections $P_i: Y \to Y_i$ such that
$$P_i S(t) = S(t) P_i, \qquad t \geq 0.$$

(H_3) $P_i X \subset X$ and $S(t) Y \subset X$ for $t > 0$.

(H_4) $S(t)$ can be extended to a group on Y_1.

(H_5) There exist constants $M \geq 0$, $N \geq 0$, $\alpha > 0$, $\beta > 0$, $0 \leq \gamma < 1$ such that
$$|S(t) P_1 x|_X \leq M e^{\alpha t} |x|_X \quad \text{for } t \leq 0, \qquad x \in X,$$
$$|S(t) P_2 x|_X \leq M e^{-\beta t} |x|_X \quad \text{for } t \geq 0, \qquad x \in X,$$
$$|S(t) P_2 y|_X \leq (M t^{-\gamma} + N) e^{-\beta t} |y|_Y \quad \text{for } t > 0, \qquad y \in Y.$$

Let $F: X \to Y$ be a given function and consider the integral equation in X:

(A.13) $$x(t) = S(t) x_0 + \int_0^t S(t - \tau) F(x(\tau)) \, d\tau.$$

For the linear semigroup $S(t)$, the subspace X_1 of X is the unstable manifold of zero and the subspace X_2 is the stable manifold of zero. If we suppose that $F \in C^k(X, Y)$, $k \geq 0$, and

(A.14) $$F(0) = 0, \qquad DF(0) = 0,$$

then one can use the contraction mapping principle as in the proof of Theorem A.2 to obtain the Lipschitz local unstable manifold $W^u_{\text{loc}}(0)$ and local stable manifold $W^s_{\text{loc}}(0)$ of the zero solution of the integral equation (A.13). These manifolds are actually C^k as stated in the following result.

THEOREM A.7. *Suppose* (H1)–(H5) *and* (A.14) *are satisfied. If* $f \in C^k(X, Y), k \geq 1$, *then* $W^u_{\text{loc}}(0), W^s_{\text{loc}}(0)$ *are* C^k-*manifolds. If, in addition,* $F = F_\lambda$ *depends on a parameter* λ *in an open subset* V *of a Banach space* Λ *and* $F_\lambda(x)$ *is* C^k *in* x, λ *with* $F_\lambda(0) = 0$, $D_x F_\lambda(0) = 0$, *then the manifolds* $W^u_{\text{loc}, \lambda}(0)$, $W^s_{\text{loc}, \lambda}(0)$ *are represented by a* C^k-*function of* x, λ.

As mentioned earlier, Theorem A.6 can be applied to the equations considered in this book (except for §4.9.6). For example, for the case in which $S(t)$ is an analytic semigroup with generator $-A$ (see §4.2), the two Banach spaces X, Y are respectively X^α, X in the notation of §4.2. For the damped hyperbolic equations considered in §§4.7 and 4.8, $X = Y = H_0^1(\Omega) \times L^2(\Omega)$.

For applications, one must also consider center manifolds and the manifolds near an equilibrium point which have a specified exponential behavior either as $t \to +\infty$ or as $t \to -\infty$. Such results are obtained by splitting the spectrum of the linear semigroup e^{At} by the circle $|z| = e^{\eta t}$ and then using weighted supremum norms so that the equilibrium point appears to be a saddle point for the linear operator in this norm (see Ball [1973a], Henry [1983], Vanderbauwhede and van Gils [1987], and Chow and Lu [1987b]).

As a final remark, we mention that results similar to the above ones are valid for fixed points of maps. In this case, the integrals in (A.8) are replaced by sums.

References

A. Acker and W. Walter, *The quenching problem for nonlinear parabolic differential equations*, Ordinary and Partial Differential Equations (Proc. Fourth Conf., Univ. Dundee, Dundee, 1976) (W. N. Everett and B. D. Sleeman, eds.), Lecture Notes in Math., vol. 564, Springer-Verlag, 1976, pp. 1–12.

____, *On the global existence of solutions of parabolic differential equations with a singular nonlinear term*, Nonlinear Anal. **2** (1978), 499–504.

S. B. Angenent, J. Mallet-Paret, and L. A. Peletier, *Stable transition layers in a semilinear boundary value problem*, J. Differential Equations **67** (1987), 212–242.

S. B. Angenent, *The Morse-Smale property for a semilinear parabolic equation*, J. Differential Equations **62** (1986), 427–442.

D. Aronson, M. G. Crandall, and L. A. Peletier, *Stabilization of solutions of a degenerate nonlinear diffusion problem*, Nonlinear Anal. **6** (1982), 1001–1022.

Z. Artstein, *Topological dynamics of ordinary differential equations and Kurzweil equations*, J. Differential Equations **23** (1977a), 224–243.

____, *The limiting equations of nonautonomous ordinary differential equations*, J. Differential Equations **25** (1977b), 184–202.

____, *Collectively limit sets and the limiting behavior of dynamical systems*, Preprint.

____, *Uniform asymptotic stability via the limiting equation*, J. Differential Equations **27** (1978), 172–189.

Z. Arstein and M. Slemrod, *Trajectories joining critical points*, J. Differential Equations **44** (1982), 40–62.

A. V. Babin and M. I. Vishik, *Regular attractors of semigroups of evolutionary equations*, J. Math. Pures Appl. **62** (1983), 441–491.

____, *Attracteurs maximaux dans les equations aux derivees partielles*, College de France Seminar, 1984, Pitman, Boston, 1985.

____, *Unstable invariant sets of semigroups of nonlinear operators and their perturbations*, Russian Math. Surveys **41** (4) (1986), 1–46.

J. Ball, *Saddle point analysis for an ordinary differential equation in a Banach space and an application to buckling of a beam*, Nonlinear Elasticity (R. W. Dickey, ed.), Academic Press, 1973a, pp. 93–160.

____, *Stability theory for an extensible beam*, J. Differential Equations **14** (1973b), 399–418.

____, *Initial boundary value problems for an extensible beam*, J. Math. Anal. Appl. **42** (1973c), 61–90.

____, *Strongly continuous semigroups, weak solutions and the variation of constants formula*, Proc. Amer. Math. Soc. **63** (1976), 370–373.

H. T. Banks and J. A. Burns, *An abstract framework for approximate solutions to optimal control problems governed by hereditary equations*, International Conf. on Differential Equations (H. A. Antosiewicz, ed.), Academic Press, 1975, pp. 10–25.

R. Bellman and K. Cooke, *Differential difference equations*, Academic Press, 1963.

J. E. Billotti and J. P. LaSalle, *Periodic dissipative processes*, Bull. Amer. Math. Soc. (N.S.) **6** (1971), 1082–1089.

L. Block, *Homoclinic points of mapping on the interval*, Proc. Amer. Math. Soc. **72** (1978), 576–580.

F. E. Browder, *On the spectral theory of elliptic differential operators*. I, Math. Ann. **142** (1961), 22–130.

W. E. Brumley, *On the asymptotic behavior of solutions of differential difference equations of neutral type*, J. Differential Equations **7** (1970), 175–188.

P. Brunovsky and B. Fiedler, *Connecting orbits in scalar reaction diffusion equations*, Preprint no. 329, Univ. Heidelberg, 1986.

J. Carr, *Applications of centre manifold theory*, Springer-Verlag, 1981.

R. C. Casten and C. J. Holland, *Instability results for reaction diffusion equations with Neumann boundary conditions*, J. Differential Equations **27** (1978), 266–273.

S. C. S. Ceron, *Compartamento assintotico de equações e sistemas hiperbolicos: soluções periodicas forcadas e convergencia para equilibria*, Inst. Mat. Ciencas de São Carlos, Univ. São Paulo, São Carlos, Brasil, 1984.

N. Chafee, *Asymptotic behavior for solutions of a one dimensional parabolic equation with Neumann boundary conditions*, J. Differential Equations **18** (1975), 111–135.

____, *Behavior of solutions leaving the neighborhood of a saddle point*, J. Math. Anal. Appl. **58** (1977), 312–325.

N. Chafee and E. F. Infante, *A bifurcation problem for a nonlinear parabolic equation*, J. Appl. Anal. **4** (1974), 17–37.

M. Chipot and J. K. Hale, *Stable equilibria with variable diffusion*, Contemp. Math. **17** (J. A. Smoller, ed.), Amer. Math. Soc., Providence, R.I., 1983, pp. 209–213.

S.-N. Chow and D. Green, Jr., *Some results on singular delay-differential equations*, Chaos, Fractals and Dynamics (P. Fischer and W. R. Smith, eds.), Marcel Dekker, 1986, pp. 161–182.

S.-N. Chow and J. K. Hale, *Strongly limit compact maps*, Funkcial. Ekvac. **17** (1974), 31–38.

S. N. Chow and K. Lu, C^k center unstable manifolds, Preprint, 1987a (to appear in Proc. Roy. Soc. Edinburgh Sect. A, 1988).

____, Invariant manifolds for flows in Banach spaces, Preprint, 1987b (to appear in J. Differential Equations, 1988).

S.-N. Chow and J. Mallet-Paret, Integral averaging and bifurcation, J. Differential Equations **26** (1977), 112–159.

B. Coleman and V. Mizel, On the general theory of fading memory, Arch. Rational Mech. Anal. **29** (1968), 18–31.

G. Cooperman, α-condensing maps and dissipative processes, Ph.D. Thesis, Brown University, Providence, R.I., 1978.

C. Conley, The gradient structure of a flow. I, RC3932, Math. Sci. IBM, Yorktown Heights, New York, 1972.

J. Constantin, C. Foiaş, and R. Temam, Attractors representing turbulent flows, Mem. Amer. Math. Soc. **53** (1985).

____, On the large time Galerkin approximation of the Navier-Stokes equations, SIAM J. Numer. Anal. **21** (1984), 615–634.

E. Conway, D. Hoff, and J. Smoller, Large time behavior of solutions of systems of reaction diffusion equations, SIAM J. Appl. Math. **35** (1978), 1–16.

C. Dafermos, Semiflows generated by compact and uniform processes, Math. Systems Theory **8** (1975), 142–149.

○ K. Deimling, Nonlinear functional analysis, Springer-Verlag, 1985.

O. Diekmann, Perturbed dual semigroups and delay equations, Dynamics of infinite dimensional systems (S.-N. Chow and J. K. Hale, eds.) Springer-Verlag, 1987, pp. 67–74.

A. Douady and J. Oesterle, Dimension de Hausdorff des attracteurs, C. R. Acad. Sci. Paris Sér. A-B **290** (1980), 1135–1138.

J. D. Farmer, Chaotic attractors of an infinite dimensional dynamical system, Phys. D **4** (1982), 366–393.

R. Fennel and P. Waltman, A boundary-value problem for a system of nonlinear functional differential equations, J. Math. Anal. Appl. **26** (1969), 447–453.

B. Fiedler and J. Mallet-Paret, Connections between Morse sets for delay-differential equations, Div. Appl. Math., Brown Univ. LCDS/CCS 87-32, 1987.

P. Fife, Boundary and interior transition layer phenomena for pairs of second order equations, J. Math. Anal. Appl. **54** (1976), 497–521.

P. Fife and L. A. Peletier, Clines induced by variable selection and migration, Proc. Roy. Soc. London B **214** (1981), 99–123.

W. E. Fitzgibbon, Strongly damped quasilinear evolution equations, J. Math. Anal. Appl. **79** (1981), 536–550.

C. Foiaş and G. Prodi, Sur le compartement global des solutions non stationaires des equations de Navier-Stokes en dimension 2, Rend. Sem. Mat. Univ. Padova **39** (1967), 1–34.

C. Foiaş, G. Sell, and R. Témam, Inertial manifolds for nonlinear evolution equations, Preprint, 1987 (to appear in J. Differential Equations, 1988).

A. V. Fursikov, *On some problems of control*, Dokl. Akad. Nauk SSSR **252** (1980), 1066–1070.

G. Fusco, *A system of ODE which has the same attractor as a scalar parabolic PDE*, J. Differential Equations **69** (1987), 85–110.

G. Fusco and J. K. Hale, *Stable equilibria in a scalar parabolic equation with variable diffusion*, SIAM J. Math. Anal. **16** (1985), 1152–1164.

G. Fusco and W. M. Oliva, *Jacobi matrices and transversality*, RT-MAP-8602, Univ. de São Paolo, Inst. Math. Stat., 1986. See also Dynamics of infinite dimensions (S.-N. Chow and J. K. Hale, eds.) Springer-Verlag, 1987, pp. 249–256.

R. Gardner, *Global continuation of branches of nondegenerate solutions*, J. Differential Euations **61** (1985), 321–334.

J. M. Ghidaglia and R. Témam, *Proprietes des attracteurs associes a des equations hyperboliques non lineares amorties*, C.R. Acad. Sci. Paris **300** (1985), 185–188.

____, *Attractors for damped nonlinear hyperbolic equations*, J. Math. Pures Appl. **66** (1987), 273–319.

____, *Regularity of solutions of second order equations and their attractors*, Ann. Scuola Norm. Sup. Pisa Cl. Sci. (4) (1987) (to appear).

J. K. Hale, *Ordinary differential equations*, 1st ed. Wiley, 1969, and 2nd ed. Kreiger, 1978.

____, *Stability of linear systems with delays*, Stability Problems (Centro Internaz. Mat. Estivo (C.I.M.E.), I Ciclo, Bressanone, 1974), Edizioni Cremonese, Rome, 1974, pp. 21–35.

____, *Theory of functional differential equations*, Springer-Verlag, 1977.

____, *Large diffusivity and asymptotic behavior in parabolic systems*, J. Math. Anal. Appl. **118** (1986), 455–466.

____, *Asymptotic behavior and dynamics in infinite dimensions*, Res. Notes in Math. (Hale and Martinez-Amores, eds.), vol. 132, Pitman, Boston, Mass., 1985, pp. 1–41.

J. K. Hale and J. Kato, *Phase space for retarded equations with infinite delay*, Tôhoku Math. J. **21** (1978), 11–41.

J. K. Hale, J. P. LaSalle, and M. Slemrod, *Theory of a general class of dissipative processes*, J. Math. Anal. Appl. **39** (1972), 177–191.

J. K. Hale, X.-B. Lin, and G. Raugel, *Upper semicontinuity of attractors for approximation of semigroups and partial differential equations*, LCDS #85-29, Div. Appl. Math., Brown Univ., Providence, R.I., 1985, J. Math. Comp. (to appear).

J. K. Hale and X.-B. Lin, *Examples of transverse homoclinic orbits in delay equations*, Nonlinear Anal. **10** (1986), 693–709.

J. K. Hale and O. Lopes, *Fixed point theorems and dissipative processes*, J. Differential Equations **13** (1973), 391–402.

J. K. Hale, L. Magalhães, and W. Oliva, *An introduction to infinite dimensional dynamical systems—geometric theory*, Springer-Verlag, 1984.

J. K. Hale and P. Massatt, *Asymptoptic behavior of gradient-like systems*, Dynamical Systems. II (A. R. Bednarek and L. Cesari, eds.), Academic Press, 1982, pp. 85–101.

J. K. Hale and G. Raugel, *Upper semicontinuity of the attractor for a singularly perturbed hyperbolic equation*, J. Differential Equations (1988a) (to appear).

____, *Lower semicontinuity of the attractor for gradient systems*, Preprint, 1988b.

J. K. Hale and C. Rocha, *Bifurcation in a parabolic equation with variable diffusion*, Nonlinear Anal. **9** (1985), 479–494.

____, *Varying boundary conditions with large diffusivity*, J. Math. Pures Appl. **66** (1987a), 139–158.

____, *Interaction of diffusion and boundary conditions*, Nonlinear Anal. **11** (1987b), 633–649.

J. K. Hale and K. P. Rybakowski, *On a gradient-like integro-differential equation*, Proc. Roy. Soc. Edinburgh **92A** (1982), 77–85.

J. K. Hale and K. Sakamoto, *Shadow systems and attractors in reaction-diffusion equations*, Preprint, 1987.

J. K. Hale and J. Scheurle, *Smoothness of bounded solutions of nonlinear evolution equations*, J. Differential Equations **56** (1985), 142–163.

J. K. Hale and G. R. Sell, *Inertial manifolds for gradient systems*, Preprint, 1987.

J. K. Hale and N. Stavrakakis, *Compact attractors for weak dynamical systems*, Applicable Anal., 1987 (to appear).

J. K. Hale and N. Sternberg, *Onset of chaos in differential delay equations*, submitted, J. Comp. Phys., 1987.

J. K. Hale and J. Vegas, *A nonlinear parabolic equation with varying domain*, Arch. Rational Mech. Anal. **86** (1984), 99–123.

A. Haraux, *Two remarks on dissipative hyperbolic problems*, Collège de France Seminaire, 1984, Pitmann, 1985.

____, *Semilinear hyperbolic problems in bounded domains*, Math. Rep., Harwood Academic Publishers, New York, 1988.

V. an der Heiden and H.-O. Walther, *Existence of chaos in control systems with delayed feedback*, J. Differential Equations **47** (1983), 273–295.

D. Henry, *Geometric theory of semilinear parabolic equations*, Lecture Notes in Math., vol. 840, Springer-Verlag, 1981.

____, *Invariant manifolds*, Dep. de Mat. Applicada, Universidade de São Paulo, São Paulo, Brazil, 1983.

____, *Some infinite dimensional Morse-Smale systems defined by parabolic differential equations*, J. Differential Equations **59** (1985), 165–205.

M. Hirsch. C. Pugh, and M. Shub, *Invariant manifolds*, Lecture Notes in Math., vol. 583, Springer-Verlag, 1977.

W. A. Horn, *Some fixed point theorems for compact mappings and flows on a Banach space*, Trans. Amer. Math. Soc. **149** (1970), 391–404.

M. C. Irwin, *On the stable manifold theorem*, Bull. London Math. Soc. **2** (1970), 196–198.

M. Ito, *A remark on singular perturbation methods*, Hiroshima Math. J. **14** (1984), 619–629.

A. F. Izé and J. G. dos Reis, *Contributions to stability of neutral functional differential equations*, J. Differential Equations **29** (1978), 58–65.

S. Jimbo, *Singular perturbation of domains and semilinear elliptic equation*, J. Fac. Sci. Univ. Tokyo, 1988 (to appear).

C. Jones, *Stability of traveling wave solution of the Fitzhugh-Nagumo equation*, Trans. Amer. Math. Soc. **286** (1984), 431–469.

G. Karakostas, *Causal operators and topological dynamics*, Ann. Mat. Pura Appl. (4) **131** (1982), 1–27.

H. Kawarada, *On solutions of initial boundary problem for* $u_t = u_{xx} + (1/(1-u))$, Publ. Res. Inst. Math. Sci. **10** (1975), 729–736.

S. N. S. Khalsa, *Finite element approximation of a reaction diffusion equation*, Preprint, 1985.

K. Kishimoto and H. F. Weinberger, *The spatial homogeneity of stable equilibria of some reaction diffusion systems on convex domains*, J. Differential Equations **58** (1985), 15–21.

P. E. Kloeden, *Asymptotically stable solutions of the Navier-Stokes equations and its Galerkin approximations*, Proc. Centre Math. Anal. Austral. Nat. Univ. (Trudinger and Williams, eds.), vol. 8, 1984, pp. 137–150.

____, *Asymptotically stable attracting sets in the Navier-Stokes equations*, Bull. Austral. Math. Soc. **34** (1986), 37–52.

P. E. Kloeden and J. Lorenz, *Stable attracting sets in dynamical systems and in their one-step discretizations*, SIAM J. Numer. Anal. **23** (1986), 986–995.

A. A. Kotsiolis and A. P. Oskolkov, *On the limit behavior and the attractor for the equations of motion for Oldroit fluids*, Zap. Nauchn. Sem. Leningrad Otdel. Mat. Inst. Steklov. (LOMI) **152** (1986), 97–100; *On the dynamical system generated by the equations of motion of Oldroit fluids*, Ibid., **155** (1986), 98–104.

S. G. Krein, *Linear differential equations in a Banach sapce*, Trans. Math. Monographs, vol. 29, Amer. Math. Soc., Providence, R.I., 1971.

H. Kurland, *Monotone and oscillatory equilibrium solutions of a problem arising in population genetics*, Contemp. Math., vol. 17, (J. A. Smoller, ed.) Amer. Math. Soc., Providence, R.I., 1983, pp. 323–342.

O. A. Ladyzenskaya, *The mathematical theory of viscous incompressible fluid*, Gordon and Breach, New York, 1963.

____, *A dynamical system generated by the Navier-Stokes equation*, Zap. Nauchn. Sem. Leningrad Otdel. Mat. Inst. Steklov. (LOMI) **27** (1972), 91–115.

____, *The boundary value problems of mathematical physics*, Appl. Math. Sci., vol. 49, Springer-Verlag, 1985.

_____, *On the determination of minimal global B-attractors for semigroups generated by boundary value problems for nonlinear dissipative partial differential equations*, (Russian) Steklov Math. Inst. Report E-3-87, Leningrad, 1987.

J. P. LaSalle, *Stability theory and invariance principles*, Dynamical Systems (Proc. Internat. Sympos., Brown Univ., Providence, R.I., 1974) (Cesari, Hale, and LaSalle, eds.), vol. 1, Academic Press, 1976, pp. 211–222.

N. Levinson, *Transformation theory of nonlinear differential equations of the second order*, Ann. of Math. (2) **45** (1944), 724–737.

X.-B. Lin and A. F. Neves, *A multiplicity theorem for hyperbolic systems*, Preprint, 1986.

X.-B. Lin and G. Raugel, *Numerical approximation of attractors in Morse-Smale dynamical systems generated by parabolic PDE's*, Preprint, 1986.

J. L. Lions and E. Magenes, *Problèmes aux limites nonhomogènes et applications*, Dunad, Paris, 1968.

P. S. Lomdahl, O. H. Soerensen, and P. L. Christiansen, *Soliton excitation in Josephson tunnel junctions*, Phys. Rev. B (3) **25** (1982), 5737–5748.

O. Lopes, *Stability and forced oscillations*, J. Math. Anal. Appl. **55** (1976), 686–698.

_____, *On the structure of the spectrum of a linear time periodic wave equations*, Rel. Int. No. 273, Univ. Est. de Campinas, Inst. Mat., Campinas, S.P., Brazil, 1984.

_____, *Perturbation of the spectrum of evolution operators*, Preprint, J. Operator Theory, 1988 (to appear).

O. Lopes and S. S. Ceron, *Existence of forced periodic solutions of dissipative semilinear hyperbolic equations and systems*, Ann. Mat. Pura Appl. (4) (submitted 1984).

M. C. Mackey and L. Glass, *Oscillation and chaos in physiological control systems*, Science **197** (1977), 287–289.

J. Mallet-Paret, *Negatively invariant sets of compact maps and an extension of a theorem of Cartwright*, J. Differential Equations **22** (1976), 331–348.

_____, *Morse decompositions for delay differential equations*, J. Differential Equations **72**, 1988, (to appear).

J. Mallet-Paret and G. Sell, *Inertial manifolds for reaction-diffusion equations in higher space dimensions*, IMA Preprint no. 331, Univ. of Minnesota, 1987.

R. Mañe, *On the dimension of the compact invariant sets of certain nonlinear maps*, Lecture Notes in Math., vol. 898, Springer-Verlag, 1981, pp. 230–242.

R. H. Martin, *Nonlinear operators and differential equations in Banach spaces*, Wiley, 1976.

P. Massatt, *Some properties of α-condensing maps*, Ann. Mat. Pura Appl. (4) **125** (1980), 101–115.

_____, *Stability and fixed points of point dissipative systems*, J. Differential Euations **40** (1981), 217–231.

_____, *Limiting behavior for strongly damped nonlinear wave equations*, J. Differential Equations **48** (1983), 334–349.

_____, *Attractivity properties of α-contractions*, J. Differential Equations **48** (1983a), 326–333.

H. Matano, *Convergence of solutions of one-dimensional semilinear parabolic equations*, J. Math. Kyoto Univ. **18** (1978), 224–243.

_____, *Asymptotic behavior and stability of solutions of semilinear diffusion equations*, Publ. Res. Inst. Math. Sci. **15** (1979), 401–458.

_____, *Nonincrease of the lap number of a solution for a one-dimensional semilinear parabolic equation*, J. Fac. Sci. Univ. Tokyo Sect. 1A Math. **29** (1982), 401–441.

M. Miklavčič, *Stability for semilinear equations with noninvertible linear operator*, Pacific J. Math. **118** (1985), 199–214.

R. K. Miller and G. Sell, *Topological dynamics and its relation to integral equations and nonautonomous systems*, Dynamical Systems (Proc. Internat. Sympos., Brown Univ., Providence, R.I., 1974) (Cesari, Hale, and LaSalle, eds.), vol. 1, Academic Press, 1976, pp. 223–249.

X. Mora, Personal communication, 1984.

_____, *Finite dimensional attracting invariant manifolds for damped semilinear wave equations*, Nonlinear Partial Differential Equations. II (J. L. Diaz and P. L. Lions, eds.), Longman, Essex, England, 1987.

X. Mora and J. Sola-Morales, *Existence and nonexistence of finite dimensional globally attracting invariant manifolds in semilinear damped wave equations*, Dynamics of Infinite Dimensional Systems (Chow and Hale, eds.), Springer-Verlag, 1987, pp. 187–210.

_____, *Diffusion equations as singular limits of damped wave equations*, Preprint (1987).

S. Murakami, *Perturbation theorem for functional differential equations with infinite delay via limiting equations*, J. Differential Equations **59** (1985), 314–335.

S. Murakami and T. Naito, *Fading memory spaces and stability properties for functional differential equations with infinite delay*, Preprint, 1987.

A. F. Neves, H. S. Ribeiro, and O. Lopes, *On the spectrum of evolution operators generated by hyperbolic systems*, J. Functional Analysis **67** (1985), 320–344.

K. Nickel, *Gestaltaussagen über Lösungen parabolischer Differentialgleichungen*, J. Reine Angew. Math. **211** (1962), 78–94.

B. Nicolaenko, B. Scheurer, and R. Témam, *Some global dynamical properties of the Kuramoto-Sivashinsky equations: nonlinear stability and attractors*, Phys. D **16** (1985), 155–183.

Y. Nishuira and H. Fujii, *Stability of singularly perturbed solutions to systems of reaction-diffusion equations*, KCU/ICS 85-05, Inst. Computer Science, Kyoto Sangyo. Univ., Preprint, 1985.

R. Nussbaum, *The radius of the essential spectrum*, Duke Math. J. **38** (1970), 473–488.

_____, *Some asymptotic fixed point theorems*, Trans. Amer. Math. Soc. **171** (1972), 349–375.

_____, *Periodic solutions of analytic differential functional differential equations are analytic*, Mich. Math. J. **20** (1973), 249–255.

J. Palis, *On Morse-Smale dynamical systems*, Topology **8** (1969), 385–405.

J. W. Palmer, *Liapunov stability theory for nonautonomous functional differential equations*, Ph.D. Thesis, Brown University, Providence, R.I., 1978.

A. Pazy, *Semigroups of linear operators and applications to partial differential equations*, Springer-Verlag, 1983.

V. Pliss, *Nonlocal problems in the theory of oscillations*, Academic Press, 1966.

J. Quandt, *On the Hartman-Grobman theorem for maps*, J. Differential Equations **64** (1986), 154–164.

C. Rocha, *Generic properties of equilibria of reaction diffusion equations with variable diffusion*, Proc. Roy. Soc. Edinburgh Sect. A **101** (1985), 45–56.

_____, *Examples of attractors in scalar reaction-diffusion equations*, J. Differential Equations, 1988 (to appear).

F. Rothe, *Dynamics of a scalar nonlinear diffusion problem*, Preprint, 1987.

P. Rutkowski, *Approximate solutions of eigenvalue problems with reproducing nonlinearities*, Z. Angew. Math. Phys. **34** (1983), 310–321.

B. N. Sadovskiĭ, *Limit compact and condensing operators*, Uspekhi Math. Nauk **27** (1972), 81–146.

K. Sakamoto, *Compact attractors for the Kuramoto-Sivashinsky equation*, Div. Appl. Math., Brown University, LCDS/CCS #87-6, 1987.

M. Schechter, *Principles of functional analysis*, Academic Press, 1971.

K. Schmitt, R. C. Thompson, and W. Walter, *Existence of solutions of a nonlinear boundary value problem via the method of lines*, Nonlinear Anal. **2** (1978), 519–535.

K. Schumacher, *Existence and continuous dependence for functional differential equations with unbounded delay*, Arch. Rational Mech. Anal. **67** (1978), 315–335.

A. C. Scott, F. Y. F. Chu, and S. A. Reible, *Magnetic-flux propagation on a Josephson transmission line*, J. Appl. Phys. **47** (1976), 3272–3286.

I. Segal, *Nonlinear semigroups*, Ann. of Math. (2) **78** (1963), 339–364.

G. Sell, *Lectures on topological dynamics and differential equations*, Van-Nostrand-Reinhold, Princeton, N.J., 1971.

_____, *Linear differential systems*, Lecture Notes, Univ. of Minnesota, 1975.

J. Sijbrand, *Properties of center manifolds*, Trans. Amer. Math. Soc. **289** (1985), 431–469.

J. Smoller, *Shock waves and reaction-diffusion equations*, Grundlehren Math. Wiss., vol. 258, Springer-Verlag, 1983.

J. Sola-Morales, *Global instability and essential spectrum in semilinear evolution equations*, Math. Ann. **274** (1986), 125–131.

J. Sola-Morales and M. Valencia, *Trend to spatial homogeneity for solutions of semilinear damped wave equations*, Seccio Mat., Univ. Autonoma de Barcelona, Preprint, 1986.

O. J. Staffans, *On weighted measures and a neutral functional differential equation*, Volterra and Functional Differential Equations, Lecture Notes in Pure and Appl. Math., vol. 81, Marcel Dekker, New York, 1982, pp. 175–183.

R. Témam, *Navier-Stokes equations, theory and numerical analysis*, 2nd ed., North-Holland, 1979.

____, *Navier-Stokes equation and nonlinear functional analysis*, CBMS-NSF Regional Conf. Ser. in Appl. Math., vol. 41, SIAM, Philadelphia, Pa., 1983.

____, *Attractors for the Navier-Stokes equation*, Nonlinear Partial Differential Equations and Their Applications, Collège de France Seminar VII, Pitman, Boston, 1984, pp. 272–292.

____, *Infinite dimensional dynamical systems in mechanics and physics*, Springer-Verlag, 1988 (to appear).

A. Vanderbauwhede and S. van Gils, *Center manifolds and contractions on a scale of Banach spaces*, J. Funct. Anal. **72** (1987), 209–224.

J. M. Vegas, *Bifurcations caused by perturbing the domain in an elliptic equation*, J. Differential Equations **48** (1983), 189–226.

____, *Irregular variations of the domain in elliptic problems with Neumann boundary conditions*, Contributions in Nonlinear PDE. II, Pitman, Boston, 1986.

W. Walter, *Differential and integral inequalities*, Springer-Verlag, 1970.

H.-O. Walther, *Stability of attractivity regions for autonomous functional differential equations*, Manuscripta Math. **15** (1975), 349–363.

____, *Homoclinic solution and chaos in* $\dot{x}(t) = f(x(t-1))$, Nonlinear Anal. **5** (1981), 775–788.

G. Webb, *Compactness of bounded trajectories of dynamical systems in infinite dimensional spaces*, Proc. Roy. Soc. Edinburgh Sect. A **84** (1979a), 19–33.

____, *A bifurcation problem for a nonlinear hyperbolic partial differential equation*, SIAM J. Math. Anal. **10** (1979b), 922–932.

____, *Existence and asymptotic behavior for a strongly damped nonlinear wave equation*, Canad. J. Math. **32** (1980), 631–643.

____, *Theory of nonlinear age-dependent population dynamics*, Marcel-Dekker, New York, 1985.

H. Weinberger, *A first course in partial differential equations*, Blaisdell, Waltham, Mass., 1965.

J. C. Wells, *Invariant manifolds of nonlinear operators*, Pacific J. Math. **62** (1976), 285–293.

E. M. Wright, *A nonlinear differential-difference equation*, J. Reine Angew. Math. **194** (1955), 66–87.

E. Yanagida, *Stability of stationary distribution in space-dependent population growth processes*, J. Math. Biol. **15** (1982), 401–441.

T. Yoshizawa, *Stability theory by Liapunov's second method*, Publ. Math. Soc. Japan, Math. Soc. Japan, Tokyo, 1966.

____, *Stability theory and the existence of periodic solutions and almost periodic solutions*, Springer-Verlag, 1975.

T. J. Zelenyak, *Stabilization of solutions of boundary value problems for a second order parabolic equation with one space variable*, Differential Equations **4** (1968), 17–22.

Index

α-contraction, 14, 37
    examples of an, 15, 16, 37
α-limit set, 8, 35
age-dependent populations, 155ff.
    attractor of, 159
    semigroup of, 158, 159
asymptotically smooth, 4, 11, 36
    collectively, 21, 41
    process, 42
attract, 9
attractor,
    approximation of the, 170
    capacity of the, 27, 28
    definition of an, 1, 17, 39
    existence of an, 5, 19, 39, 40, 46
    global, 1, 17, 39
    Hausdorff dimension of an, 27, 28
    local, 17, 160
    lower-semicontinuity of the, 171
    smoothness of flow on the, 56, 57
    upper-semicontinuity of the, 22, 41, 52, 163, 165, 167
beam equation, 148
    attractor of a, 149
    periodically forced, 151
β-condensing, 14
β-contraction,
    definition of, 14, 36
    examples of a, 15
capacity, 27
    finiteness of, 27
completely continuous, 13, 36
    process, 42
continuity of the attractor, 53, 173
dimension, Hausdorff, 26
    finiteness of, 27, 28
dissipative,
    bounded, 1, 16, 38
    compact, 16, 38
    in two spaces, 28ff., 54ff., 120ff., 148
    locally compact, 16, 38
    point, 1, 16, 38
distance from B to A, 17

$\delta(B, A)$, 17
equilibrium point,
    definition of, 39
    existence of, 40
    hyperbolic, 49
    stability of, 47, 48
equivalent, 25
evolutionary equations,
    abstract, 120ff.
    sectorial, 71ff.
    semigroups of, 73, 120, 121, 122
    strongly damped, 146
fixed point,
    existence of a, 23, 25
    hyperbolic, 25
global attractor,
    definition of, 1, 17
    existence of a, 5, 19, 39, 40, 42
gradient system,
    definition of a, 49
    attractor of a, 51, 52
Hausdorff dimension, 26
    finiteness of, 27, 28
hyperbolic, 25
inertial manifold, 169, 176
invariant set,
    definition of, 9, 36
    isolated, 11, 38
    maximal compact, 17, 38
Kuramoto-Sivashinsky equation, 153
    attractor of the, 154
lower-semicontinuity, 171ff.
maps,
    asymptotically smooth, 4, 11, 36
    α-contracting, 14
    β-condensing, 14, 37
    β-contracting, 14, 37
maximal compact invariant set,
    definition of, 17, 38
    existence of, 17, 39
measure of noncompactness, 13
    Kuratowskii, 13
Morse decomposition, 68

Morse-Smale system, 26, 53
Navier-Stokes equation, 107ff.
  attractor of the, 111
  periodic, 112
neutral functional differential equations, 113ff.
  attractor in C of, 116, 117, 165
  attractor in $W^{1,\infty}$ of, 117
  definition of, 113
  periodic, 117
  singularly perturbed, 165
  smoothness of solutions of, 116
  the semigroup of, 115
nonlinear diffusion, 154
orbit,
  positive, 8, 35
  negative, 8, 35
  complete, 8, 35
parabolic equation,
  attractor of a, 90, 91, 92
  bifurcation in a, 87ff.
  gradient structure of a, 77
  Morse-Smale, 84
  ODE equivalent to a, 94ff.
  quenching in a, 105ff.
  singularly perturbed, 99
  stability of equilibria of a, 84, 97, 100, 102
  transition layers in a, 83
  transversality in a, 83
periodic point, 26
periodic trajectory, 43, 117, 132, 139, 145, 151
process, 41
  almost periodic, 45
  asymptotically autonomous, 45
  asymptotically smooth, 42
  autonomous, 42, 45
  completely continuous, 42
  periodic, 42, 45
  periodic trajectory of a, 43
  quasiperiodic, 45
  trajectory of a, 42
reaction diffusion equations, 161
  attractors of, 161, 164
  shadow system of, 163
retarded functional differential equations, 61ff.
  attractor of, 63
  gradient, 65
  infinite delay, 145
  periodic, 70

  semigroup of, 61
  smoothness of solutions of, 63
  with negative feedback, 67
sectorial evolutionary equations, 71ff.
  existence of the attractor of, 74
  semigroup of, 73
  smoothness of solutions of, 74
sectorial operator, 71
  fractional powers of a, 71
  semigroup of a, 73
skew product flows, 43ff.
spectrum, essential, 14
stability,
  asymptotic, 10, 38
  definition of, 10, 38
  of attractors, 20, 25, 26, 53
  uniform asymptotic, 10, 38
stable manifold, 26, 49, 179ff.
transmission lines, 151
  attractors in, 152
transversality, 83
unstable manifold, 26, 49, 179ff.
upper-semicontinuity,
  definition of, 21
  of attractors, 22, 41, 52, 163, 165, 167
wave equation,
  attractor of, 125, 128, 130, 132, 134, 138, 139, 143
  bifurcation in a, 129
  inertial manifold of a, 169
  linear, 124
  linearly damped, nonlinear, 125, 139
  nonlinearly damped, nonlinear, 130, 134
  periodically forced, 132, 138, 145
  singularly perturbed, 130, 167, 169
ω-limit set, 1, 8, 35